Performance of Bolting Materials in High Temperature Plant Applications

ORGANISING COMMITTEE

A Strang (Chairman)
GEC ALSTHOM Large Steam Turbines, Rugby, UK

RD Conroy
Parsons Power Generation Systems Ltd, Newcastle-upon-Tyne, UK

H Everson
UES Steels, Stocksbridge, UK

DJ Gooch
National Power PLC, Swindon, UK

KH Mayer
MAN Energie, Nürnberg, Germany

RD Townsend
ERA Technology, Leatherhead, UK

INTERNATIONAL LIAISON COMMITTEE

Professor I Artinger
Technical University of Budapest, Hungary

Dr P Auerkari
VTT Metallilaboratorio, Espoo, Finland

Dr R Blum
I/S Fynsvaerkt, Odense, Finland

Dr V Foldyna
Vitkovice Research Institute, Ostrava, Czech Republic

Dr B Scarlin
ABB Power Generation, Baden, Switzerland

Dr V Viswanathan
EPRI, Palo Alto, CA, USA

rmance of
g Materials
in
emperature
Applications

:d by A Strang

Large Steam Turbines, Rugby, UK

ence Proceedings
–17 June 1994
York, UK

Organised by the
erature Materials Committee
rials Engineering Division of
Institute of Materials

CRC Press
Taylor & Francis Group
Boca Raton London New York

shed 1995 by The Institute of Materials

Published 2019 by CRC Press
    Taylor & Francis Group
oken Sound Parkway NW, Suite 300
Boca Raton, FL 33487-2742

5 by Taylor & Francis Group, LLC
nt of Taylor & Francis Group, an Informa business

First issued in paperback 2019

n to original U.S. Government works

N 13: 978-0-367-44890-5 (pbk)
N 13: 978-0-901716-72-9 (hbk)

the Taylor & Francis Web site at
p://www.taylorandfrancis.com

nd the CRC Press Web site at
http://www.crcpress.com

rary Cataloguing in Publication Data
    Available on application

ngress Cataloging in Publication Data
    Available on application

e world's largest and most powerful steam turbine –
erating at Chooz B Power Station in France.

Contents

SESSION 4   *Austenitic Bolting Steels*   *201*

SESSION 5   *Nickel Based Bolting Alloys*   *257*

Foreword

This is the second international conference on high temperature creep resistant materials, organised by the High Temperature Materials Committee of the Institute of Materials, to be held in York. The first dealt with the general subject of rupture ductility in creep resistant steels and addressed the compositional, heat treatment and microstructural factors responsible for controlling this important material property. This conference is concerned with a similar topic in addressing the application and performance of bolting materials in high temperature plant. In such applications bolts are often required to operate for long periods of time under high duty plant operational conditions. Performance and material reliability are important factors in the selection of suitable alloys for high temperature plant applications. For these applications bolting materials are required to have high creep and stress relaxation resistance in combination with good creep rupture ductility and tolerance to the presence of notches. Methods of achieving the optimum combination of these properties are essential to ensure the safe and reliable performance of bolted joints in high duty plant applications.

In a series of 33 invited and contributed papers presented in 8 sessions, this Conference considers the design, materials supply and control, material testing and selection, data analysis and life management factors which affect the performance and reliability of high temperature bolting in a wide variety of high temperature plant applications. It is hoped that this Conference has not only provided an opportunity to debate the wide range of factors affecting the performance of bolting materials in service but also, by extending our knowledge and understanding of their behaviour, will lead to the development of new and more reliable materials for the design of future advanced high temperature plant.

I wish to thank my colleagues on the Organising Committee for their hard work and expertise in formulating such a stimulating technical programme for this Conference and on their behalf express our thanks to Lisa Davies of the Institute of Materials Conference Department for her enthusiasm and sterling efforts in making this meeting such a success. Finally I would like to express my thanks to Professor F B Pickering for his expertise, enthusiastic support and willingness to undertake the unenviable task of reviewing all of the papers and presenting us with his technical overview of the state of the art of bolting materials, their performance and future on the last day of the Conference.

A Strang
Organising Committee Chairman

SESSION 1

Design and Performance

Chairman: A Strang

GEC ALSTHOM Large Steam Turbines, Rugby, UK

1

Design Considerations for High Temperature Bolting

J. BOLTON

GEC ALSTHOM Large Steam Turbines, Rugby, UK

1. Introduction

The object of the present paper is to preface this conference, which will inevitably focus on the technology of high temperature bolting materials, with a discussion of the design considerations of high-temperature bolted joints. The two major considerations in design are relaxation behaviour and the prevention of failure. This paper concentrates on relaxation behaviour, making only limited reference to failure considerations which are the subject of the following paper to this conference by Dr R. D. Townsend. Calculation of the relaxed strength of a bolted joint often involves the designer in some manipulation of the available relaxation data. The type of material modelling employed is discussed, illustrating the additional materials data which the designer finds useful.

2. Mechanism of bolt relaxation

Figures 1 and 2 show typical bolted joint designs in modern high-temperature, high-pressure steam turbine plant. Figure 1 shows the arrangement of a cover joint for an HP steam chest and Fig. 2 shows the arrangement of a VHP cylinder horizontal joint. Pressure and temperature conditions for the HP steam chest are 80 bar and 580°C. Maximum pressure and temperature conditions for the VHP inner casing are 245 bar and 555°C. In both cases the sealing face is a metal-to-metal joint and the effectiveness of the seal depends upon the flatness of the joint surfaces and the clamping pressure exerted by the tension in the bolts. The tensile strain in joint bolts at the beginning of service life may be established by heating the bolt through a central hole and screwing the nut down whilst it is hot, or by hydraulic means, and control of the initial strain level may be achieved by micrometer measurement of the bolt extension. In less demanding applications the pre-tension may be adequately controlled by the measurement of applied tightening torque or by the measurement of nut rotation during tensioning. Ultrasonic methods also exist for indirectly measuring the applied strain. Typical initial design strains are in the region of 0.15–0.18% hot. The fundamental design criterion for the bolts is a very simple one, that the gradual extension of the bolts due to creep at operating temperature does not reduce the clamping force to the point that the contact pressure on the joint face falls below the level at which the joint leaks. The mechanism of bolt stress relaxation is that the initial elastic strain imposed by tightening is gradually converted to permanent creep strain. It is a convenient simplification to regard the clamped flange itself as rigid, having no elastic flexibility and accumulating no creep strain, so that the bolt

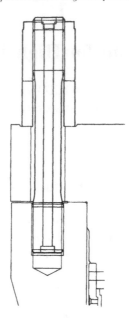

Fig. 1 *Bolted steam chest cover for a modern large steam turbine (HPSV chest, M72).*

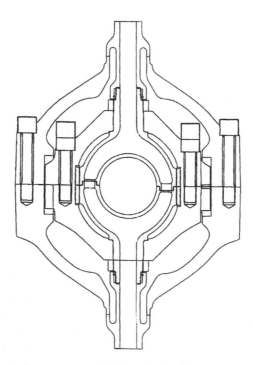

Fig. 2 *Bolted casing joint for a modern large steam turbine (VHP cyl., M140).*

may be considered to be subject to a total (elastic plus creep) strain which is invariant with time. Design practices for flanged joints will normally ensure that the sectional area of the flange in compression is large in comparison to the sectional area of the bolts in tension. In such cases, and provided that the creep strength levels of bolt and flange are not grossly different, both the elastic and creep strains in the flange will be relatively low. The compressive creep strain in the flange will be compensated to some extent by its elastic recovery as the bolt stress relaxes and, if these effects are reasonably balanced, the net effect will be as for a rigid flange. Under these circumstances the conversion of elastic strain to creep strain is expressed by the equation

$$-\frac{1}{E} \cdot \frac{d\sigma}{dt} = \frac{d\epsilon_c}{dt} \tag{1}$$

where s = bolt stress, t = time, E = elastic modulus and ϵ_c = creep strain.

If creep strain accumulation in the bolt material is represented, very simply, by an equation of the form

$$\epsilon_c = Bs^n t \tag{2}$$

where the coefficient B and exponent n (which has a value greater than 1) are material constants at a given temperature, then eqn (1) gives the following expression for the relaxation of stress with time:

$$\sigma / \sigma_i = \left[\frac{1}{1 + EB(n-1)t\sigma_i^{n-1}} \right]^{1/(n-1)} \tag{3}$$

Equation (3) describes a relaxation curve as shown in Fig. 3, which is broadly similar to the observed experimental relaxation curves of bolting materials — in that the rate of relaxation of stress decreases with time. It may be noted that the coefficient B in eqn (2) can be inferred from available creep data and that it is the creep strength of the bolt material, not its

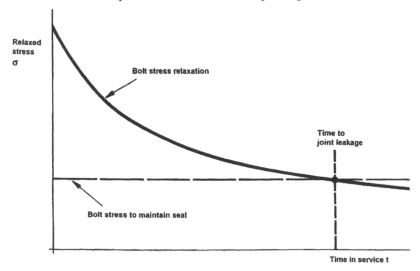

Fig. 3 *Relaxation of bolt stress in service.*

rupture strength, which governs its relaxation strength. Moreover, since relaxation strains of interest are only in the region of 0.05–0.10%, it is creep strength at low strain that is important.

The dashed horizontal curve in Fig. 3 represents the bolt stress calculated to be required to maintain a pressure seal, illustrating the basic design problem that the gradual decline in residual stress limits the operating life of the joint before re-tightening becomes necessary. It is a considerable burden to a plant owner to have to re-tighten joint bolting, especially where an outer casing must be opened to gain access to an inner casing joint, and there is an obvious incentive to extend re-tightening times to match the times at which overhauls become necessary for other reasons. There has been a long-established practice in large steam turbines of designing for re-tightening cycles of 30 000 h but, as target periods between overhauls have been extended in the interests of greater operational efficiency, it is now common to design for a retightening interval of 50 000 h. This presents the practical problem of the need to gather relaxation data over longer duration and it is helpful to have some analytical basis for extrapolating observed relaxation curves. Equation (3) is of little assistance in this respect since the shape of the calculated relaxation curve diverges considerably from that observed, as shown in Fig. 4. The data plotted is for 1%CrMoVTiB steel (Durehete 1055) at 550°C and for this comparison the values of the material constants in eqn (2) have been chosen to give a coincident relaxed strength at 30 000 h.

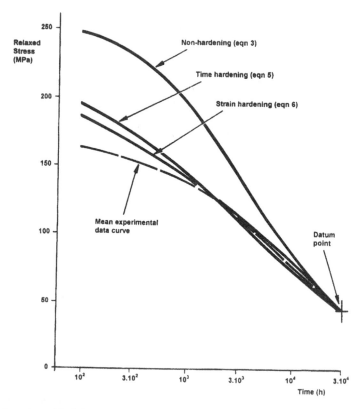

Fig. 4 *Relaxation with different hardening models (D1055, 550°C).*

3. Relaxation Models with Simple Hardening

A better basis for extrapolated relaxation strength is obtained with a modified material model, introducing a non-linear time dependency to the representation of conventional creep strain data.

$$\epsilon_c = B\sigma^n t^{\,m} \tag{4}$$

where the exponent m has a value less than 1 corresponding to conventional creep data for primary and secondary creep in which the rate of strain accumulation diminishes with time.

The strain rate equation can then be written either in terms of time-hardening as

$$\frac{d\epsilon_c}{dt} = mB\sigma^n t^{m-1} \tag{5}$$

or, in terms of strain-hardening, as

$$\frac{d\epsilon_c}{dt} = mB^{1/m}\sigma^{n/m}\epsilon_c^{(m-1)/m} \tag{6}$$

Relaxation curves calculated from hardening relationships display a much improved correspondence to observed relaxation, as illustrated in Fig. 4.

Such relationships are useful for extrapolating available data to longer times or to cycles of repeated strain as discussed in the following section. Their use is, however, dependent on judicious selection of the values of the exponents n and m. If available, uniaxial creep data may be used as the basis of determining suitable values of n and m. Constant load uniaxial creep data plotted in the form log creep strain vs log time produce curves typically of the form of Fig. 5 for creep resistant bolting materials. If the strain dependence on stress and time were truly as in eqn (4) then these curves would form a set of straight, parallel lines of slope m. It is clear from Fig. 5 that in fact both n and m are variable and the selection of pseudo-constant values for analytical purposes should be done with due regard to the times and strains of interest.

If curves as in Fig. 5 are available, the limitations of constant n and m values are avoidable by stepwise integration of the relaxation as shown in the figure, where stress is assumed constant in each time increment and decrements of stress are made on strain hardening assumptions. In creep resistant steels it is observed that the relationship between creep strain ϵ_{ct}, and stress σ_t at any given time is satisfactorily represented by the relation

$$\epsilon_{ct} = \frac{\epsilon_D(\sigma_{Rt}/\sigma_{Dt} - 1)}{(\sigma_{Rt}/\sigma_t - 1)} \tag{7}$$

where σ_{Rt} is the rupture strength at the time of interest and σ_{Dt} is the stress to produce datum strain ϵ_D (e.g. 0.2%) in the same time. Equation (7) is equivalent to defining a value of n which is variable with stress, such that $n = 1$ at low stress and tends to infinity at the rupture stress

$$n = \frac{1}{1 - \sigma_t/\sigma_{Rt}} \tag{8}$$

Hence, where only the time variation of rupture strength and creep strength for a datum strain are known, a consistent set of curves with the characteristics of Fig. 5 may be generated and stepwise integration of the relaxation may be carried out as described above.

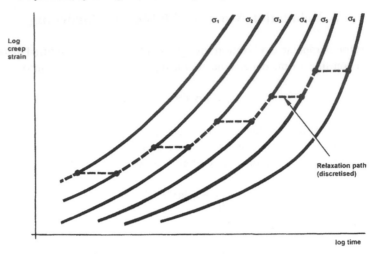

Fig. 5 *Typical creep strain vs time data with schematic relaxation path.*

The material models described here are very simple models constructed for the mathematical convenience of the design engineer rather than for precision in representing material behaviour. The paper to this conference by Osgerby and Dyson discusses more sophisticated models constructed on the basis of the physical behaviour of the materials.

4. Successive Relaxation Cycles

Design procedures are normally based upon single cycle relaxation data but consideration of the second relaxation cycle, following re-tightening to the original elastic strain, is of some interest from the design point of view. It is natural to expect that the hardening effects apparent in single cycle relaxation will lead to an enhanced relaxation strength in the second cycle. Values of the exponent m in eqn (4) are observed from conventional creep data (at appropriate strains and times) to be commonly in the region of 1/3. Equation (6) thus implies that the creep strain rate varies as the inverse square of the creep strain, indicating a strong tendency to strain hardening. Third and subsequent cycles might manifest a further enhanced relaxation strength as creep strain accumulates, until the accumulated creep strain reaches the point at which tertiary creep intervenes. It is rare for repeated-cycle testing to be carried out for relaxation times of practical interest, but repeated relaxation tests for relatively short cycles show the expected effect very clearly, as in Fig. 6.

In theory this presents a number of interesting design possibilities, although these are difficult to exploit in practice.

(a) The first-cycle re-tightening period could be shorter than the second and subsequent periods. The practical drawbacks are that, firstly the plant operator is no less interested in the first-cycle time than in subsequent cycles. Secondly, the replacement of any bolts after the first cycle, through accidental damage, would invalidate the assumption of prior hardening.

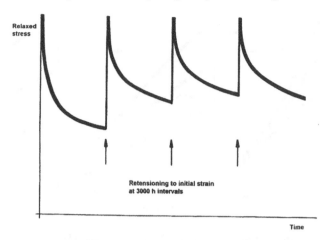

Fig. 6 *Increase of residual stress in repeated relaxation cycles (D1055, 550°C).*

(b) The second-cycle pre-strain could be reduced, extending the bolt life without reducing the re-tightening period. The drawback, again, is that any bolt replacement invalidates the assumptions.

(c) The in-service relaxation strength of bolts could be enhanced by prior creep straining. Unfortunately there exists no practical means of carrying out the prior straining of large bolts.

One way of extending the re-tightening period is to raise the level of initial strain. Obviously there is a limit to initial strain beyond which further strain will merely cause the bolt to stretch plastically when hot without increasing the initial hot elastic strain. Moreover, if re-tightening periods are extended in theory but not in fact then the only consequence of increased initial strain is a higher rate of strain accumulation in the bolts and a reduced bolt life. It is therefore important to take a realistic view of achievable periods between overhauls.

Different material models diverge considerably in predicting the response of materials relaxed strength to increased initial strain. A non-hardening model (eqn (2)) predicts little gain in relaxed strength as a result of increased strain, as shown in Fig. 7(a). The inference has sometimes been drawn that initial strains may be reduced in order to extend bolt life with little or no reduction in the re-tightening period. However, time-hardening and strain-hardening behaviour models predict a much stronger dependence of relaxed strength on initial strain, as in Figs 7(b) and 7(c). Relaxation data gathered more recently demonstrates a roughly linear relationship over the range of strain tested, similar to Fig. 7(b).

5. Exhaustion of Bolt Life

The failure of high temperature bolts in steam turbine plant is an uncommon occurrence nowadays, but there have been periods in the history of the power industry when failures

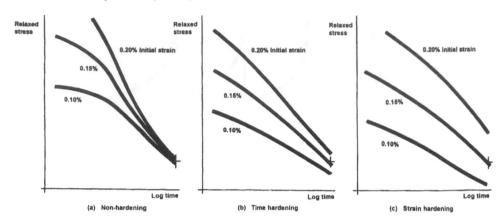

Fig. 7 Influence of initial strain with different hardening models (D1055, 550°C).

were unacceptably frequent. The failure of bolts securing steam valve covers has been known to result in the ejection of the covers, with significant consequential damage. Known failures have initiated in the first engaged thread of the screwed ends of the bolts, as indicated in Fig. 8(a). Elastic analysis shows that there are two stress components of similar magnitude at the root of the first engaged thread, one due to the load reacted on the first thread itself and the other due to load reacted on 'downstream' threads but concentrated in passing by the notch at the first thread. The thread load tends to concentrate heavily at the first thread, as indicated in Fig. 8(b), owing to the relationship between thread core, nut core and thread-form flexibility. For a typical turbine casing bolt thread geometry the elastic thread load concentration reaches about four times the average for the whole of the thread. The effect of creep is to redistribute the load with a beneficial reduction of load in the first thread, but there remains a significant concentration of stress and strain at the root of the first engaged thread.

Consequently, it is important that bolting materials have sufficient ductility to absorb the locally concentrated strain without fracture and to allow the thread deformations which relieve the severe elastic concentration of load. It is not possible to state a ductility requirement in simple terms of the elongation at rupture of a plain uniaxial specimen, but the experience of creep fractures in bolts suggests that problems are encountered where rupture elongations fall below about 5%. Bolting materials are sometimes tested in notched rupture, but standard notched specimens are not closely representative of bolts despite the superficial resemblance of a circumferential notch to a thread. In a notched specimen the section is a minimum at the notch, whereas the shank of a bolt is normally waisted at least to the core diameter of the thread. A greater degree of waisting has often been introduced in practice for the express purpose of restricting the stress level in the threaded region. Notched specimen rupture life might therefore be less than the life of a bolt at the same net section stress. On the other hand, bolt threads are subjected to direct loading on the thread flanks which is not present in a notched specimen.

One method that has been used to discriminate between the tolerance of different bolting materials to accumulated thread strain is accelerated, repeated relaxation cycling of full size bolts. This method simulates bolt service conditions very accurately apart from the high

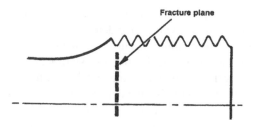

(a)   Location of observed bolt fractures

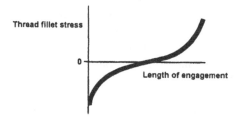

(b)   Stress distribution in threads due to bolt load

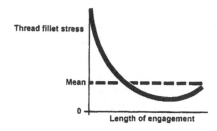

(c)   Stress distribution due to differential strain

Fig. 8 *Stresses in bolt threads.*

initial strain and short re-tightening period employed to accelerate the tests. This type of testing and the results will be discussed in the paper by Holdsworth and Strang.

Within the power industry practical procedures for determining bolt life have been drawn up on the basis of limiting the notional level of strain accumulated at the first engaged thread. In principle these procedures consist of numerically combining the effect of a series of physical factors, each of which contributes to strain accumulation, into an overall penalty factor of life reduction. The factors considered are:

(1) Material;
(2) Temperature of application;
(3) Nominal applied strain;
(4) Bolt form (degree of waisting);
(5) Temperature gradient, and
(6) Number of re-tightening cycles.

The roles of the first four of these are self-evident. The importance of the fifth factor has emerged from failure investigations which have demonstrated the sensitivity of bolt life to axial temperature gradients. These failures occurred in steam chest securing bolts where the difficulties of, or inadequacies in, the application of external insulation resulted in significant thermal gradients along the length of the bolts. Such a gradient can be shown, by calculation using material models incorporating the dependence of strain rate on temperature, to result in a much higher hot end strain than would occur if the whole of the bolt were at the maximum temperature. The stress relaxation which would have occurred at a uniform maximum temperature occurs much more slowly and the hot end strain is much increased. For the materials affected at the relevant nominal temperature, an increase of hot end strain of two to threefold can be calculated for a temperature differential of 100°C, although this varies depending on what model is used for material hardening. Temperature gradients along cover bolts also imply temperature gradients through the covers, leading to thermal distortion and the superimposition of bending loads on the bolts. This further reduces their safe life in relation to bolts at uniform temperature.

The last of the listed factors, the number of re-tightening cycles, also makes a self-evident contribution to strain accumulation. Within this factor allowance can be made for the duration of individual cycles and for the reduced strain accumulation due to strain hardening after the first relaxation cycle.

The designer can mitigate the degree of strain concentration to some extent by attention to the detail design of the threaded connection. Taper threads are beneficial in this respect, since the load on the first threads can be reduced and the load on the succeeding, lower stressed threads can be increased. The effect is equivalent to introducing a differential strain between the bolt and nut threads which superimposes a thread load distribution as shown in Fig. 8(c), off-setting the original concentration at the first thread as in Fig. 8(b).

6. Differential Expansion between Bolt and Thread

Where the expansion coefficients of bolt and flange materials differ significantly, additional design problems arise. Important material combinations in large steam turbine design include:

(1) 12%CrMoVNb steel bolts in 1%CrMoV steel flanges, where the bolts expand less than the flange between room temperature and operating temperature. Consequently the hot bolt strain exceeds the cold strain by *ca.* 0.06%, depending on temperature and flange geometry, which is large in relation to typical initial strains.

(2) Nimonic bolts in 9%CrMo steel flanges, where the bolts expand more than the flange. Consequently the hot bolt strain is less than the cold strain by about 0.06%, depending on temperature and flange geometry.

In both cases the differential expansion strain is initially absorbed by elastic strain changes in the bolt and the flange, and is shared between bolt and flange in a proportion dictated by their elastic stiffness. For a differential expansion strain $\Delta T \Delta \alpha$, the strain change in the bolt is given by

$$\Delta T \Delta \alpha \frac{L}{AE} \frac{C_B C_F}{C_B + C_F} \qquad (9)$$

where

    L = Length of assembly over which there is a difference in coefficient of expansion $\Delta \alpha$.
    A = cross sectional area of bolt shank.
    E = Hot elastic modulus of bolt.
    C_B = Hot elastic stiffness of bolt (load per unit increase in length).
    C_F = Hot elastic stiffness of flange.

The stiffness C_B and C_F include the influence of the bolt threaded lengths and the nuts. Since C_B is approximately equal to AE/L it can be seen that if, for example, the flange stiffness is three times the bolt stiffness then about 75% of the differential strain is absorbed by the bolt and 25% by the flange. Their relative stiffness are generally such that 70–80% of the differential strain is absorbed by the bolt and 20–30% by the flange. For a joint design based on the relaxation strength of the bolt from a selected initial hot strain it is necessary to compensate for the expansion by adjusting the cold strain (or by other means) to give the required strain when hot.

In the case of bolts expanding less than the flange, if the cold strain is reduced then the joint must be checked to verify that after relaxation in service adequate joint tightness is retained when cold. This could present problems, necessitating initial strains higher than required for joint sealing when hot.

In the case of bolts expanding more than the flange, if the cold strain is increased to compensate it must remain within the cold elastic limit of the bolt. This may impose an unacceptable limit on the achievable hot strain and necessitate the introduction of compensating sleeves between the flange and the nut. Austenitic sleeves compensate adequately for the differential expansion between Nimonic and 9%CrMo steel, but not without introducing problems of space in the design of the joint. The creep resistance of the austenitic sleeve also becomes a factor in the relaxation strength of the assembly and it can be difficult to obtain sleeve material, at large diameters, of sufficient strength so as not to reduce the strength of the assembly.

Furthermore, differential thermal strain between a bolt thread and the interlocking thread in a casing (or in a nut) will cause redistribution of thread load as in Fig. 8(c). Therefore the optimum thread taper for materials with differing expansion coefficients is not the same as the optimum for materials with matching coefficients.

Thus a number of design problems are introduced by a mismatch in thermal expansion between bolt and flange. From a design viewpoint a bolt material with a matching expansion coefficient is much to be preferred.

7. Effect of Flange Creep

Whilst it is a convenient simplification to suppose that any flange creep will be offset by flange elastic recovery, and therefore does not influence bolt relaxation, it is by no means certain that this will always be so. The greater the disparity between the creep resistance of the flange and the bolt, the more likely it is that flange creep will reduce the relaxation strength of the joint to significantly below the bolt relaxation strength determined by conventional testing.

The effect can be illustrated by considering a simple bolt and flange arrangement where the sectional area of the flange under compression is R times the sectional area of the bolt in tension. Local deformations in the threads and the regions of contact between nuts and flange are assumed negligible so that joint relaxation is controlled only by bolt tensile creep under instantaneous stress a and flange compressive creep under instantaneous stress σ/R. Assuming that strain accumulation in the bolt and flange materials is governed by simple relationships of the form of eqn (2), such that the flange material accumulates creep strain K times faster than bolt material at the same stress

$\epsilon_{CB} = B\sigma^n t$ for the bolt

$\epsilon_{CF} = KB(\sigma/R)^n t$ for the flange

then the joint relaxation can be expressed as

$$\frac{\sigma}{\sigma_i} = \left[\frac{1}{1 + EB(n-1)t\sigma_i^{n-1} \dfrac{R^n + K}{R^{n-1}(1+R)}} \right]^{1/(n-1)} \tag{10}$$

This reduces to eqn (3) for R equal to infinity, for which condition both elastic and creep strains in the flange are negligible. According to this equation, the influence of the area ratio R and creep rate ratio K are as shown in Fig. 9 for the case that a rigid flange gives a relaxed stress equal to 60% of the initial stress in a given time and $n = 1.5$. It can be seen that for equal creep strengths ($K = 1$) and equal areas ($R = 1$) the result is identical to that obtained for a rigid flange. This is because the compressive creep in the flange is a mirror image of the tensile creep in the bolts and the total (elastic plus creep) strain in both remains constant. For $K = 1$ and $R > 1$ the elastic recovery of the flange more than compensates for the flange creep and the relaxation strength of the assembly is calculated to be slightly enhanced.

For a flange which is moderately weak in creep relative to the bolt ($K = 3$), the relaxation strength of the assembly is slightly reduced over a practical range of R values ($3 < R < 8$), but only by an amount that might be within normal design margins. However, when the bolt creep strength greatly exceeds the flange creep strength ($K = 10$, $K = 30$) the relaxation strength of the assembly can fall well below that of the bolt and shows a significant variation with the flange to bolt area ratio.

The differences between the creep strengths of superalloy bolting and steel flanges are sufficiently great for considerable strength reductions to be observed in model testing, where the flange is simulated by a hollow cylinder. This type of test is described in the paper by Mayer and König. Test data shown in Fig. 10 illustrate the considerable reduction in the relaxed strength of Nimonic 80A bolts in 9%CrMo model flanges relative to conventional data for Nimonic 80A. While this type of data provides the designer with a clear indication of where significant weakening may occur due to flange creep, it also presents some prob-

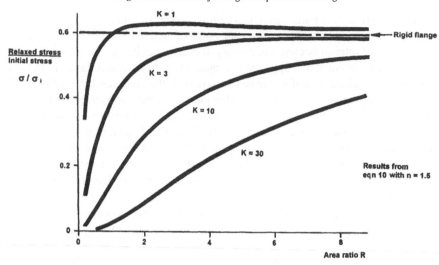

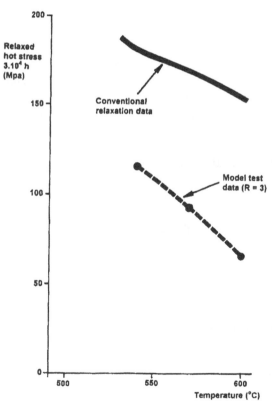

Fig. 9 *Expected influence of flange creep on bolt relaxation.*

Fig. 10 *Comparison of conventional and model test data (Nimonic 80A bolts with mod. 9%CrMo sleeve).*

lems of interpretation. It is hardly practical to test over a wide range of flange to bolt area ratio, yet design applications will cover a wide range of area ratio and some basis of extrapolation to the case of interest becomes necessary. It is also inevitable that the loads carried by the bolt and flange are equal and opposite in a model test. In an actual flange joint the effect of the pressure contained within the vessel is to relieve the compressive load in the flange and to, very slightly, increase the tensile load in the bolt. Where model tests indicate a large influence of flange creep, the designer will wish to discount any excessive influence due to inexact simulation of the service condition. This leads the designer into complex calculations involving the simultaneous relaxation of the bolt and the flange. Such calculations necessitate the use of mathematical models of the form of eqns (2)–(8), the values of the constant coefficients and exponents being determined from available conventional relaxation and creep data.

A closely associated problem arising from a large difference between the flange and bolt creep strengths is that the region of the flange directly under the nut may be overloaded and may deform locally, leading to loss of relaxed strength of the assembly. This is compensated in design by the use of nuts of larger than normal outside diameter, but this necessitates an increased pitch between bolts and consequently reduces the total available joint closing force.

8. Conclusion

The purpose of this paper was to present the materials specialists assembled for this conference with a practical perspective on how available bolting data is actually used and manipulated for the purposes of high temperature joint design. In conclusion the following short list summarises the practical problems that the designer would wish the materials specialist to keep in mind.

(i)   Extrapolation of relaxation data to longer duration.

(ii)   Extrapolation of relaxation data to higher or lower initial strains.

(iii)   Material tolerance to accumulated concentrated strain in threads.

(iv)   Problems of differential expansion between bolt and flange.

(v)   Problems of difference in creep strength between bolt and flange.

(vi)   Interpretation of model assembly relaxation tests.

2

Performance of High Temperature Bolting in Power Plant

R. D. TOWNSEND

ERA Technology Ltd, Leatherhead, UK

1. Introduction

Bolts are used in the power industry for the many joints which need to be separated after service for maintenance and repair and also as a means of attachment for components such as cabling, thermocouples and thermal insulation. In both types of application the bolts are used at all temperatures but it is only the application and performance of bolts operating above 370°C where creep becomes a potential degradation mechanism that are considered in this paper. Typically in a modern steam turbine there may be up to 800 bolts and studs operating in the creep range in sizes between 40–150 mm dia. and 300–1000 mm long. On nuclear plant several thousand small bolts of 10–25 mm dia. and up to 100 mm long may be used for thermocouple, instrumentation attachments and insulation.

The main design requirements for bolts operating under these conditions are:

• to maintain a steam tight joint throughout the service life of the plant

• to avoid bolt fracture since this might lead to catastrophic failure of the joint.

As an aspect of historical record these design requirements have not always been realised and there are many examples of steam leaks occurring in service and even examples of catastrophic failures. It is important to appreciate however, that there may be significant differences between the design faults and metallurgical factors which result in these different joint failure modes. Typical bolting situations for turbine and nuclear applications are illustrated schematically in Figs. 1 and 2. Apart from the obvious differences in size and perhaps environmental situation, aspects which are likely to affect bolt integrity are that both types are likely to operate in a temperature gradient and for nuclear attachments there are often several plies or washers along the bolt shank [1]. Under normal circumstances both types of assembly are tightened at ambient temperatures and then raised to the operating temperature. If this is in the creep range then the initial elastic stress in the assembly is progressively converted to creep strain thereby reducing the effective load on the bolt (Fig. 3). Clearly, the relaxed load σ_R after a period of service will depend on a number of factors including:

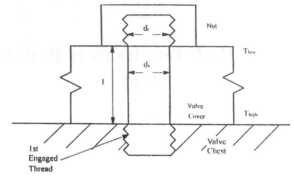

Fig. 1 *Stud bolt in turbine valve chest.*

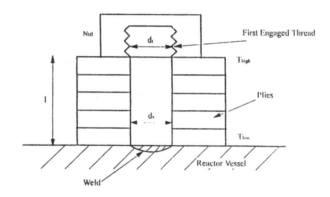

Fig. 2 *Attachment bolt in nuclear assembly.*

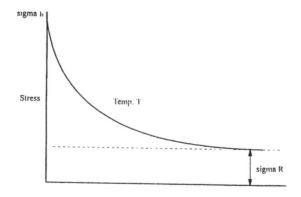

Fig. 3 *Schematic stress relaxationcurve.*

• the initial load applied at ambient temperature;

• the mismatch in thermal coefficients of expansion between the bolting material and flange (or ply) materials;

• the tensile and creep properties of the bolt and flange (ply) materials; and

• operating time and temperature.

In idealised situations, creep strain accumulation will occur along the entire length l of the bolt shank and within corresponding regions of the flange or plies. In more realistic situations however, this does not pertain and for factors which will become apparent later the strain is often concentrated in the area of the first engaged threads. Under such circumstances, considerable strain concentration can build up in this region and eventually this can lead to exhaustion of the local strain tolerance of the material and ultimately to premature bolt failure. Figure 4 is a schematic illustration of the build-up of creep strain in the shank and first engaged regions as a function of operating time and (for turbine applications) multiple tightening events. Also shown is the variation of the fracture strain tolerance ε_T of the material which will often decrease with increasing time due to the build-up of creep damage in the material and to in-service embrittlement. Clearly, in the situations shown, failure would not occur in the shank portion of the bolt but would occur in the region of the first engaged thread, at time t_2, after a single tightening, or at time t, after multiple tightenings. In practice, the factors which can and indeed (in service) have given rise to enhanced thread strain and to premature bolt failures are:

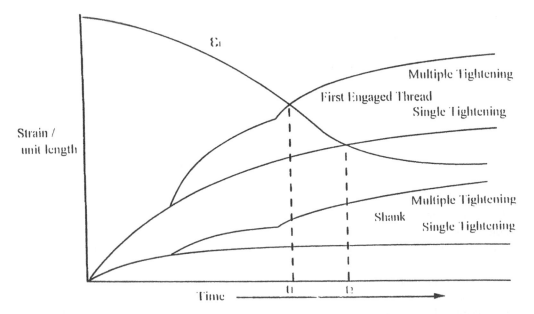

Fig. 4 *Schematic illustration of creep strain accumulation per unit length of shank and first engaged thread.*

• multiple re-tightening events;

• bolt designs for which the thread diameter D_T is less than the shank diameter D_S which exacerbates the geometrical notch aspects of the thread and reduces the load bearing area;

• operating conditions which give rise to temperature gradients along the length of the bolt so that creep strain accumulation occurs much faster in the hottest region of bolt (often coincident with the site of the first engaged thread);

• the establishment of additional loads on the bolts by: superimposed bending loads;

• contraction of the bolt material in service expansion of the flange or ply region in service; and

• in-service embrittlement of the bolting material which reduces the strain tolerance, ε_T.

In practice, bolting failures in service often occur due to a combination of these factors and the Case Studies chosen in this paper are intended to illustrate these interactions. For turbine applications they include:

• failures in ferritic bolts caused by poor design leading to the establishment of temperature gradients along the bolts and in-service embrittlement;

• failures in a nickel base bolting material due to in-service embrittlement and lattice contraction due to ordering (giving rise to bolt shrinkage) and for nuclear applications;

• the consequences of oxide jacking in gas cooled reactors; and

• bolt failures due to irradiation embrittlement.

2. Bolting Failures in Nuclear Applications

High temperature bolting failures are relatively rare in nuclear applications simply because the most common reactor types, the water cooled reactors, do not function with structural components operating in the creep range. This statement, however, does not pertain to gas cooled reactors and bolting failures have occurred in these systems. An important aspect to note is that bolts used within reactor pressure vessels, ducting and steam generator units have often been used for permanent fixtures rather than on components which require to be separated after a period of service for maintenance as in many turbine applications. The use of bolting in this manner in nuclear situations has often been used for simplicity in construction or to avoid welding on critical components which could give rise to distortion. The first two case studies presented in this paper are associated with bolting failures in gas cooled reactors.

2.1. Case Study 1: In-Service Embrittlement of Alloy 800 Cover Plate Bolts

During inspection of a hot gas duct in the helium cooled THTR-300 nuclear reactor at Hamm-Uentrop in West Germany, some 33 fractured bolt heads were found on the floor of the duct [1]. The bolt heads had become detached from modified Alloy 800 bolts each 130 mm long and 16 mm dia. These bolts were a small fraction of the 2574 bolts used to retain and clamp an insulation pack assembly of a perforated plate of modified Alloy 800 comprising a stack of 10 layers of X2CrNi19 11 steel and eight layers of X10CrNiNb18 9 steel (Fig. 5). The service temperatures of the insulation pack ranged from 680–810°C and the failed bolts were centrally located in each cover plate compressing the insulation pack with a force of about 70 kg.

In all cases the failures occurred in the transition from the shank to the conical bolt head which finite element analysis identified as the region of highest stress concentration. The fracture surfaces indicated the failures were due to intercrystalline, brittle creep failures and showed no evidence of fatigue. Detailed metallography showed there was negligible oxidation between the insulation layers nor on other mating surfaces and no evidence of in-service carburisation.

In examining the causes of failure it was clear that the centrally mounted failed bolts would have sustained higher thermally induced displacement than bolts at the corners of the cover plates and hence subjected to higher bending loads. Additionally differential expansion of the insulation pack and the bolts themselves would have increased the axial tensile

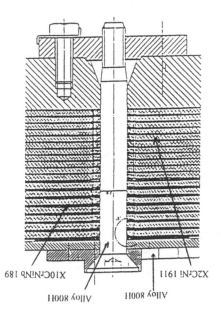

Fig. 5 *Cross-section of THTR-300 hot duct insulation package.*

stress along the bolts. However, it was also evident that the prime cause of failure was the severe in-service embrittlement due to thermal neutron irradiation which drastically decreased the tensile ductility to less than 10% of the original ductility (Fig. 6).

It would appear therefore that the enhanced stresses on the central bolts due to bending and differential expansion culminated in an enhanced strain accumulation during stress relaxation at temperature and caused total exhaustion of the strain tolerance of the material in its irradiation embrittled condition, thus promoting bolt fracture. In this particular incident it was fortunate that the integrity of the THTR reactor was not seriously compromised because of considerable redundancy in the design. Analysis of the situation by Schuster and Schubert [1] indicated that failure of a few centrally mounted bolts could be tolerated without seriously jeopardising plant integrity.

2.2. Case Study 2: Oxide Jacking in Gas Cooled Reactors

Economically and technologically one of the most important failures in bolting occurred during early operation of the UK CO_2 cooled Magnox reactors. Here the problem was *in situ* straining of small ($^1/_2$ and $^5/_8$ in. dia.) carbon steel bolted assemblies by oxide jacking between the interfaces of washers or plies. Whilst carbon steels oxidise protectively in carbon dioxide at atmospheric pressure, under the original reactor operating conditions (maximum coolant gas temperatures approaching 400°C and pressures 10–30 bar) rapid breakaway linear rates of oxidation occur after an incubation period of protective oxidation. This behaviour is shown schematically in Fig. 7 (Meadowcroft) [2]. The time to breakaway can be several years at Magnox operating conditions and has been found to depend on a range of factors including the silicon content of the steel, its surface finish and the moisture and carbon monoxide concentrations in the gas. The breakaway oxide thus formed is porous and continues to grow at interfaces even when the gap between such interfaces appears full

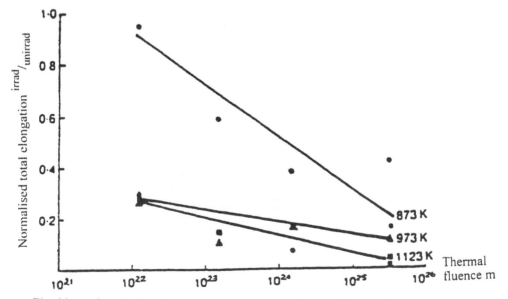

Fig. 6 Loss of tensile ductility by thermal neutron irradiation (Alloy 800 H).

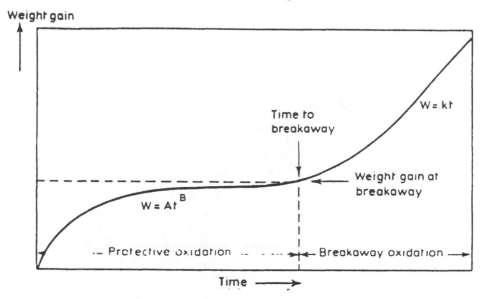

Fig. 7 *Schematic — onset of breakaway oxidation in CO_2.*

of oxide. The volume of oxide produced is greater than that of the metal consumed in its formation (the linear Pilling–Bedworth ratio ie ratio of oxide thickness to metal loss for breakaway oxide on mild steel is 2.4) and growth of the oxide causes adjacent components to be forced, or jacked, apart. A comparison of unoxidised and oxidised bolted assemblies with multiple plies is shown in Figs. 8(a) and (b). In this case the oxidation was performed in the laboratory in high pressure CO_2 as was the fractured sample shown in Fig. 9 [3]. Note the failure occurred, as in the majority of turbine bolt failures, in the first engaged thread.

Since most of the bolted joints in the reactors contain mild steel bolts and plies the assessment procedures were first devised to predict strain in bolts of this material. Clearly it has been necessary to predict the strain distributions along the bolts. The major effect of oxide growth is to promote axial bolt strain and the same forces also compress the plies, which acts as an alleviation to bolt strain. In some cases, there is a tendency to extract the bolt from the nut which is also an alleviation on bolt strain. Examination of sectioned samples Figs. 9 and 10 [3, 4] did indicate that most of the deformation extends the unmeshed region of the bolt (the free threads) while some extends the meshed (engaged) threads and small extension of the unthreaded shank also occurs. The relative cross-sectional area of the shank and threads is clearly important in this situation. The greatest deformation is found where the bolt enters the nut near the first free thread. The effective change in section here acts to concentrate the stress as confirmed by finite element stress analysis Fig. 11 [3]. A schematic of the components of axial strain based on such analyses is shown in Fig. 12 [3]. Under these conditions deformation is strain controlled and takes place by creep and failure is determined by the ductility of the material or more appropriately, the notched ductility.

Temperature influences both oxidation and mechanical behaviour. A reduction in temperature reduces the oxide growth rates and hence oxide jacking and bolt strain. It should be noted, however, that paradoxically, this actually leads to an increase in the load on the bolts since the activation energy for oxide growth is less than for creep. But since fracture of the

(a)   (b)

Fig. 8 *Multi-interface samples: (a) before, and (b) after oxidation in high pressure CO_2.*

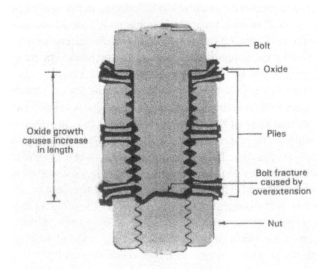

Fig. 9 *Section of multi-interface sample fractured after oxidation in high pressure CO_2 in the region of first engaged thread.*

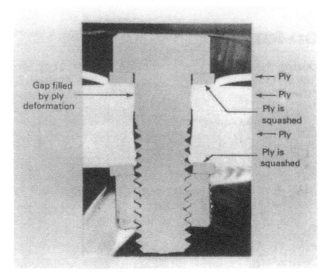

Fig. 10 *Section of multi-interface sample after oxidation in high pressure CO_2 showing distortion of plies and shank.*

bolts depends on bolt ductility reducing the temperature does in fact reduce the overall strain and hence does achieve the desired effect of extending the life of bolted joints. For this reason the major action taken when the phenomenon was first encountered in the Magnox reactor gas circuits in the late 1960s was to reduce the gas temperature. This has obviously reduced the efficiency of this reactor system and reduced the output but has not caused reductions in the design lives.

Over the years, a substantial amount of work has been conducted in oxidation studies, statistical analyses of in-reactor samples and theoretical analyses of the jacking phenomenon (Rowlands *et al.* [5]). These have culminated in procedures agreed between the Nuclear Utilities and the Nuclear Installations Inspectorate for use in statutory Magnox Oxidation Safety Assessments. Analyses show that the lifetime of important structural joints can now be considerably extended, with clear benefits for the future economic and safe operation of Magnox reactors.

3. Bolting Failures in Turbine Applications
3.1. Design Requirements

In contrast to many nuclear applications bolting used in turbine applications is often required to be slackened and retightened after periods of service to allow for essential maintenance of the component. The consequences of this requirement in design have been covered extensively in the previous paper by Bolton [6]. Here only a few essentials of the design aspects are covered to provide background to the case studies selected.

To prevent steam leakage of a joint it is necessary that the load applied by the tensile loads within the bolts at all times exceeds the steam load on the flange. The actual number of the bolts, their cross sectional area and distribution within a joint are dictated by the load

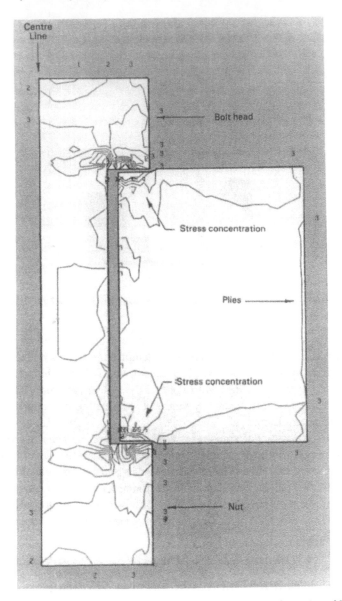

Fig. 11 *Finite element stress analysis showing stress concentration in the region of first engaged thread.*

required, the properties of the bolts, the operating conditions of time and temperature and geometrical considerations.

An example of failure to meet this criterion was found recently by ERA engineers during a life assessment investigation of a power plant in the Far East (Fig. 13) [7]. On opening a steam chest the chest cover bolts were significantly distorted with necking up to 15% clearly visible to the naked eye. On investigation the $1^1/_4$Cr1MoV1Ni25V bolts had been subjected to a number of retightening campaigns using an uncontrolled technique such that elastic

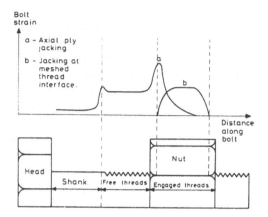

Fig. 12 *Schematic representation of components of oxidation strain in a bolt resulting from ply and meshed interface jacking.*

Fig. 13 *Steam chest valve cover bolts showing necking in bolt shank.*

strains much greater than 0.15% were achieved. Clearly, the bolts were not strong enough to withstand this type of mistreatment.

At high temperatures, creep within the bolts causes stress relaxation of the load and the elastic strain(s) introduced by the initial tightening of the bolts are progressively converted to creep strains, thereby reducing the effective load on the joints. The final load thus depends on the initial load at temperature, the creep properties of the flange and bolting materials and the operating times and temperature. At the end of an operating regime, it is this reduced load which must still exceed the steam loading.

In practice the initial load on an individual bolt at the operating temperature cannot be measured easily. Bolts are therefore tightened to a prescribed strain and the load at temperature can then be estimated provided the tensile properties of the bolt and flange materials are known and allowance made for any differences in thermal expansion between the two materials. Matching of the thermal coefficients of expansion is in fact an important consideration in joint design, too large a mis-match could lead to slackening of the joint if the bolt has a higher expansion coefficient than the flange or excessive tightness if the reverse situation pertained.

The strain to which the bolts are initially tightened depends on the properties of the bolting and flange materials and on the service requirements. The usual strain of 0.15% represents a compromise between a sufficiently high load (to maintain joint tightness) and excessive plastic strain accumulation during service. In designing for relaxation at temperature, it is normally assumed that the flange remains rigid and that the strain distribution along the bolt is uniform and that the entire bolt operates at the maximum temperature within the joint. These assumptions lead to an overestimation of the amount of stress relaxation and hence in terms of the final load on the joint a common conservative design.

The temperature distribution along the bolt depends on the temperature of the steam and the degree of lagging. In well lagged joints, the metal temperature is assumed to be the same as the steam temperature, but in poorly lagged joints considerable temperature gradients can pertain along the bolts and we will see later this gives rise to considerable non-conservatism in the design.

Joint lives are normally equated in design to an idealised situation in which overhauls, and therefore joint dismantling, takes place at fixed intervals of 30 000 h and between which no re-tightening is carried out. It is then assumed that the joints are reassembled using the same bolts and that up to six re-tightening campaigns may be required during the life of a turbine. Clearly the creep relaxation of a bolt re-used in this way must continue to meet the design requirement such that the residual stress on the bolt does not cause the applied load on the flange to fall below the steam load. In practice of course a joint may experience many more re-tightenings than six if the source of the operating periods are significantly less than 30 000 h and for continued use of the bolts it would be necessary to assess the effects of this on joint integrity.

Typical compositions of ferritic bolting steels and the various national standards pertaining to them are given in Table 1(taken from the paper by Everson, Orr and Dulieu [8]) which also tabulates their heat treatments and properties. It should be noted that these cover the same range of materials as used in the UK (Table 2). For convenience the UK Central Electricity Generating Board classified these materials into eight groups according to composition and properties (see Table 2). The stress-relaxation behaviour of these groups are illustrated in Fig. 14 [9] and it can be seen that the Nickel-base alloy Alloy 80A Group 8 has the best resistance to stress relaxation followed by the 12CrMoV alloy Group 7 and Durehete 1055 material Group 6 with the CrMoV Group 2 and CrMo Group I materials being very much weaker. As a consequence the maximum operating temperatures for the lower grouped steels are restricted as indicated in Table 3, taken from the CEGB Generation Operation Memorandum GOM85 [10].

Table 1. *Commonly used ferritic alloy bolting steels (in non-UK specifications)*

Specn.	Grade	Typical Composition					Heat Treatment
		C	Cr	Mo	V	Others	°C
USA	B7	0.4	1	0.2	—	—	$T \geq 593$
A193-84a	B7M	0.4	1	0.2	—	—	$T \geq 620$
	B16	0.4	1	0.6	0.3	—	$T \geq 650$
Germany	21CrMoV57	0.2	1.2	0.7	0.3	—	930 OQ + 720
DIN 17240	40Cr1MoV47	0.4	1.2	0.6	0.3	—	930 OQ + 700
	X22Cr1MoV121	0.2	12	1	0.3	0.5 Ni	1050 AC + 680
France	25 CD 4	0.25	1	0.2	—	—	880 OQ + $\geq$ 600
NF A 35 558	20 CDV 5.07	0.2	1.3	0.7	0.25	—	930 OQ + $\geq$ 700
	Z20 CDNbV11	0.2	11	0.7	0.2	0.4 Nb	1120 AC + $\geq$ 670
Russia	20KhIIFITR	0.2	1.2	1	0.9	Ti, B	1000 Q + 700
Australia	Comsteel 029	0.2	1	1	0.6	Ti, B	990 OQ X 700

Table 2. *UK ferritic bolt steel designations*

SES Trade Name	CEGB Code	BS 1506: 1986 Type No.	Nominal Composition %				
			C	Cr	Mo	V	Others
Durehete 900	GP 1 (1Cr–Mo)	631–850	0.4	1	0.5	—	—
Durehete 950	GP 2 (CrMoV)	671–850	0.4	1	0.5	0.25	—
—	GP 3 (3CrMoV)	—	0.3	3	0.5	0.75	—
—	GP 4 (MoV)	—	0.2	—	0.5	0.25	—
Durehete 1050	GP 5 (1CrMoV)	—	0.2	1	1	0.75	—
Durehete 1055	GP 5 (1CrMoV)	861–820	0.2	1	1	0.75	Ti, B

Table 3. Basic material lives of common UK bolting materials

	Temperature					
	370–400°C	401–454°C	455–485°C	485–515°C	516–538°C	539–560°C
Durehete 1055	350 000	350 000	300 000	250 000	200 000	175 000
Durehete 950	350 000	300 000	200 000	150 000	—	—
Durehete 900	300 000	250 000	150 000	—	—	—
Esshete 1250	350 000	350 000	300 000	250 000	250 000	200 000
316	250 000	250 000	250 000	250 000	250 000	150 000
Nimonic	350 000	350 000	350 000	350 000	350 000	300 000

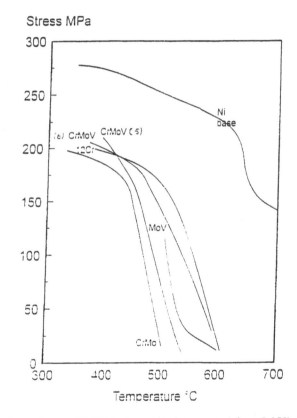

Fig. 14 Comparison of mean 30 000 h stress of bolting materials at 0.15% strain.

3.2. Service Experience

Statistical data on operating experience with high temperature bolting is difficult to find in the open literature because most utilities have kept such information confidential. Some information is available from the Central Electricity Generating Board and this indicates that a significant increase in bolt failure rates occurred in the UK during the 1960s as the steam temperatures and pressures increased and as more plant was brought into operation

[9]. This led to the reconstitution of the UK bolt collaborative committee with representatives from the CEGB and the major UK turbine manufacturers.

This committee concentrated on factors involved in bolt fracture since the CEGB operating experience indicated that steam leakage from joints was neither particularly prevalent nor as serious as bolt fracture, particularly as a leaking joint can be re-tightened without great inconvenience.

The committee found from surveys conducted into the bolt failure incidents that fractures were generally found during maintenance. While many incidents only involved a single bolt in a given joint, other joints contained several fractures indicating more than random cause. For all such incidents as much information as possible was collected on the location of cracking, joint design, operating and maintenance histories and bolt materials. The main causes identified were:

Material related — in that cracks were frequently found in materials with low creep rupture ductility or notch sensitive materials.

Design factors — the incidence of cracking was more prevalent in studs and invariably occurred in the first engaged thread particularly at the hotter end. The incidence of cracking was also affected by the actual bolt geometry (shape).

Tightening procedures — many of the cracked bolts had been subjected to uncontrolled tightening procedures and had probably been overstrained.

The committee concluded that the situation could be improved by tackling all three factors in the establishment of the CEGB Code of Practice for bolt replacement [11] which was the precursor to the present GOM 85 discussed in the papers by Bolton [6] and Jones [12]. The publication of this Code of Practice immediately resulted in a sigruficant reduction in the number of bolt failures on CEGB plant Fig. 15 [9]. However, as indicated in the following case studies, these same factors are still dominant in determining bolt integrity and service lives.

3.3. Case Study 3: Catastrophic Failure of Ferritic Steel Bolts

Figure 16 is a photograph of the damage caused by the failure of 24 Durehete 1055 stud bolts on a steam chest at a CEGB station in 1979. The incident was remarkable not only for the violence of the event, the chest cover and associated valve gear was blown 80 ft through the roof of the turbine hall, but also because of the large number of design parameters (faults) and adverse metallurgical parameters which contributed to the failure. At the time of the failure the unit had operated for 54 000 h and the stud bolts on the steam chest had been subjected to five tightening operations. According to GOM 85 the effective bolt lives were thus:

Hours in services plus Tightening Penalty = 54 000 h plus (5 × 15 000 h) = 129 000 h

In this case the Design Factor $KD = 1$ and the Strain Factor $= \frac{\varepsilon}{0.15} = 1$, since each re-tightening operation on all bolts had been performed (within the permitted tolerance) to

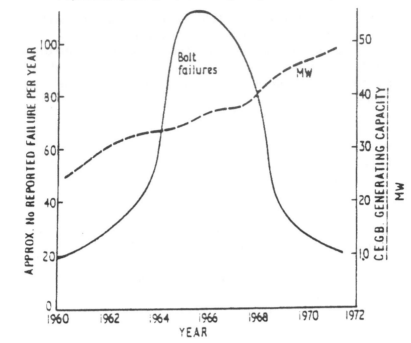

Fig. 15 *Interrelationship between bolt failures and CEGB generating capacity.*

Fig. 16 *Catastrophic failure of 24 low alloy steel bolts on a steam valve body (1979).*

0.15% strain. It should be noted that since the GOM 85 Basic Material Life for D1055 up to 570°C = 150 000 h the stud bolts in the chest were due for replacement at the next outage.

Visual examination of the 24 fractured stud ends retained in the chest suggested that the failures were initiated by creep cavitation and cracking in the region of the first engaged threads (Fig. 17). Most of the fractures were relatively flat and typical of bolt creep failures. However, six fractures exhibited areas of cup and cone shear failures but with some creep cracking at the perimeters. It was noted that the initiation site of failure in all studs occurred on the outside of the stud pitch circle. From examination of the studs in situ, it was clear that the final failure event was associated with fracture of only six bolts and that the remainder had in fact fractured well before then. Oxide thickness measurements made on the fracture surfaces allowed estimates of the crack ages. Assuming parabolic oxidation kinetics $\chi^2 = kt$, where χ is the oxide thickness in cm, t the time and k the parabolic rate content = 1.65×10^{-12} cm^2 s^{-1} [13], crack initiation times were determined on some studs to have occurred between 14 000–15 000 h prior to the final failure which compared well with the total operating hours (15, 143) since the last inspection. Complete fracture of these same studs was calculated to have occurred at over 1000 h prior to the final failure. In most cases the oxide was thicker (indicating older cracks) on the outer portion of the stud fracture surfaces.

Hardness measurements on the nut end of the studs indicated all but two had been within the required specification range 250–320 HV for D1055 but longitudinal hardness traverses indicated significant hardness gradients along the studs. In some cases the hardness dropped by 30–40 HV points from the nut end to the fractured end (Fig. 18). These observations were consistent with the studs having operated with a longitudinal temperature gradient, later determined in some cases to have been as high as 40–70°C.

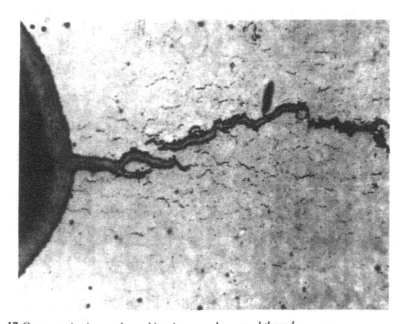

Fig. 17 Creep cavitation and cracking in second engaged thread.

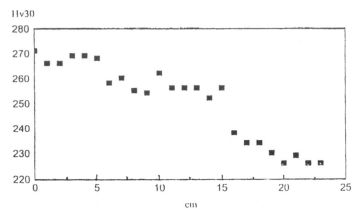

Fig. 18 *Hardness gradient along failed studs.*

Since fracture initiated in all studs on the outer stud pitch circle, it was clear that bending of the studs could have contributed significantly to the failure process. Bending of studs in joints of this type occurs when the valve cover operates at significantly lower temperatures than the valve body and hence sustains a smaller thermal expansion when at service temperatures. Measurements made on unfractured studs in a similar chest indicated permanent out of plane deflection of up to 0.075 in. suggesting that the creep strain due to bending could be as high as 0.12% per tightening and this of course would have been superimposed on the relaxed creep strain from the original tightening tensile strain of 0.15%.

On this basis of these observations, it was concluded that the operating condition which had contributed most significantly to this failure, and not accounted for in the original design, was the severe temperature gradient in the valve cover chest. This occurred primarily because of the difficulty (almost inability) to properly lag the valve cover due to interference by the valve gear mechanism. The presence of the temperature gradient contributed to failure in two significant ways:

(i) by the superimposition of a bending stress in addition to the original tensile stress thereby increasing (possibly doubling) the total elastic strain relaxed during each operating period, and

(ii) by concentrating the creep strain (acquired during relaxation of the studs in service) in the hottest region of the studs, the first engaged thread located just below the flange cover.

Although the lack of proper lagging and consequential temperature gradients in the valve cover were major contributing factors to this failure it was clear that other factors were also significant. The oxide dating of cracks had indicated that the studs had in fact failed in three different batches/phases. Detailed metallurgical investigations indicated that early cracking occurred in studs with coarse grained bainitic structures, of high hardness and significant residual element content. In contrast, the last studs to fail were fine grained, much lower hardness and had much lower residual element contents.

observations by Gooch and Kimmins [14] and King
h Laboratories demonstrated the effect of grain size on
dness on creep crack endurance (Fig. 20), and residual
Fig. 21).

ack

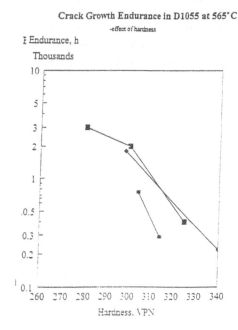

Fig. 20 Effect of hardness on creep crack
endurance.

crack

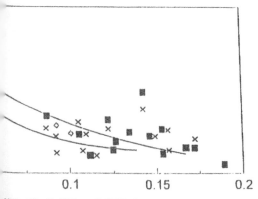

'R'=(P+2.43As+3.57Sn)

Ultimately, this failure incident and the subsequent investigations following it, led to significant improvements in the understanding of bolt performance, the need to control operating regimes more carefully and improvements in the metallurgical design and performance of D1055. Specific improvements arising from the incident were:

(1) Better control of lagging on valve chests and similar components.

(2) The introduction of a penalty within the GOM to allow for the effect of temperature gradients the issue of a new GOM, GOM 85B.

(3) Improvements to the metallurgical specification for D1055 with improvements in heat treatment to control grain size and hardness and of composition to lower the residual element content and reduce creep embrittlement. Following these developments significant improvements have been made in steel making practice by the use of secondary melting techniques such as VAR (Vacuum Arc Remelting) and VIM (Vacuum Induction Melting) for primary melting [16].

These techniques however only reduce the levels of residuals with high vapour pressures. To achieve an overall low residual element content an important stage of the steelmaking process is to control scrap selection [17].

3.4. Case Study 4: Lattice Shrinkage and Embrittlement in Alloy 80A

Alloy 80A (Ni–20Cr–2.4Ti–1.4Al) has been used extensively as a bolting material in the UK [18] in 100–660 MW turbines, and there is also significant experience in Germany in 220–860 MW sets and in France in 250–600 MW sets [19]. The applications for Alloy 80A have been for valve covers, steam strainers, cylinders, loop pipe flanges and nozzle plates in sizes up to 1000 mm long and 115 mm dia. Steam pressures have been up to 165 bar and the bolts have operated at temperatures in the range 450–550°C.

Altogether there have been over 14 000 Alloy 80A bolts used in the UK and some 6 500 used elsewhere and operating times now exceeding 175 000 h. In general the service experience has been excellent with 74 reported failures in the UK and less than 0.4% of failures in total [18, 19]. Of these approximately 50% have been attributed to intergranular fast fracture after in-service embrittlement, 33% to stress corrosion cracking and the remainder to creep attributed to over-tightening and high strain fatigue in one specific location [19]. Apart from the early creep failures, it was generally found that the failed bolts had actually shortened in length during service and the material had apparently embrittled exhibiting low impact energies. A further significant aspect of the failures in the UK was that most of these occurred on CEGB oil fired plant with an operating temperature of 540°C or in joints on coal fired stations which were operating below this temperature.

A relevant factor to these observations is that Alloy 80A undergoes ordering reactions, short range order (SRO) of Ni and Cr atoms at temperatures below 550°C and at temperatures below a critical temperature $T_c = 525$–530°C, the SRO may transform to long range order (LRO) with the formation of Ni_2Cr. The important aspect about these transformations is that they give rise to lattice contraction, 0.03% for SRO [14] and ~0.11% for LRO after 30 000 h at 450°C [20]. These reactions, particularly LRO, are extremely sluggish with an incubation period of > 10 000 h and may take upto 30 000 h for completion.

The significance of these reactions to the use of Alloy 80A as a bolting material, is the way in which they influence the stress relaxation properties of the material. Thus at temperatures above 550°C Alloy 80A displays normal stress relaxation behaviour as the initial elastic strain, 0.15%, is converted to plastic strain due to creep [21] (Fig. 22). At lower temperatures, in the ordering range, there are two competing processes: (a) lattice contraction due to ordering which gives an increase in stress during a 'stress relaxation' test and (b) a decrease in stress due to creep, giving rise to the complex behaviour shown in Fig. 22 at 500 and 550°C. At 450°C the ordering reactions dominate giving rise to the large pick-up in stress shown. These observations have been extensively studied by Nath *et al.* [18–22]. It should be noted, however, that the actual increase in stress realised in service may be made less than that shown in Fig. 22 due to creep of the flange material as observed by Mayer and Konig in model bolt tests [19, 23]. Nevertheless, it is considered that a contributory factor to the Alloy 80A bolt failures in service is an increase in stress on the bolts due to lattice contraction arising from atomic ordering.

In addition to lattice contraction the other common feature to most Alloy 80A bolt failures has been embrittlement. It is considered [18] that the combination of these two phenomena is largely responsible for the failures due to intergranular fast fracture and that embrittlement also increases the susceptibility to stress corrosion cracking. Again, work

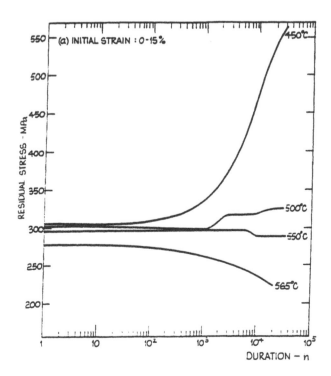

Fig. 22 *Stress relaxation in Alloy 80A.*

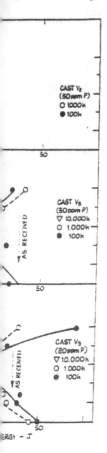

CAST V₂
(50 ppm P)
O 1000h
● 100h

CAST V₈
(30 ppm P)
▽ 10.000h
O 1.000h
● 100h

AS RECEIVED

CAST V₃
(20 ppm P)
▽ 10.000h
O 1.000h
● 100h

AS RECEIVED

50

ENERGY - J

Fig. 23 *Variation of impact energy with ageing — Alloy 80A.*

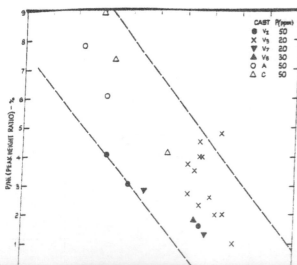

CAST	P(ppm)
● V₁	50
✕ V₃	20
▼ V₇	20
▲ V₈	30
O A	50
△ c	50

P/HI (PEAK HEIGHT RATIO) - %

by Nath *et al.* [18] has found Charpy impact values in failed ex-service bolts in the range 5–17 J whereas in unfailed bolts the impact values were in the range 25–48 J (Fig. 23). Fracture surfaces in the former displayed brittle intergranular cleavage with no evidence of ductile micro void formation on boundary carbides. Materials with higher impact values displayed transgranular/ductile intergranular fracture with micro voids associated with grain boundaries. Ageing studies on the more brittle materials demonstrated that the embrittlement phenomenon was reversible and Auger Electron Spectroscopy revealed phosphorous segregated to grain boundaries. In new casts of Alloy 80A the kinetics of embrittlement decreased significantly with decreasing bulk phosphorus content over the range 50–20 ppm P (Fig. 24) [18].

A variety of laboratory investigations [24–26] has now established that Alloy 80A is susceptible to stress corrosion cracking in certain environments. These include sulphuric and hydrochloric acid solutions and water with circulating SO_2 but seemingly not in non acidic chloride (with or without circulating SO_2) nor concentrated sodium hydroxide solution.

Failures in service attributed to SCC have invariably sulphur contamination as a common feature on fracture surfaces and particularly those which had been lubricated with molybdenum disulphide, which laboratory tests now indicate promotes the formation of acidic environments. Evidence is also available to indicate that embrittlement in service due to P segregation also increases the susceptibility to SCC in this material.

4. Conclusions

The general experience with high temperature bolting has been good in that most bolts have survived the required operating regimes. Those failures which have occurred can often be attributed to several quite different factors involving design and metallurgical aspects and operational or maintenance procedures.

The metallurgical aspects which increase the susceptibility to bolt failures are those which promote low ductility and notch sensitivity, i.e. in-service embrittlement due to the build-up of creep damage, temper-embrittlement by phosphorus and sulphur and embrittlement by residual element deposition on grain boundaries. Other aspects of metallurgical importance are short and long range ordering phenomena which lead to bolt contraction in nickel base alloys or oxidation effects increasing the loading on the bolts.

Important design aspects are the existence of temperature gradients which can cause strain accumulation in specific regions of the bolts and additional loads due to bending. Bolt geometry and differential thermal expansion between bolt and flange materials are also important design factors.

Operational/maintenance factors affecting bolt integrity include the number of re-tightening events, uncontrolled tightenings, service time and temperature.

Bolt integrity is thus assured provided the design, metallurgical and maintenance engineers are fully aware of these complex interactions.

5. Acknowledgements

This paper is published by permission of ERA Technology Ltd. The Author wishes to thank those who assisted in the preparation of the paper including Dr G. Breitbach of KFA, Dr B. Nath of National Power and Dr I. R. McLauchin of Nuclear Electric.

References

1. H. Schuster and F. Schubert, Contribution to Nuclear Corrosion Working Party, Stockholm, Helsinki, 14–16 June, 1989

2. D. B. Meadowcroft, Corrosion control in power plants. In *Characterisation of High Temperature Materials 6 — Surface Stability* (T. N. Rhys-Jones, Ed.) Institute of Metals, London (1989).

3. I. R. McLauchlin, Effect of oxidation on integrity of bolted structures, *Met. Sci. Technol.*, 1990, **6**, 39.

4. I. R. McLauchlin, J. D. Morgan, L.H. Watson and M. R. Wooton, Methods of assessing the effects of interface oxide growth in Magnox and AGR plant, *Nucl. Energy*, 1986, **25**, (5), 287–292.

5. P. C. Rowlands, J. C. P. Garrett, L. A. Popple, A. Whittaker and A. Hoaksey, The oxidation performance of Magnox and AGR reactor steels in high pressure CO_2, *Nucl. Energy*, 1986, **25**, (4), 67–275.

6. J. Bolton, Keynote paper: Design and materials requirements for high temperature bolting, this volume, pp.1–14.

7. J. A. White, ERA Confidential Report, 1983.

8. H. Everson, J. Orr and D. Dulieu, Low alloy ferritic bolting steels for steam turbine applications: the evolution of the Durehete steels. *EPRI-ASM Int. Conf. on Advances in Material Technology for Fossil Power Plants*, Chicago, IL, USA, September 1987.

9. G. D. Branch, J. H. M. Draper, N. W. Hodges, J. B. Marriott, M. C. Murphy, A. I. Smith and L. H. Toft, High temperature bolts for steam power plant, *Int. Conf. on Creep and Fatigue in Elevated Temperature Applications*, Sheffield, UK and Philadelphia, PA, USA. Inst. Mech. Engineers, London, Publication B, 1973, pp 1921–1929.

10. CEGB GOM 85B 1986. The selection, care and replacement of threaded fasteners operating at temperatures above 370°C.

11. CEGB Private Document. High temperature studs and bolts — Part 1: Code of Practice for Replacement Procedure. Part 2: — Code of Practice for Controlled Tightening, Dec 1968.

12. G. T. Jones, Review of high temperature bolt tightening removal and replacement procedures. Keynote Paper, this volume, pp.331–342.

13. L. W. Pinder, *Corros. Sci.*, 1981, **21**, 749–763.

14. S. T. Kimmins and D. J. Gooch, *Met. Sci.*, 1993, **17**, 519.

15. B. L. King, *Phil Trans. R. Soc., Lond.*, 1980, **A295**, 235.

16. A. Strang and S. R. Holdsworth, *Int. Conf. on Rupture Ductility of Creep Resistant Steels*, Inst. of Metals, York, 1990, p.224.

17. H. Everson, J. Orr, D. Dulieu and D. Burton, *Int. Conf. on Rupture Ductility of Creep Resistant Steels*, Inst. of Metals, York, 1990, p.235.

18. S. M. Beech, S. R. Holdsworth, H. G. Mellor, D. A. Miller and B. Nath, An assessment of alloy 80A as a high temperature bolting material for advanced steam conditions: *EPRI-ASM Conf. on Advances in Material Technology for Fossil Power Plants*, Chicago, IL, USA, 1987.

19. K. H. Mayer and H. König, High temperature bolting of steam turbines for improved coal fired power plants, *2nd Int. Conf. on Improved Coal Fired Power Plants*, Palo Alto, CA, USA, Nov. 1988.

20. E. Metcalfe, B. Nath and A. Wickens, *Mat Sci. Eng.*, 1984, **67**, 157–162.

21. E. Metcalfe and B. Nath, *Int. Conf on Phase Transformations*, York, UK. Publ. Inst. Metallurgists, London, 1979.

22. A. Marucco, E. Metcalfe and B. Nath, *Int Conf. on Solid–Solid Phase Transformations*, Pittsburgh, PA, USA, Metall. Soc. AIME, 1981.

23. K. H. Mayer and H. König, *Int. Conf. on Advances in Material Technology for Fossil Power Plants*, Chicago, IL, USA, 1987, ASM/EPRI.

24. K. H. Mayer and K. H. Keienburg, *Inst. Mech. Eng. Creep Conf.*, Sheffield, UK, 1980.

25. J. Parker and T. E. Parsons, Private communications, CEGB (OED), Bedminster Down, UK.

26. D. A. Miller, L. Miles and J. A. Roscow, Private Communications, CEGB (OED), Bedminster Down, UK.

SESSION 2

Materials Testing and Data Analysis

Chairman: KH Mayer
MAN Energie, Nürnberg, Germany

3

Uniaxial Stress Relaxation Testing — a UK Perspective

P. R. McCARTHY and A. STRANG*

ERA Technology Ltd., Leatherhead, Surrey, UK
*GEC ALSTHOM Large Steam Turbines, Rugby, Warwickshire, UK

1. Introduction

Stress relaxation testing has a long history in the UK, the first tests being performed in the early 1900s. The first part of this paper reviews the developments of the past 90 years, both in terms of technique development and the refinement of testing machines during this period.

The second half of the paper explores the practical realisation of long term (>10 000 h) stress relaxation testing. The accuracy and stability needs of the measurement and loading systems are addressed, as are the testpiece specific parameters which greatly influence the success, or otherwise, of the test itself.

2. The History of Stress Relaxation Testing

The development of stress relaxation testing over the past 90 years is examined below, focusing upon how the technique has evolved over this period. Arbitrary ten year segments have been chosen within this period, the key advances from each decade being identified.

2.1. The 1900s

The earliest reference to stress relaxation testing which the authors have discovered is a paper published by Troughton & Rankine [1] in the early years of the twentieth century. This reported creep and stress relaxation of lead wires one metre in length at ambient temperature. The load was applied using a tank of water, elongation was monitored using an optical telescope and load adjustment effected via a siphon (Fig. 1). The siphoned water was weighed and thus load–time data were derived. The design of the experiment was both simple and elegant, the data generated subsequently being fitted to an equation of the form:

$$W = a\text{-}k \log(t + 1)$$

where W is the load at time t, a is the initial load and k is a constant.

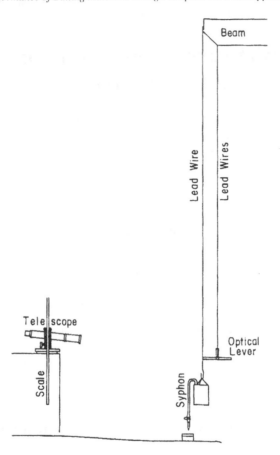

Fig. 1 *Troughton & Rankine's stress relaxation test configuration [1].*

Examples of the stress relaxation curves generated by Troughton & Rankine are presented as Fig. 2, complete with their curve fit equations.

Also included in this paper was a schematic diagram of an automated stress relaxation test configuration, this is reproduced as Fig. 3. In this machine the testpiece A is loaded by a tank of water which, as it stretches, makes an electrical contact at point E. This completes an electrical circuit which, via relays R_1 and R_2, opens a solenoid valve at K. Water is thus siphoned off until the gauge length recovers to the original length and the circuit is broken. The siphoned off water is collected in tank T which is supported by a spring. Movement of the tank due to ingress of water causes pointer V to move about pivot U, resulting in the generation of a load vs time trace being recorded on the motorised drum W. This system can therefore be viewed as the first automated stress relaxation test machine.

2.2. The 1920s

In the 1920s the turbine makers were assessing the suitability of materials for bolting service at elevated temperatures. Bailey, of the Metropolitan-Vickers Electrical Company, was one

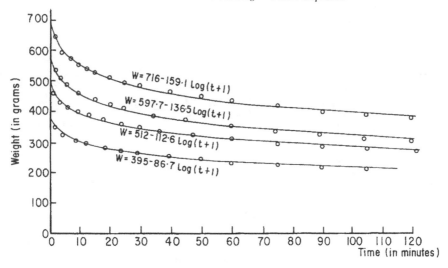

Fig. 2 *Data as generated by Troughton and Rankine [1].*

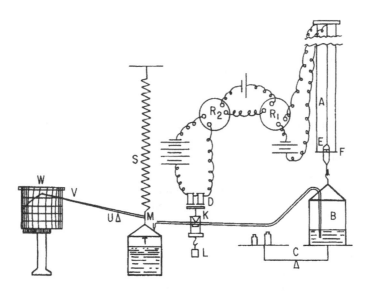

Fig. 3 *Schematic for an automated stress relaxation system by Troughton and Rankine [1].*

of those active in this field and reported his findings [2] on ranking tests for bolt materials. The procedure involved subjecting a double notched testpiece (Fig. 4) to temperature and load for given durations. After this exposure the threaded areas were machined away, leaving a double notched impact testpiece. The testpiece was impact tested at ambient temperature, with one groove level with the anvil of the Izod impact machine. This test generated an impact energy by which the material could be assessed; the unbroken notch was available for subsequent metallographic examination.

The results from a series of tests were used to establish an impact energy vs time under

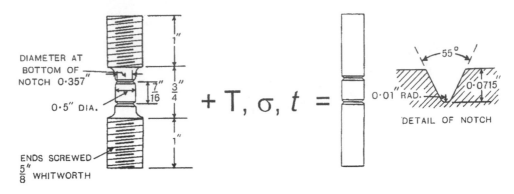

Fig. 4 *Testpiece configuration as used by Bailey [2].*

stress relationship for each material; this behaviour was subsequently used to rank materials. This analysis, coupled with service experience, was used to eliminate those materials unsuited to service as bolts.

2.3. The 1930s

During this period attention was firmly focused upon the science and engineering of flanged joints, the UK collaborative effort being co-ordinated by the Flange Bolting Working Party, under the aegis of the Institution of Mechanical Engineers. Their remit covered:

(a) the testing of flanged joints with various types of seal;
(b) the influence of bolt pitch upon flange performance under internal pressure;
(c) the behaviour of packing materials;
(d) creep tests on bolt materials;
(e) notched bar impact tests;
(f) full scale bolted flange joints subjected to temperature and pressure;
(g) retightening of flange joints and bolt lifing; and
(h) creep relaxation tests on model flanges.

At this time stress relaxation behaviour was assessed by examining the behaviour of the flanged joint/bolts/seals as a complete system, mimicking as closely as possible the real life behaviour of jointed assemblies. In order to do this a special loading apparatus was developed by NPL. with extremely high stiffness which, in conjunction with a furnace and loading system, enabled one quarter size scale model flanges to be tested (Figs. 5, 6). The mode of operation was to mount flanges in the test rig and load them hydraulically. The relative movement of the flanges, due to relaxation, was transmitted to mirror extensometers and recorded, the system having a claimed resolution of 3 micro-inches. The extensometer was also used to drive an hydraulic screw pump, thereby adjusting the hydraulic pressure and hence the load on the flanges (Fig. 6). This could be construed as the first servo-hydraulic, closed loop control, test machine. For further details the relevant reference [3] should be addressed.

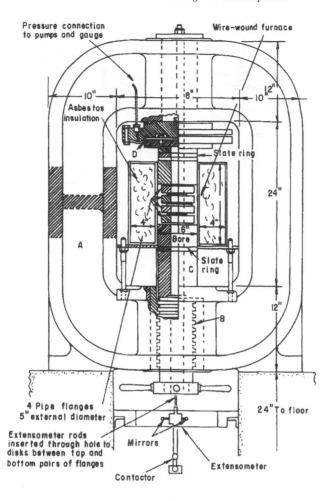

Fig. 5 Apparatus for creep relaxation tests on model flanges [3].

2.4. The 1940s

In the second half of this decade attention focused upon the examination of stress relaxation testing from an engineering standpoint, led by Herbert and others at NPL. This was facilitated by developments in test equipment undertaken by Tapsell *et al.*, who addressed both test machine and extensometry requirements.

The great advance in test machine technology at this time was the development of an underslung double lever creep machine with an adjustable poise weight [4]. This design was subsequently developed by Denison to become the ubiquitous T45 machine, the workhorse of those involved in stress relaxation testing in subsequent decades. A key factor of their design was the easy adjustment of the test load which, coupled with the wide load range (0.05–5 tons), made these machines ideal for stress relaxation testing. The availability of these machines, coupled with the tried and tested Martens Mirror extensometer [5], enabled long term (>10 000 h) stress relaxation data generation to become feasible.

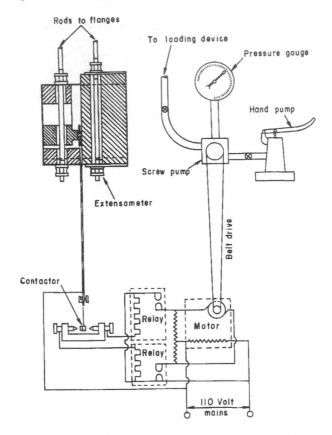

Fig. 6 *Automatic pressure control system for creep relaxation apparatus [3].*

2.5. The 1950s

Throughout the 1950s the turbine makers were utilising higher steam temperatures and pressures to increase power output and efficiency. The advent of 1% CrMoV steels for rotors meant that the existing bolting materials were entering service regimes where creep became a design consideration. In the UK this led to the development of Durehete 900, 950, 1050 and 1055, the numbers signifying the service temperature in °F. To match these materials developments there was a corresponding need for stress relaxation data to form part of the design database; this resulted in the commencement of large data generation programmes by the UK bolting material producers and users. The extent of this data was controlled by the utility inspection interval of 25 000 h, hence tests to 30 000 h became the norm.

During this period significant effort was expended to develop automated load adjustment systems for stress relaxation testing. Two principal approaches were used, the first, developed by Herbert and Armstrong [6], involving the use of a modified Martens type extensometer, the second by Draper [8] used a mirror extensometer coupled with a pair of photo-cells. The Herbert and Armstrong system (Fig. 7) operated via an extension arm to the

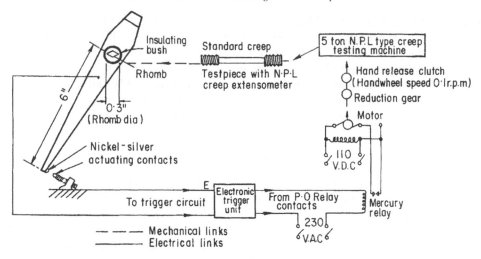

Fig. 7 *Automatic strain control system [7].*

Martens type extensometer, as the testpiece stretched a contact closed. This contact triggered a motor control system which adjusted load by moving the poise weight of a Denison T45 machine, thereby maintaining constant strain. The Draper system (Fig. 8) utilised a light source, the mirrors on the Martens extensometer and a pair of photo-cells. As the testpiece elongated the light beam was deflected, causing the beam to move from its null position and fall onto one of the photo-cells. This triggered the motor control which, as in the Herbert and Armstrong system, caused load to be adjusted and the strain to be maintained at the required level.

Both systems were reported to work well; however to obtain best performance they required careful setting up. In addition the gearing of the load adjustment motor had to be such

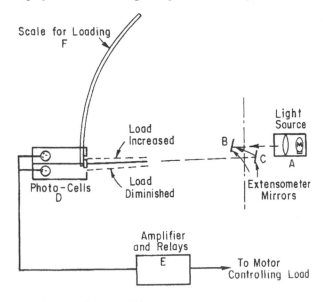

Fig. 8 *Automatic strain control system [8].*

that it would not enter into load hysteresis due to motor over-run upon removal of the drive signal; Herbert & Armstrong overcame this problem by the use of a d.c. shunt motor.

During this period other types of testing were still undertaken to assess suitability for bolt service, i.e. plain and notched stress rupture, impact, etc., however the prime requirement of designers was the stress relaxation behaviour of the materials. Other effects, such as re-loading or re-tightening were examined by workers such as Draper [6]; the resulting data was used to generate guidelines for bolt re-use after inspections.

2.6. The 1960s

The work commenced in the 1950s continued into the 1960s, driven by continuing bolt design development to address the needs of larger machines and higher inlet temperatures. The bolting material producers continued to 'fine tune' their materials, the designers required corresponding data for design. This need was addressed by the material producers, turbine makers and the National Engineering Laboratory at East Kilbride in a collaborative data collection exercise. A consequence of this activity was the need for a testing standard to specifically address stress relaxation. This led to the production of BS 3500: Part 6: 1969 [9], paralleling similar activity by ASTM [10].

2.7. The 1970s

The major advances in stress relaxation testing during the 1970s were associated with instrumentation, in particular the development of high stability electronic systems for strain measurement. The availability of conditioning units for Linear Variable Displacement Transducers (LVDTs) and capacitance transducers opened the doors to computerised strain monitoring and the replacement of mirror based optical extensometers. The development of high stability conditioning units enabled a strain of 0.15% on a 100 mm gauge length to be controlled to the ±1% level demanded by BS 3500: Part 6: 1969 over long durations. Such units also contributed to the development of closed loop servo machines for stress relaxation testing.

2.8.   The 1980s to the Current Day

Developments during the 1980s were focused upon extension of inspection intervals to 50 000 h, precipitating an examination of data extrapolation methods and a reassessment of the existing database. Testing continued to generate 30 000 h data on materials produced using 'cleaner' production methods, with the testing technology continued to evolve in parallel. Yet further improvements in electronics arose, providing greater long term stability and reliability. Such developments enabled automated stress relaxation machines to be developed by several laboratories in the UK, which are in use at the current time.

3. The Practicalities of Stress Relaxation Testing

Having examined the historical development of stress relaxation testing, it is appropriate to review the achievement of good testing practice; in this context the comments herein are limited to long term (>10 000 h) testing.

In order to perform high quality stress relaxation tests the following aspects require detailed attention:

(i)   definition of the test requirements;
(ii) testpiece design;
(iii) test machine performance;
(iv) extensometry, and
(v)   data requirements.

These five areas, and their interactions, are reviewed below.

3.1. Test Requirements

The test requirements are defined in the British Standard BS 3500: Part 6: 1969, which addresses stress relaxation testing. It is currently under review by a BSI working party, whose brief is to incorporate the developments of the past 25 years into the current document, possibly expanding it to incorporate more recent techniques such as re-tightening simulation and the re-start of tests.

The three most important parameters in any stress relaxation test are the required strain, the test temperature and load.

(a) *Strain*
The strain must be controlled to ±1% of the specified level; for a typical test this would be 0.15%. Typical testpieces have gauge lengths of ~100 mm, hence this must be controlled to ±1.5 μm over very long periods, an exacting requirement.

(b) *Temperature*
Test temperature needs to be maintained at a constant level and it is important to have as uniform a temperature distribution along the testpiece as possible. This is necessary as relaxation rates are very temperature sensitive. Hence thermal gradients will result in a relaxation rate profile along the gauge length; this is to be avoided as the resulting data will be of a poor quality.

(c) *Load*
Test load must be continuously adjustable, hence those active in stress relaxation testing of bolting materials in the UK use the Denison T45 machines. It is possible to use servo-hydraulic or servo-electric machines to perform such work, however the economics of operation favour the 'gravity driven' equipment.

3.2. Testpiece Design

Stress relaxation is one of a small, select group of tests where the size of the available material does not constitute the sole arbiter in testpiece design. Indeed, with stress relaxation the converse is true: the discrimination of the transducery and the required number of 'digits' which constitute the strain control band for the test combine to define the required gauge length.

Testpiece design also needs to address extensometry attachment as well as gripping. The former is usually effected via ridges on the testpiece, by pointed screws in a triform arrangement or via clamps at the testpiece shoulders; the latter requiring an effective gauge length. Load is transmitted to the testpiece via threaded ends, the testpiece being linked to the machine loading bars using threaded couplings.

3.3. Test Machine

The test machine can be of the servo or lever type. The machine needs to be rigid, well aligned and have a very stable load application system. Servos achieve this via their closed loop control systems, lever machines via gravity.

Automatic load adjustment is provided on servo machines from the extensometry control loop. Most lever machines are manually adjusted, however some have been modified to automatically adjust the load. It is important that any load control system is not affected by hysteresis within the strain control system. It can arise where the control limits are too tight, leading to motor over-run and hunting between limits.

3.4. Extensometry

The efficacy of the extensometry is intimately related to the performance characteristics of the transducery, which translates the physical displacement on the testpiece into a control voltage. Ideally the transducers should be unaffected by changes in temperature, humidity and voltage fluctuations over long periods, i.e. up to 4 years. They should have frictionless bearings, be immune to electromagnetic fields created by the furnace and have excellent discrimination, linearity and repeatability.

The extensometer, to which the transducers are attached, needs to be made from suitable materials; guidance is available [11]. The extensometer design needs to address both the attachment to the testpiece as well as the mounting method for the transducers. All extensometers should be of the 'double sided' type, which enable bending of the testpiece to be assessed.

Finally, the locality of the extensometer needs to be shielded from draughts, it is also necessary to control the ambient temperature in order to minimise temperature generated scatter.

3.5. Data Requirements

The data requirements are clearly set out in the BS 3500: Part 6: 1969. It is important that not only the stress–time data pairs be recorded but also all of the pertinent machinery details, including the calibration traceability. This is particularly so when data is to be exchanged between laboratories for analysis.

4. Concluding Remarks

Stress relaxation testing has a long history which includes a continuing trend of innovation and incorporation of advances in relevant technologies. The prime aim of all laboratories active in this field is to generate high quality data for use by designers and engineers; by focusing upon the detailed practicalities of stress relaxation testing this will continue to be achieved.

5. Acknowledgements

The permission of the Directors of ERA Technology and GEC ALSTHOM Large Steam Turbines in allowing the publication of this paper is gratefully acknowledged. The assistance of the Information Centres of both organisations in obtaining the historical references is appreciated, as is the kind help of Mr Maurice Day in providing information about the post-war activities of NPL in this field of testing.

References

1. A.Troughton and O. Rankine, 'On the stretching and torsion of lead wire beyond the elastic limit', *Phil. Mag.*, 1904, **8**, pp. 538–556.

2. R. W. Bailey, 'The mechanical testing of materials', *Proc. Inst. Mech. Eng.*, May 1928, pp. 417-452.

3. H. J. Tapsell, 'Second Report of the Pipe Flanges Research Committee', *Proc. I. Mech. E.*, 1939, **141**, (5), pp. 433-471.

4. R. A. Beaumont, *Mechanical Testing of Metallic Materials*, Pitman, pp. 146–149.

5. *ibid.*, pp. 151–156.

6. J. H. M. Draper, 'Relaxation testing of steam turbine bolt materials with simulated re-tightening', *Iron & Steel*, December 1961, pp. 622–625.

7. D. C. Herbert and D. J. Armstrong, 'The development of a strain or displacement actuated electronic trigger for high temperature stress relaxation testing and other purposes', *Metallurgia*, 1952, **45**, 267–270.

8. J. H. M. Draper, 'Creep relaxation testing: tests at constant strain and decreasing load', *Engineering*, 1955, **179**, 564–565.

9. BS 3500: Part 6: 1969, 'Methods for creep and rupture testing of metals. Part 6. Tensile stress relaxation testing', British Standards Institution, 16pp.

10. ASTM E328-67T, 'Standard methods for stress relaxation testing for materials', ASTM, Philadelphia, PA, USA.

11. M. S. Loveday and T. B. Gibbons, 'A code of practice for the use of Ni–Cr base alloy extensometers for measurement of creep strain', ISBN 0 946754 08X, National Physical Laboratory.

4

Testing of the Relaxation Strength of Bolted Joints

H. KÖNIG and K. H. MAYER

MAN Energie, Nuremberg, Germany

ABSTRACT

For about 50 years the results of relaxation tests on models of heat-resistant bolted joints have been used for sizing the heat-resistant bolted joints of steam turbines. Whereas the uniaxial relaxation test in the testing machine determines the behaviour of the bolt material, the testing of bolted joint models provides close-to-practice sizing parameters at low cost. The results not only encompass the influence of the bolt, flange and nut materials but also the geometrical influence, such as the local deformation on force transmission through the thread and at the nut-to-flange contact faces as well as global deformation of the flange. Equally close-to-practice is the type of preloading at room temperature with subsequent thermal loading to test or operating temperature. Relaxation tests on bolted joint models are of particular significance when, for instance, very high temperature-resistant bolt materials, such as Alloy 80A, are combined with ferritic flange materials. The results then also yield details on the influence of the negative creep effect of Alloy 80A and the reduction of pre-load due to the different thermal expansion coefficients of the bolt and flange materials, which can, practically, be further reduced by an austenitic sleeve.

The report provides in-depth details on test methods and on a large number of long-term tests which have been carried out by the authors and by joint research work over the past 30 years on homogeneous and non-homogeneous bolted joints.

1. Introduction

The sizing of bolted joints at elevated temperatures in steam turbines have for about 50 years relied on the results of relaxation tests made with bolted joint models [1–8, 11, 12] (Fig. 1). Whereas the relaxation tests in the relaxation test machine is used to determine the relaxation behaviour of the bolt material, the bolted joint model supplies results for a close-to-practice determination of design parameters, i.e. it is used to determine system properties. For this reason, it is obviously desirable to construct the models analogously to the feasible design and to test them under close-topractice conditions.

Close-to-practice design means: similar material combinations with specimens from real components and bar materials with real cross-sectional dimensions.
Close-to-practice testing means: similar temperature and pre-stress ranges.
Close-to-practice results means: taking into consideration the effects of the bolt, flange and nut materials, local deformations at the force application faces in the thread and at the nut-to-flange contact faces and global deformations of the bolt and flange.

1937	Robinson *et al.*	General Electric
1942	Wellinger, Keil	MPA Stuttgart
1965	Weigand, Beelich	IfW Darmstadt
1966	Erker, Klotzbücher, Mayer	MAN Nuremberg
1978	**EGKS-Report**	
	Jakobeit	ABB Mannheim
	Keienburg	Siemens Mülheim
	Mayer	MAN Energie Nuremberg
	Gulden	Krupp Siegen
	Schmidt, Huchtemann	Thyssen Krefeld
1992	**VDI-Report**	
	Löser	IfW Darmstadt
1994	**AIF/VGB-Report**	
	Maile, Purper, Hänßel	MPA Stuttgart

Fig. 1 *History of bolt model testing.*

2. Relaxation Testing Procedures

The following procedures have found acceptance in determining the relaxation behaviour of bolt materials and bolted joints (Fig. 2).

2.1. The Isothermal Relaxation Test

This is used exclusively to determine the isothermal relaxation behaviour of the bolt material (Fig. 2, top). Procedure: Heating to test temperature, then application of total strain. Continuous measurement of residual stresses [9, 15, 17].

2.2 Relaxation Test with Bolts Including Nuts in the Relaxation Machine

Determination of the relaxation behaviour of the bolt/nut combination, i.e. procedure same as in the case of the isothermal relaxation test. In addition to the relaxation of the bolt material, this test also determines set effects in the threads [5].

2.3. The Anisothermal Relaxation Test

This is used to determine the anisothermal relaxation behaviour of the bolt material. Procedure: Application of total strain at room temperature (RT), heating to test temperature. Continuous measurement of residual stresses starting from RT [14].

2.4. Relaxation Test with Bolted Joint Models

These are designed to determine the relaxation behaviour of the bolted joint, i.e. the bolt, nut and flange system, including expansion sleeves of studs screwed into the casing lower half with bolted joints made of similar and dissimilar materials [6, 8, 9].

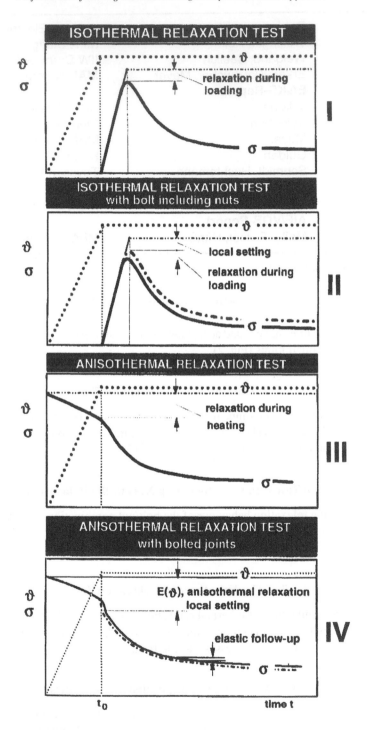

Fig. 2.

3. Bolted Joint Models
3.1. Bolted Joint Models of Similar Materials

3.1.1. Size of models

Tests for the low-cost determination of the long-term relaxation behaviour of bolt materials rely on models whose individual components consist of the bolt material. Moreover, these models feature a high flange-to-bolt area ratio. In the case of the models used in Germany, this ratio is between 8 and 9 (Fig. 3) [20].

3.1.2. Testing procedure

3.1.2.1. Pre-loading. The pre-load is applied free of torsion effects in a fixture by tightening up a nut. To reduce the setting effects, pre-loading is carried out twice at 20% increased pre-load before the desired pre-load is reached during a third tightening operation. It will be recalled that real bolted joints are also subjected to a 20% increased preload during the internal pressure test of the casings.

Model I   **Model II**

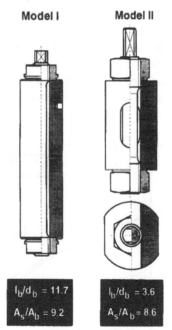

| $l_b/d_b = 11.7$ | $l_b/d_b = 3.6$ |
| $A_s/A_b = 9.2$ | $A_s/A_b = 8.6$ |

- **Determination of elastic strain of the bolt:**

    Model I : by Δl of bolt. Reference length = 153.3 mm
    Model II : by two strain gauges $\underline{\varepsilon_{el}}$

- **Calculation of elastic stress according:**

$$\sigma_{el} = \varepsilon_{el} \cdot E_{stat}$$

Fig. 3.

3.1.2.2. Long-term exposure at elevated temperatures. Following the application of the pre-load, the models are heated in an forced-air-draft furnace at 100 °C h⁻¹ to the test tempera-ture and removed after different times from the cooled-down furnace. The accuracy of the temperature measurement is within ±3 °C.

3.1.2.3. Unloading. After applying the strain gauges through the lateral openings in the case of model II or, respectively, after measuring the total length of the bolt prior to unloading in the case of model I, unloading of the bolt is effected by mechanically destroying a nut. On model II, this is done by drilling three holes and rupturing the nut by means of a screw driver and on model I by cutting two grooves on opposite sides [3, 11, 12].

3.1.3. Determination of relaxation strength

The elastic residual strain in the bolt shaft is determined.

3.1.3.1. Model I. In the case of model I, the change in bolt length is measured in a measuring device with a dial indicator before and after applying the load and, respectively, before and after unloading. As mentioned earlier, it is necessary for determining the strain by calcula-tion to calibrate the model once with the aid of strain gauges applied to the bolt shaft.

3.1.3.2. Model II. In the case of model II, this is effected directly by means of two strain gauges applied 180° apart. For this purpose, two lateral openings are required in the sleeve.

The elastic stresses in the bolt shaft are then calculated by multiplication with the static modulus of elasticity. For a definition of the static modulus of elasticity (see Fig. 4). Gener-ally, this is determined in the relaxation or tensile testing machine (loading rate about 1–3 N mm² s⁻¹) or taken from existing standards. The differences compared to the dynamic modu-

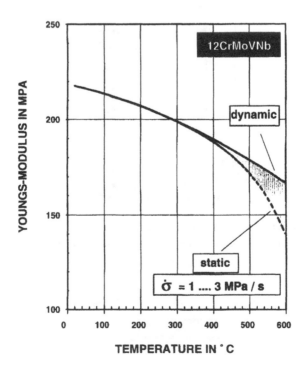

Fig. 4.

lus of elasticity tend to increase as the temperature rises, the dynamic modulus of elasticity being invariably greater than the static modulus.

3.1.4. Number of models

Each model provides one test point per test period. For this reason, about 5 models are used per test temperature for long-term tests (e.g. 1000, 3000, 10 000, 20 000 and 30 000 h). The use of several specimens offers an advantage in that major inhomogeneities in the material, if any, can be detected from the trend of the curves.

3.2. Bolted Joint Models of Dissimilar Materials

3.2.1. Close-to-practice model test

The models to determine the close-to-practice long-term relaxation behaviour are constructed close to design, i.e:

* similar cross-sectional and length ratios $A_s/A_b = 3.5$;
* similar physical design of the models, i.e. for instance with expansion sleeve or for instance casing bottom half (screw-in-piece) instead of second nut;
* similar material pairings;
* similar pre-load at temperature — similar temperature range.

A few of the model variants used for research projects are shown in Fig. 5 [13].

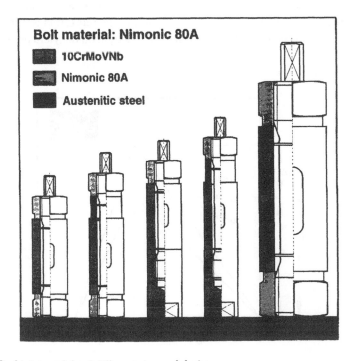

Fig. 5 Bolted joint models of different size and design.

3.2.2. Determination of relaxation strength

Measurements are made analogously to those on models I and II. In calculating the pre-stress or residual stress at T from the strains measured at RT, the influence of the different moduli of elasticity, cross-sectional areas and thermal expansion coefficients has to be taken into account [9–11].

The influence that can be subdivided into a thermal element and a mechanical element can be seen from Fig. 6. The pre-loading graph included in the illustration shows that, depending on the selection of the materials, there may be a decrease or increase in pre-load as the temperature rises, depending on thermal expansion. This, in turn, has to be taken into consideration in selecting the level of the pre-load at RT.

4. Mechanisms in Bolted Joints
4.1. Bolted Joints of Similar Materials

In bolted joints using similar materials for the bolt, flange and nuts, the pre-load is applied at room temperature as in the case of casing and flange joints. Relaxation and deformation processes as outlined schematically in Fig. 7, tend to occur already during the subsequent heating [9, 10, 18]. The effects involved are given below.

4.1.1. During heating

• Loss of pre-load due to decrease in modulus of elasticity as the temperature rises (Fig. 7(a)).

• Loss of pre-load and total strain by set and plastic flow in the threads and the contact faces (Fig. 7(b)). This is a substantial difference compared to the relaxation test in the relaxation machine with smooth specimens. It depends to a great extent on the yield point, pre-load and temperature.

• Anisothermal relaxation.

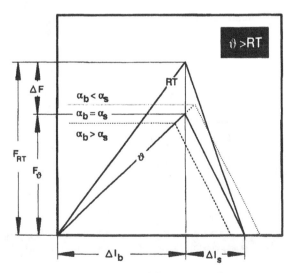

Fig. 6 *Estimation of stresses in bolted joint models at elevated temperatures.*

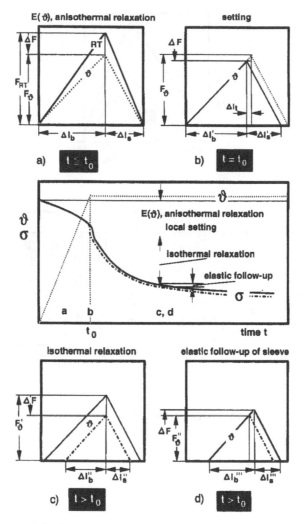

Fig. 7 *Mechanisms in bolted joint models.*

4.1.2. At constant temperature
• Isothermal relaxation (Fig. 7(c)).
• Elastic follow-up of flange F due to decrease in preload (Figs. 7(d) and 8) [9, 16, 18, 19].

Of course, this will occur in every case of a decrease in pre-load and counteracts the loss of pre-load due to relaxation. It depends to a marked extent on the flange-to-bolt area ratio and the creep strength of the flange and bolt materials, and also represents a significant difference compared to the isothermal test with the smooth specimen in the machine.

The effects described also occur in real bolted joints. The conditions during erection, thermal exposure and long-term stressing are practically comparable if one ignores the relatively small additional stresses resulting from the internal pressure and the thermal stresses during start-up and shut-down.

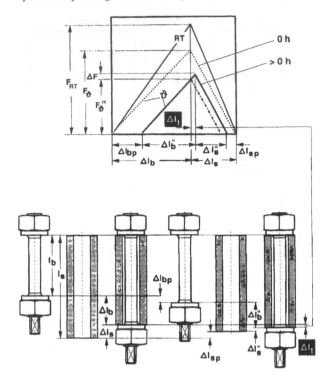

Fig. 8 *Relaxation and elastic follow-up in bolted joints.*

5. Results of Relaxation Testing

The results of the following examples should not be generalised, irrespective of the test conditions and the material examined.

5.1. Documentation

Figure 9 shows a typical pattern of isothermal relaxation curves of a 1CrMoV heat determined at 425, 480 and 540°C with the bolted joint model II. In the documentation, the model design, materials and modulus of elasticity should also be recorded.

5.2. Results of Similar Materials

5.2.1. Comparison of small and large bolted joints with uniaxial specimen test of 1CrMoV
A comparison of the results of tests made with small bolted joint models with those obtained with a large bolted joint model, with the smooth specimen in the relaxation machine and the results of residual stresses in real bolted joints of the same material is given in Fig. 10 [18].

Entered into the common scatter band of specimens and model joints have been the results of a large and a small bolted joint model with bolts originating from the same heat. It will be noted that, in this example, the large model has higher residual stresses than the small model. However, the differences are still within the scatter inherent in the materials and test

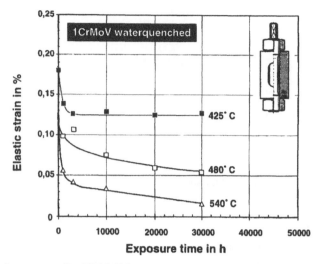

Fig. 9 *Relaxation curves of a 1CrMoV heat.*

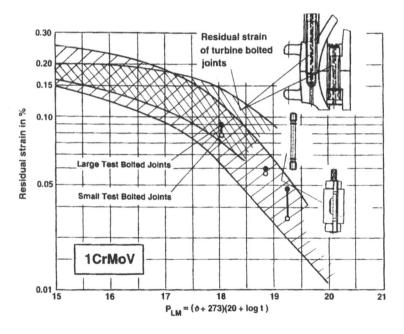

Fig. 10 *Relaxation of bolted joints of different sizes and specimen.*

conditions. In the case of the models, differences are primarily caused by differences in the elastic follow-up, redistribution of stresses and in setting effects. Even higher residual stresses on average were found in real bolted joints (second scatter band). Uncertainties in the measurement, temperature gradients and casing deformations probably are responsible for the higher residual stress values of the bolted casing joints.

A comparison between the mean curves of small models and relaxation specimens of

the specimens tend to result in somewhat higher residual

olt lengths of 1CrMoV
h from 70 to 140 mm resulted in the same residual strains
g. 12) [3].

l uniaxial specimen test of 12CrMoV
elt shows that neither at 500 nor at 550°C were there any
idual stress values of the specimen and model (Fig. 13) [12].
ble from the 12CrMoVNb bolt steel at 550°C. This is in
carried out with the 1CrMoV steel, where the results always
or the models.

leeve cross sections of 12CrMoVNb and 1CrMoV
oints and the 1CrMoV bolted joints showed no, or only a
he flange cross section $(A_xA_b > 3)$ on the elastic residual
°C/100 h, Fig. 14 [3].

ut strengths of 1CrMoV
t material decreases, the elastic residual strain also tends to

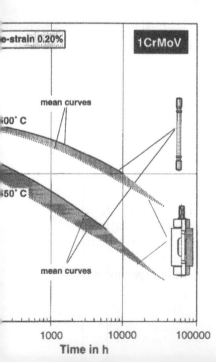

e-strain 0.20%   1CrMoV

mean curves

00° C

50° C

mean curves

1000   10000   100000
Time in h

Fig. 15 [6]. It is evident that with increasing tempera-

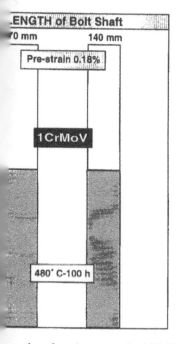

on the relaxation strength of *1CrMoV bolt steel.*

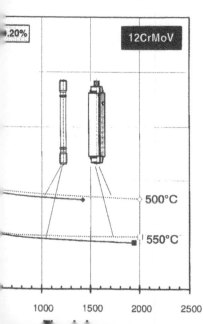

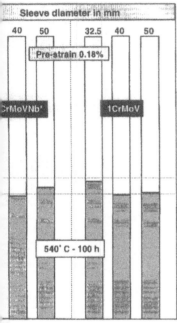

axation strength version

eter on the elastic residual strain.

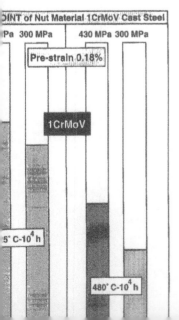

5.3. Results of Dissimilar Materials

5.3.1. Model/test specimen comparison of Nimonic 80A bolt, nuts 1CrMoV sleeve
Due to the negative creep encountered up to about 550°C in the case of Nimonic 80, different residual stress values will be obtained with specimens compared to those in models of the combination of a Nimonic 80A bolt and 1CrMoV sleeve (Fig. 16) [8]. This practical example serves well to bring out the advantage and usefulness of the model tests compared to the tests with specimens which they certainly do not make unnecessary having regard to the direct use of the test results for the sizing of real bolted joints.

5.3.2. Small model/large model comparison of Nimonic 80A bolt, nuts/T91 sleeve
The comparison of this material combination between 540 and 600°C does not reveal any significant differences (Fig. 17) [13]. Presumably this is due to the fact that in the case of bolt models the nuts as well as the bolts were made of Nimonic 80A and that the pre-load reducing setting effects in the threads and in the contact faces are independent of model size.

It is possible that the results are dominated by the extreme difference in strength between the bolt and flange material, with the result that the minor influences of the elastic follow-up of the flange, as well as of the different setting, are exceeded.

6. Conclusion

The bolted joint model permits a low-cost and close-topractice determination of long-term design criteria that can be directly used for heat-resistant bolted joints made of similar and dissimilar materials for steam turbines.

The isothermal short-time relaxation tests done in the relaxation machine using smooth

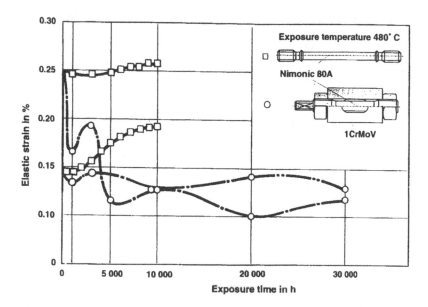

Fig. 16 *Influence of test method on relaxation strength.*

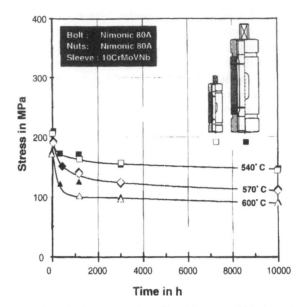

Fig. 17 *Comparison of residual stresses of small and large model bolt testing.*

specimens are not rendered superfluous as a result, but rather represent a necessary and meaningful complement to the model tests.

7. References

1. L. Robinson, 'A relaxation test on 0.35% C steel K20'. *Trans. ASME*, 1937, **59**, 451–452.
2. K. Wellinger and E. Keil, Der Spannungsabfall in Stahlschrauben bei höherer Temperatur unter Last', *Arch. Eisenhütt.*, 1942, **15**, 475–478.
3. A. Erker, E. Klotzbuecher and K. H. Mayer, 'Entspannungsversuche an warmfesten Schraubenverbindungen', MAN Forschungheft Nr. 13 (1966/67), pp. 62–76.
4. A. Erker, 'Die vorgespannte Schraubenverbindung unter Dauerbeanspruchung und Überlastung'. MAN Forschungsheft (1953), pp. 1–15.
5. H. Wiegand and K. H. Beelich, 'Relaxation bei statischer Beanspruchung von Schraubenverbindungen', *Draht-Welt*, 1968, **54**, 5, 306–322.
6. A. Erker and K. H. Mayer, 'Relaxations- und Sprödbruchverhalten von warmfesten Schraubenverbindungen', *VGB-Kraftwerkstechnik*. 1973, pp. 121–131.
7. H. Wiegand and K.H. Beelich, 'Relaxationsverhalten hochfester Schraubenverbindungen', *Konstruktion*, 1965, **17**, 315–320.
8. K. H. Mayer and H. König, VGB-Konferenz 'Forschung in der Kraftwerktechnik' (1988), Essen, 'Untersuchungen zum Relaxationsverhalten warmfester Schraubenverbindungen'.
9. H. Wiegand, K. H. Kloos and W. Thomala, *Schraubenverbindungen*, Konstruktionsbücher, Band 5, Springer Verlag, 1988.
10. K. Löser, 'Mechanische Langzeiteigenschaften von Nickelbasislegierungen für Dampfturbinenverschraubungen', VDI-Fortschrittsberichte, Nr. 255, 1992.
11. EGKS-Report, 'Relaxationsverhalten warmfester Stähle für Schrauben', Doc.-No. 6210-KF/1/101 – F6.2/74 (1978).

12. Research Project 'Langzeituntersuchungen zum Relaxationsverhalten mit und ohne Betriebsvorbeanspruchung', AIF-No. 8198/VGB-No. 79, 1994.

13. K. H. Mayer and H. Koenig, 'Hightemperature Bolting of Steam Turbines for Improved Coal-fired Power Plants', *2nd Int. Conf. on Improved Coal-fired Power Plants*, 2–4 November, 1988, Palo Alto, CA/USA.

14. M. C. Lecoco and C. Leymonie, 'Influence of thermal and mechanical cycling on the stress relaxation of a CrMoV bolt steel', *2nd Int. Conf. on Creep and Fracture of Engineering Materials and Structures*.

15. C. Tanaka and T. Ohba, 'Analysis of Reloading Stress Relaxation Behaviour with Specified Reloading Time Intervals for High Temperature Bolting Steels', *Trans. NRIM*, 1984, **26**, 1–20.

16. R. M. Goldhoff, E. E. Zwicky and H. C. Bahr, 'The Effect of Simulated Elastic Follow-up on the Relaxation Properties of Bolting Steels', ASME publication, 1967.

17. K. Wellinger and E. Keil, 'Entspannungsversuche mit einer Schraubenverbindung für 600 °C', Mitteilungen der VGB, 1951, pp. 317–320.

18. D. Tremmel, 'Bauteileinfluß auf das Relaxationsverhalten warmfester Stahle für Schrauben', VDEL-Tagung, Düsseldorf, 1978.

19. M. Boenick, 'Untersuchungen an Schraubenverbindungen', Doktorarbeit, Berlin, 1966.

20. SEP 1260 (Draft), 1994.

5

Behaviour of Durehete 1055 in Full Scale Bolt Tightening

A. ELSENDER and D. CARR*

International Research & Development Ltd, Fossway, Newcastle-upon-Tyne, UK
*Parsons Power Generation Systems Ltd, UK

ABSTRACT

Durehete bolts have been loaded, under controlled conditions, to 0.15% initial strain, soaked at 565°C for 1000 h and untightened. After twenty of these cycles (20 000 h) there was no superficial damage at the thread roots whether the bolts were tightened by hydraulic extension, torque or heat. 20 000 h represents ~ 320 000 effective operating hours which is 110% greater than the material design life. Destructive metallurgical assessment of the bolt threads after 12 000 h (≡ 30% greater than the material design life) revealed the torque tightening method to have generated slightly more creep damage than the other two methods but within acceptable levels.

1. Introduction

Bolt failures in power plant have been of concern since the 1950s leading to the formation of a bolt working party [1] to review operating procedures. Uncontrolled tightening, levels of residual elements in bolting steels and temperature gradients were identified as being instrument in many of these failures and the recommendation of the panel resulted in upgrading of bolting standards [2]. Extensometry has superseded angle of turn to control initial applied strains and electric heating still remains the preferred method offered to Parsons TG Ltd customers. The development of torque and hydraulic extension tightening methods offer the advantage of quicker times to secure bolt jointed components. However these methods, particularly torque tightening, have introduced some further elements of unquantified residual stresses/localised strain deformations. The effect of any additional residual stresses, from torque and hydraulic extension tightening on creep cavitation or cracking at the thread root regions was investigated.

1.1. Programme Details

Twelve full size bolts (4 for each tightening method) were tightened to 0.15% strain, soaked for 1000 h at 565°C, cooled and untightened. This loading/heating/unloading cycle was repeated twenty times (20 000 h at temperature) with bolts of each tightening method being retired for destructive metallographic examination after six and twelve cycles. The six, twelve and twenty cycles represent an equivalent 96 000, 192 000 and 320 000 h life calculated using a penalty of 15 000 h for each pre-stressing [3]. The 'basic material life hours' for the Durehete 1055 bolts (group 6) is shown in Table 1 to be 15 0000 h.

Table 1. *Criteria for examination and replacement basic material life hours*

°C	371/399	400/426	427/454	455/485	486/515	516/538	539/555	556/570
°F	700/750	751/800	801/850	851/905	906/960	961/1000	1001/1030	1031/1058
Group 1	300 000	250 000	200 000	150 000				
Group 2	350 000	300 000	250 000	200 000	150 000			
Group 5	350 000	300 000	250 000	Replace with Group 6 material				
Group 6	350 000	350 000	350 000	300 000	250 000	200 000	175 000	150 000
Group 8	350 000	350 000	350 000	350 000	350 000	250 000	300 000	250 000

1.2. Material and Components

The studbolts were made from Durehete 1055 material with the cap nut manufactured from Durehete 900. The bolt material was purchased to an old CEGB specification but Fig. 1 (a) shows the composition ranges are within the new specification [4] with the exception of Ni which was slightly high. The chemical composition and mechanical properties of both materials are shown in Fig. 1 (a) along with the new specification. The test bolts were 1.75" BSF, this being the minimum size that can withstand the 0.69" dia. bore necessary for extensometry and also affords test handleability. Eight standard studbolts and four with extended threads (for hydraulic extensions) were used for the programme. The bolts were tightened onto 6" dia. cylinders (representing the flange of a pressure vessel) with a 1.875" dia. bore; made from $1/2\%$ CMV thick walled pipe. The final assembly is shown in Fig. 1 (b).

MATERIAL			C	Mn	Si	S	P	Ni	Cr	Mo	V	Cu	Sn	Ti	B	HEAT TREATMENT	0.2% P.S. MPa	UTS MPa	F1 %	RofA %	HARDNESS Hv30
Specification 221 (Durehete 1055)	S	Min	.15						.9	.85	.6		.05	.001			679	849	14	55	
		Max	.25	.75	.35	.04	.04		1.3	1.1	.8			.2	.01			1003			
	C / L		.2/.2	.5/.5	.26/.26	.009/.005	.009/.011	.15/.26	1.03/.96	.95/.92	.65/.66	.19/.19	.018/.015	.09/.1	.003/.003	980°C 2hr WQ 680°C 6hr AC	821	896	20	71.7	283–299
Revised CEGB Standard 02596 1982		Min	.17	.35	.1				.9	.9	.6			.07	.001						
		Max	.23	.75	.35	.02	.02	.2	1.2	1.1	.8	.2	.02	.15	.01						
Specification 11	S	Min	.35	.4	.1				1.0	.5						875°C OQ T 650–700°C	633	849	14		
		Max	.45	.7	.35	.04	.04	.4	1.5	.9								1003			
Specification 97	S	Min	.08	.4	.1				.25	.5	.22					N 940–980°C T 660–700°C	295	460	21		
		Max	.15	.7	.35	.04	.04	.3	.5	.7	.28	.25	.03					620			
	C		.16	.49	.21	.02	.01	.28	.34	.53	.23	.18	.014								

S : NEI Parsons Specification
C : Steelmakers Cast Analysis and Properties
L : Laboratory Analysis

Fig. 1(a) *Bolting materials properties.*

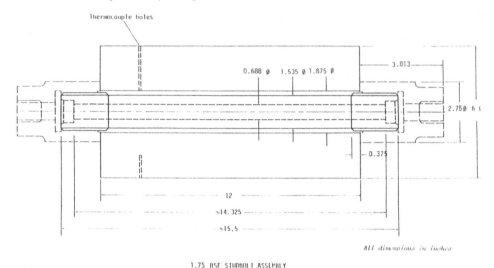

Fig. 1(b) *Studbolt assembly.*

1.3. Bolt Tightening Techniques

1.3.1. Torque tightening

The torque tightening equipment consisted of a hydraulically operated torque wrench with a 1.5" drive with a torque range of 2000–10 800 NM. The square drive was bored to facilitate the *in situ* extensometry for direct extension measurement. A purpose-built jig was manu-factured to hold the bolt assembly and supply the torque reaction. Roller bearings were employed to enable the top nut to turn and react to eliminate any bending in the bolt. The jig and assembly are shown in Fig. 2.

1.3.2. Hydraulic stretching

The hydraulic jack and adaptors were also purpose built allowing for *in situ* extensometry. The assembly jacking system is shown in Fig. 3.

1.3.3. Heat tightening

Conventional 2 kW heating elements with an 11" active stem length were supplied from a 12 kVA transformer.

1.3.4. Extensometer

The extensometer (shown *in situ*, Figs. 2 and 3) can measure absolute extension of the bolts to within 0.005 mm. Heat tightening requires re-insertion of the extensometer after the nut angle of turn has been applied and the bolt cooled. The re-insertion error is approximately 0.025 mm or 5%.

1.3.5. Metallography

The micro-structures of the first five engaged thread roots of the bolts were established by a polish–etch–polish etch process [5] suitable for detecting small amounts of creep cavitation.

Fig. 2 *Torque tightening equipment and jig with extensometer in position.*

Fig. 3 *Hydraulic extension assembly.*

The microstructure of the first engaged thread of a heat, torque and hydraulically stretched tightened bolts after 12 000 and 20 000 h temperature/load dwells are shown in Figs. 4 and 5.

1.3.6. Thread lubricant

A nickel-based anti-seize compound (Belzona) was used to lubricate the threads. The lubricant was free of residual elements deemed detrimental for use with high temperature bolts.

1.3.6.1. Tightening parameters*

Angle of turn

This angle is given by the following form:

$$\emptyset = 3.6 \, Sn \, (1.35L + 0.9D + 3)$$

where $\emptyset$ = angle of turn in degrees, S = initial cold strain %, L = active length (in.), D = nominal diameter (in.), and n = number of threads/in.

Torque

The torque required to give a strain of 0.15% is obtained from the following form [6]:

$$T = 1866.7 \left[\frac{D_o + D_i}{2} + D_m \tan\{\tan^{-1}\left(\frac{1}{\Pi D_m n}\right) + 6.432\} \right]$$

where T = torque (ft lb), A = minimum cross-section area of thread rood (in.^{2}), D_o = nut outer diameter (in.), D_m = maximum effective thread diameter of external threads (in.), n = number of thread/in.

The calculated and design torques to give 0.15% strain are ~1250 ft lb but the actual measured values on the torque gauge to achieve 0.15% strain are 1400 ft lb.

2. Results
2.1. Bolt Tightening Tests

Detailed measurements of thread pitch (averaged over 5 threads from root 3–8), thread diameter and bolt length were taken before the bolts were tested and after each thermal cycle in the unloaded state. This continued for as many thermal cycles as possible before corrosion damage precluded accurate measurements. The measurements enabled assessment of the accumulation of plastic strain within the bolt during testing.

The 'linearised' change in average thread pitch for typical bolts from each tightening method are presented in Fig. 6. The accumulation of permanent extension of the bolt gauge length after each tightening and heating cycle is presented in Fig. 7. The angle of nut rotation

---

* At the time of the project the standards for bolt tightening were in imperial units.

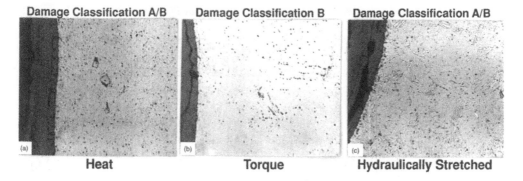

Damage Classification A/B	Damage Classification B	Damage Classification A/B
Heat	Torque	Hydraulically Stretched

Fig. 4 Photomicrographs of thread roots after 12 000 h. (a) Heat × 500; (b) Torque × 500; (c) Hydraulically stretched × 500.

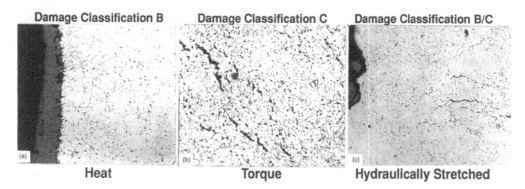

Damage Classification B	Damage Classification C	Damage Classification B/C
Heat	Torque	Hydraulically Stretched

Fig. 5 Photomicrographs of thread Roots After ~20 000 h. (a) Heat × 500; (b) Torque × 500; (c) Hydraulically stretched × 500.

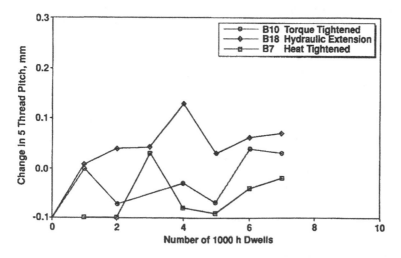

Fig. 6 Strain accumulation in the threads of torque, heat and hydraulic tightened bolts.

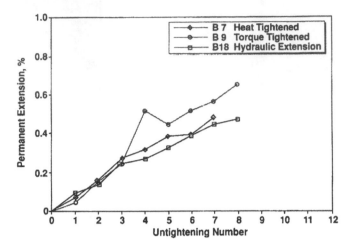

Fig. 7 *Accumulation of permanent strain during testing for heat, torque and hydraulic tightened bolts.*

and the over extension required for the heat tightened and hydraulically extended bolts during tightening are plotted in Figs. 8 and 9 respectively. Residual shank stress calculated from residual extensions (using an elastic load/extension curve) are plotted in Figs 10–12.

3. Discussion

The design criteria for examination and replacement of turbine bolts is based on 'basic material life hours' which has been derived from high temperature properties and failure experience gathered by the CEGB [7]. Table 1 shows the design lives of the various gauges of bolts at specified operating temperatures. The materials range from a 0.5% CrMo (Group 1) steel up to a 10–12% CrMoV (Group 7) steel with the addition of a Nimonic 80A Group 8. The Durehete 1055 bolts used in the above programme are covered by Group 6 which have a basic design life of 150 000 h at the 565°C operating temperature. Taking the penalty hours into consideration the bolts in the present programme were tested to 96 000, 192 000 and finally between 288–320 000 h.

The longest of these times represents more than double the design life for the studbolts. Under the controlled laboratory programme using all three tightening methods reported above there was no evidence of bolt failure or superficial cracking in any of the thread roots even after an equivalent 320 000 test h. This indicates that the design lives for this category of bolt is somewhat pessimistic with respect to bolt failure if bolt tightening procedures and complementary extensometry are adhered to.

An assessment of the cumulative creep damage was made on metallographic sections taken from the highly stressed regions of the first few engaged bolt threads. The creep damage has been assessed using a ranking procedure designated A, B, C and D [8]. Where Class A is the least damage and Class D the most damage. The reference photographs used in the categorisation are reproduced in Fig. 13 (p.80). The creep damage accumulation steadily

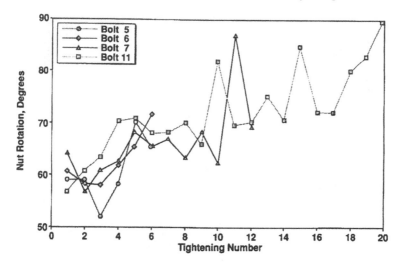

Fig. 8 *Variation of nut rotation at each tightening for heat tightened bolts.*

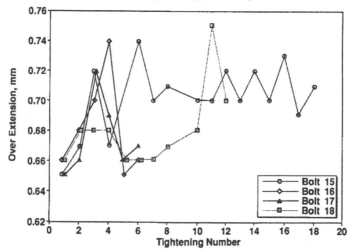

Fig. 9 *Variation of over-extension at each tightening for hydraulically extended bolts.*

increased from the 6000 h duration tests. Within the context of the test programme it appears that there is little difference, with respect to the creep damage, between the heat tightened and hydraulically extended bolts at all three test durations.

It is evident that the creep damage accumulation is greater in the thread roots of the torque tightened bolts as the test duration increases with the cavities linking to form crack like features after 20 000 h. The observation that, after 12 000 h testing or an equivalent 192 000 h effective operation, the creep damage accumulation in the torque tightened bolts is only slightly greater than that of the heat tightened and hydraulically extended bolts, indicates that within the design life of these fasteners torque tightening is an acceptable procedure.

According to GOM 85 [7] the examination on replacement of fasteners should be based

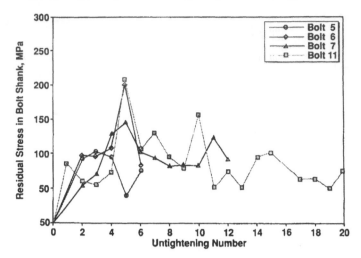

Fig. 10 *Variation in bolt shank residual stress after each untightening for torque tightened bolts.*

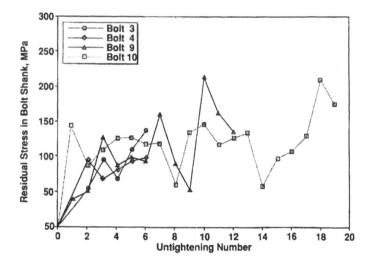

Fig. 11 *Variation in bolt shank residual stress after each untightening for heat tightened bolts.*

on sampling, metallurgical examination and NDT. For category 2 fasteners operating above 455°C the fasteners should be examined when 80% of their 'basic material life' has been expended. This would have resulted in all of the bolts in the present programme being passed for continuing service. At effective operating hours some 30% greater than the 'basic material life' (i.e. 192 000 h) all the bolts using the three differing tightening methods would have passed inspection. It is only when 'the effective operating hours' reached 110% of the 'basic material life' that damage have been recognised from metallurgical investigation. The torque tightened bolts would almost certainly be retired, with reservation as to the severity of the damage in the heat tightened and hydraulically tensioned bolts.

The accumulated plastic strains caused by the numerous pre-strainings will cause some

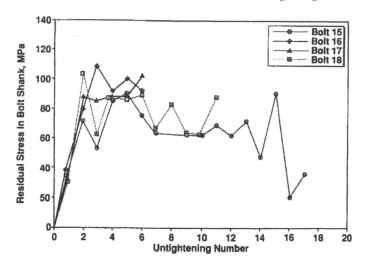

Fig. 12 *Variation in bolt shank residual stress after each untightening for hydraulically extended bolts.*

permanent change in the physical dimensions of the bolt. The measurements of thread pitch change (taken over 5 thread pitches, Fig. 6) show no evidence that any one tightening method is causing more deformation than another, although the measurements were terminated after the seventh unloading cycle because of corrosion products and thread crest damage precluding further measurements. The maximum change in thread pitch represents a strain of 1% for the heat tightened bolt from the present programme which compares to other researchers measurements [9] of 3.5% change in thread pitch for hydraulically stretched bolts after 10 tightenings and 1000 h holds at 565°C. Because of associated errors in the measurements and the changes are so small it would be imprudent to make any judgements as to the damage assessments at the thread roots or life predications of the bolt fastener from thread root measurements.

The accumulation of plastic strains over the effective gauge length of the bolts (Fig. 6(b)) shows that the heat tightened bolt would have approximately 0.8% permanent elongation when extrapolated to 10 test cycles (= 160 000 effective operating h). When extrapolated to 1% permanent strain the number of equivalent loading cycles would be 12 or 192 000 effective operating h. The corresponding permanent strains for the hydraulically tensioned and torque tightened bolts are 0.65% and 0.72% for 10 cycles and 0.76% and 0.82%, for 12 cycles respectively. The permanent elongation for the hydraulically extended bolt after 10 cycles (extrapolated) compares favourably with the permanent plastic strain in the bolt shanks of the larger diameter bolts (3", or 76.2 mm Ø) tested, to the same 0.15% pre-strain, by other researchers [9]. The advantage of being able to use permanent elongation for the monitoring of bolts are recognised by the turbine plant operators. It has been suggested that 1% permanent strain be used as a criteria for replacing high temperature bolts [10, 11]. The practice for replacing bolts after 1% permanent elongation is attributed to this being the end of the secondary creep range for bolt materials and also the relaxation strength of the next prestress cycle is distinctly lower than in the previous cycle after the 1% elongation is attained [11, 12]. Unfortunately, because physical measurements were not continued beyond the eighth

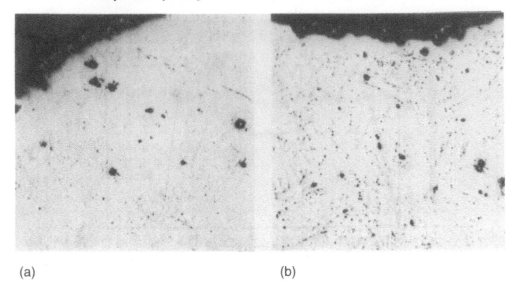

(a)   (b)

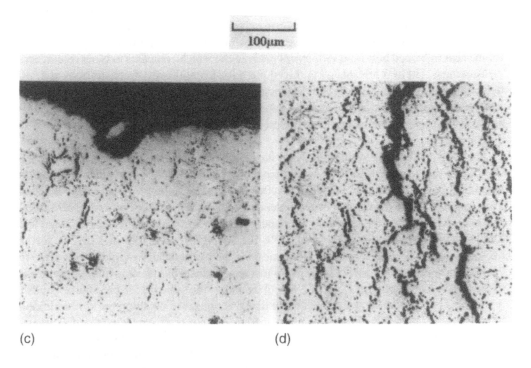

(c)   (d)

Fig. 13 *Damage classification in D1055 stud material. (a) Class A; (b) Class B; (c) Class C; and (d) Class D.*

cycle no permanent strain measurements are available for operation life times of 130 000 h or greater. In considering permanent elongation by measurements across the effective length of the bolt, oxidation across the measuring lands must be taken into consideration [13].

These estimates are of the same order of magnitude as the actual residual extension measurements inferring that the increase in extension is predominantly oxide growth. Any systematic increase in stud bolt length with time will be totally masked by the build up of oxide film during testing.

Two other parameters, 'nut rotation' and hydraulic over-extension were measured during the relevant tightening periods for the heat tightened and hydraulic extensioning methods respectively. The nut rotation to give 0.15% strain in all four heat tightened bolts was approximately 25% less than the calculated angle of 81° for the first three loading cycles and by 17% less up to the 9th loading cycle. The angle of nut rotation increased slightly as testing progressed from the 9th loading cycle but only approached the calculated 81° towards the 16th and 17th cycle. The degree of scatter in the angle of turn increased with the last few loading cycles. The trend in actual angle of turn being less than the calculated values is the reverse of the findings from previous laboratory tests by other researchers [9], where the required angles to give 0.15% strain for larger bolts was some 1.7–2.5 times the calculated angle. The conclusion from the tests conducted in [9] was that the angle of turn method was unsatisfactory in controlling the applied strains for high temperature bolting applications.

The difference between the calculated and final angle of turn (and the calculated and applied over extensions) may be due to the elastic follow-up of the flange which is dependent upon the bolt/flange CSA. A typical area ratio for close pitched cylinder bolts, i.e. bolt/flange area would be 0.23 whilst the area ratio for the tests was 0.06. This reduced bolt area would have the effect of increasing the cylinder rigidity and thus, the bolt extension 'lost' through flange compression, (i.e. load loss or follow-up) would be significantly reduced when compared to actual turbine bolting.

Hydraulic tensioning tests on larger bolts [9] have shown that residual bolt tension after several thousand hours at 565°C was similar for bolts tightened to different initial strains of 0.1, 0.15 and 0.2%. Only when nuts of a different specification (to that of the bolts) are used does the residual tension, for 0.1% initial strain, not quite reach the levels of the bolts tightened to higher initial strains (0.15 and 0.2%). In the tests at initial strains of 0.1% the residual tension after several tightenings was always greater that the residual tension after the first tightening for the 0.15% initial strain tests. The pattern of increasing residual tension for the first few tightening/heating cycles is consistent with the creep strain rates being higher and also the bedding in of the threads due to the conversion of elastic strains to plastic strains [9]. The residual stresses from the hydraulic extensioned bolts in the present programme follow the same pattern as in [9], where the steep increase at the initial cycles levels off for the remainder of the testing (Fig. 11). There is a lot more scatter in the residual stress results for heat and torque tightened bolts (Figs. 9 and 10), however from the 4 to 6 cycles the residual stress values level off at a higher value than after the first heating cycle. One of the major recommendations [9] is that after an initial tightening of say 0.15% strain, then, the bolts could be further tightened to only 0.1% strain without detriment to the jointed component or damage within the bolt threads. The same recommendation would apply to repeated tightenings of steam joints, from the results in the present project.

4. Conclusions

1. Under controlled laboratory conditions and after 12 loading/1000 h @ 565°C cycles (equivalent to 192 000 h effective operating hours), Durehete 1055 bolts tightened by hydraulic, torque and heating standard methods showed no signs of superficial thread root damage.

2. The torque tightened bolts generated slightly more creep cavitation at the thread roots than either the heat tightened or hydraulically extended bolts, but was not considered excessive considering 12 loading/heating cycles is equivalent to a period of approximately 30% greater than the basic material life of 150 000 h.

3. At an effective operating hours, some 110% greater than the basic material life, damage at the thread roots would cause retirement of the torque bolts. Reservations as to the severity of damage in the heat and hydraulically tightened bolts remain. In all three methods the damage would probably not be detected superficially or by dimensional measurements.

4. Hydraulic torque tightening has shown itself to be the quickest method of fastening bolts to 0.15% strain. When used with extensometry, to obtain the correct strain measurements, this controlled method is not detrimental to the integrity of the bolts within the design life of fasteners.

References

1. CEGB High Temperature Bolt Working Party.
2. J. S. Mitchell 'The Design Requirements for Studs, Studbolts, Bolts and Nuts', NEI Parsons R&M 70/996, 1970.
3. Design Standard Instruction D9 1006. NEI Parsons Design Standard Instruction 1985.
4. '1 Chromium–1 Molybdenum, 0.75 Vanadium Steel Bars for High Temperature Bolting', CEGB Standard 02596, GDCD Standard 2, Issue 2, 1982.
5. S. M. Beech 'An Examination of Creep Damage Development in Durehete D1055 Bolt Steels', NEI Parsons Technical Memorandum MET 82-79, 1982.
6. A. D. Batte, J. M. Clarke, J. J. O'Connor and E. Taylor 'A Re-analysis of the Single Cycle Stress Relaxation Data for 1%CrMo, 1%CrMoV (0.4%), 1%CrMoTiVB and Nimonic Alloy 80A Steam Turbine Bolt Material', NEI Parsons Memorandum. CAP 80-29, 1980.
7. 'The Selection, Care and Replacement of Threaded Fasteners Operating at Temperatures Above 370°C', CEGB GOM 85 1985.
8. S. M. Beech and N. Reid 'Examination of Durehete 1055 Reheat ESV Valve Cover Studs from Alcan Unit 1 (machine 3436)', NEI Parsons Technical Memorandum MET 83-108, 1983.
9. H. G. Mellor 'Investigation of some Factors Affecting High Temperature Bolt Life', CEGB SSD/MID/R45, 1978.
10. M. De Witte and J. Stubbe 'Failure Mechanisms and Practical Approach for the Followingup of Higher Temperature Turbine Bolts', Eurotest Conference, 'Remanent Life: Assessment and Extension', Brussels, 19–21 March, 1985.
11. K. H. Mayer and K. H. Keienburg 'Operating Experience and Life Span of Heat Resistant Bolted Joints in Steam Turbines of Fossil Fired Power Stations', IMechE Conference *Engineering Aspects of Creep*, Sheffield, 15–19 September, 1980.

12. V. G. B. Empferhlug für Schrauben in Bereich Hoher Temperaturen.
13. D. J. Gass 'Oxide Dating Techniques for High Temperature Components', NEI-IRD Report 89-4, 1989.

6

The Performance of High Purity 1CrMoVTiB Steel in Full Size Bolt Tests

S. R. HOLDSWORTH and A. STRANG

GEC ALSTHOM Large Steam Turbines, Rugby, UK

ABSTRACT

Full size bolt tests have been employed to confirm that the superior elevated temperature properties of high purity 1CrMoVTiB steel determined in conventional laboratory tests are realised in circumstances more closely representing those in large steam turbine service situations.

Long duration model bolt stress relaxation tests have been conducted on studs of similar size and design to IP turbine valve chest cover studs. Bolts have been cyclic loaded to strains of 0.15 or 0.20% in individual restraining blocks and held at 550°C for periods of 144, 1000 or 3000 h before unloading to inspect for cracking. Service assembly conditions have been reproduced by using a hydraulic loading device with strain control by means of a special extensometer inserted through the central hole of the bolt.

The high purity bolting steel was developed to have improved rupture ductility and notch tolerance properties over conventional 1CrMoVTiB for circumstances where many repeated retightenings were anticipated. The results of the full size bolt tests demonstrate the advantage of the high purity steel over its commercial equivalent and the advantage of limiting the level of tightening strain for conventional purity bolts which are likely to to be repeatedly re-tightened.

1. Introduction

During the 1970s a number of developments were directed towards improving the performance of bolting materials for steam turbine applications. One of these exploited the important observation that creep rupture strength and ductility are enhanced by restricting the residual element content in low alloy creep resistant steels [1]. Tipler's work was based on steels produced from relatively small experimental heats, but the results of a comprehensive evaluation of the properties of a high purity VIM/VAR 1CrMoVTiB bolting alloy confirmed that the same benefits were realised in material from a production size melt [2].

Long duration plain and notched specimen creep rupture and stress relaxation tests clearly demonstrated that reducing the levels of impurity elements such as S, P, As, Sb and Sn leads to significant improvements in the rupture ductility, notch tolerance and creeP crack growth resistance of 1CrMoVTiB steels, without loss in creep strength.

The superiority of the high purity VIM/VAR steel was particularly evident from the results of cyclically loaded notched stress relaxation tests, devised to simulate the cyclic retightening experienced by bolts in service. In weekly reloading tests at 550°C, the high purity alloy testpieces survived at least ten times the number of cycles to failure associated with conventional purity 1CrMoVTiB (D1055) steels.

Following on from the conventional laboratory testing programme, the performance of the high purity alloy was finally assessed in full size model bolt tests. The results of these are the subject of the following paper.

2. Materials

Bolts were manufactured from a conventional purity 1CrMoVTiB steel (CP D1055) and a high purity version of the alloy prepared using Japanese electrolytic iron, high purity alloying additions and a double vacuum melting VIM/VAR steel making route (referred to below as HPVV). The chemical compositions, heat treatment details and uniaxial tensile properties of the two steels are summarised in Table 1.

The purity of HPVV was controlled through limits set on the residual element content and by restricting R to 0.04, R being defined after [3] as:

$$R = P + 2.43 \text{ As} + 3.57 \text{ Sn} + 8.16 \text{ Sb}$$

The long term creep rupture, cyclic stress relaxation and creep crack growth properties of the two steels have been reviewed elsewhere [2].

3. Full Size Bolt Tests
3.1. Testing Arrangement

Model bolt tests were performed using the full size IP valve chest cover stud shown in Fig. 1. The studs were end-faced with Haynes 25 to minimise errors during measurement due to oxidation and wear. Overall length measurements were made using a Hytorc extensometer which located against the two end-faces through the central hole of the bolt. The relationship between end-face displacement and shank strain was established in strain gauge trials prior to the start of the main test programme.

The bolts were loaded to shank strains of 0.15 or 0.20% in individual restraining block assemblies (Fig. 2) and held at 550°C for periods of 144, 1000 or 3000 h, before unloading for fluorescent penetrant dye crack inspection. The cycle was repeated until failure or until significant plastic strain had been accumulated in the bolt shank. Bolt loading was achieved using a hydraulic pilgrim nut. This provided a convenient method of tensile bolt loading without torsion but required a degree of take-up loading at room temperature (AB and DE, Fig. 3) during loading (OABC) and unloading (DEO') procedures. This was to compensate for load equilibration when the action of the pilgrim nut was replaced by the bolt assembly nut. At the end of each hold time at temperature, the assembly was unloaded at room temperature with the extensometer in position to allow measurement of the relaxed stress (Figs. 4(a), 5(a) ,6)a)) and the plastic strain accumulated (Figs. 4 (b), 5(b), 6)b)). Final fracture almost always occurred during the load take-up operation in the unloading procedure (i.e. bolt loading to σ_{TL2}, DE in Fig. 3).

The results of twelve model bolt tests are summarised in Table 2.

Table 1. *Chemical composition, heat treatment details and tensile properties*

	CP 1CrMoVTiB	HP 1CrMoVTiB
Chemical Composition		
C	0.21	0.21
Si	0.26	0.24
Mn	0.55	0.35
Ni	0.17	0.01
Cr	0.97	1.03
Mo	0.96	1.07
V	0.62	0.70
Ti	0.10	0.08
B	0.004	0.003
S	0.015	0.007
P	0.017	0.003
Al	0.045	0.013
Cu	0.18	0.02
Sn	0.015	< 0.003
Sb	0.0037	< 0.0010
As	0.015	0.002
Pb	0.0013	< 0.0005
N_2	0.016	0.002
R	0.137	0.027
Heat Treatment Details		
Normalise		3 h 950°C, AC
Harden	3 h 980°C, WQ	3 h 980°C, WQ
Temper	6 h 690°C, AC	6 h 690°C, AC
Tensile Properties		
20°C $Rp_{0.2}$ (MPa)	920	859
Rm (MPa)	979	944
Z (%)	18	19
A (%)	65	74
550°C $Rp_{0.2}$ (MPa)	582	559
Rm (MPa)	661	630
Z (%)	16	18
A (%)	70	81

3.2. Model Bolt Performance

In the B15/0.20ε%/144 h test, a CP D1055 bolt failed during unloading at the end of the 21st cycle, having accumulated 0.53% plastic strain in the shank at the end of the 20th cycle (Fig.

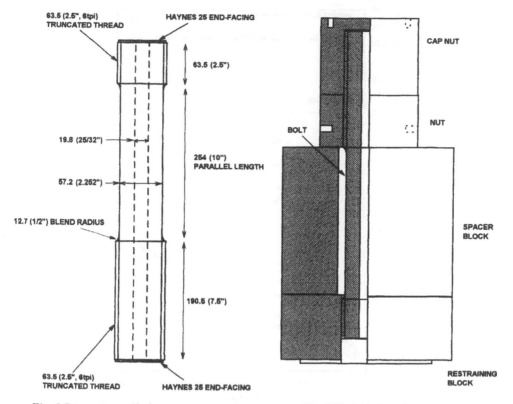

Fig. 1 *Dimensions of bolt.*

Fig. 2 *Model bolt testing arrangement.*

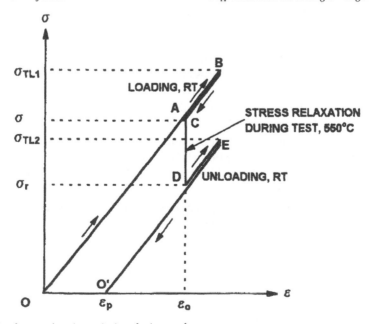

Fig. 3 *Bolt shank stress/strain variation during cycle.*

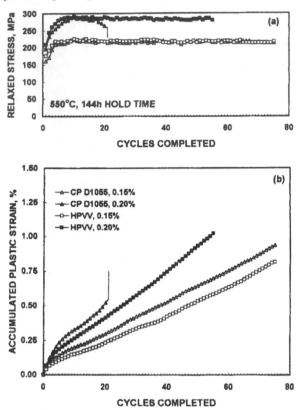

Fig. 4 *Variation of (a) relaxed stress and (b) accumulated plastic strain, with number of 144 h retightening cycles in model bolt tests at 550°C.*

4(b)). The B14 HPVV bolt subject to the same loading history endured 55 cycles before the test was discontinued. No cracking was observed after 75 cycles in bolts of either material retightened to 0.15%ε every 144 h (i.e. B13 and B12).

CP D1055 bolts retightened to 0.15 and 0.20%ε every 1000 h failed after completing 12 (B16) and 6 (B17) cycles respectively. These bolts had accumulated 0.56 and 0.55% plastic strain in the shank at the end of their penultimate cycles (Fig. 5(b)). The HPVV bolts were uncracked after 25 cycles in the B19/0.15%ε/1000 h test and 20 cycles in the B20/0.20%ε/ 1000 h test, the latter being discontinued because hole deformation prevented further insertion of the extensometer.

There was no evidence of cracking after 8 cycles in any of the bolts subject to the 0.15%ε/ 3000 h and 0.020%ε/3000 h tests (Fig. 6).

3.3. Mechanics of Fracture

CP D1055 bolt failure occurred from the first engaged thread in the base of the restraining block assembly (Fig. 7(a)). Creep cracking developed from grain boundary cavitation, forming

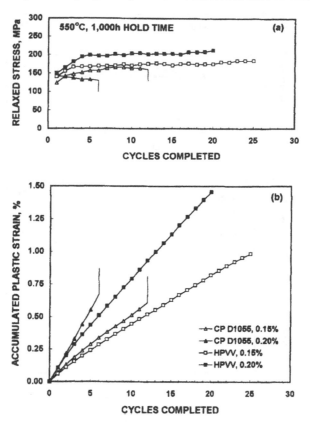

Fig. 5 *Variation of (a) relaxed stress and (b) accumulated plastic strain, with number of 1000 h retightening cycles in model bolt tests at 550°C.*

at the thread root due to the accumulation of plastic strain and leading to ductility exhaustion (Fig. 8). Grain boundary cavitation and microcrack formation were also in evidence at the root of the first engaged thread at the non fractured end of the bolt (Fig. 8) and the second engaged thread at the fractured end (Fig. 9).

Cracking extended by a stable creep crack growth process (Figs. 7–9) to a critical size which could not be tolerated during load take-up in the unloading procedure (Fig. 3). Final fracture was due to overload at room temperature (Fig. 7)

4. Discussion
4.1. High Purity 1CrMoVTiB

Previous results have shown that the high purity 1CrMoVTiB alloy exhibits superior creep rupture ductility, notch tolerance and creep crack growth resistance [2]. The performance of HPVV is equally impressive in full scale bolt tests extending out to 24 000 h at 550°C (Figs. 4–6). No HPVV bolt failures were experienced, whereas three CP D1055 bolts fractured in parallel tests.

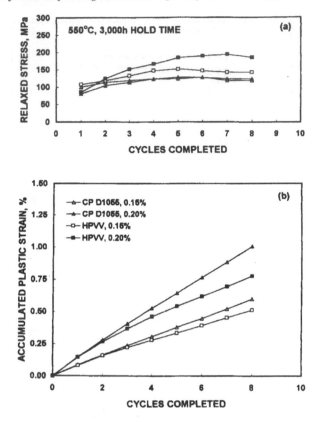

Fig. 6 *Variation of (a) relaxed stress and (b) accumulated plastic strain, with number of 3000 h retightening cycles in model bolt tests at 550°C.*

4.1. High Purity 1CrMoVTiB

The outstanding performance of the HPVV bolts is due to two factors. Firstly, the alloy is more resistant to the accumulation of plastic strain with retightening than CP D1055 (Figs. 4(b), 5(b), 6(b)). More importantly, the long time uniaxial rupture ductility of the high purity alloy is significantly greater than that of conventional purity 1CrMoVTiB. After 20 000 h at 550°C, the rupture ductility of HPVV is ≈14% compared with ≈7% for CP D1055, these values being typical for the two variants of the alloy [2, 4].

Multiaxial rupture ductility (ε'_r) may be estimated from uniaxial ductility (ε_r) using the relationship [5, 6]:

$$\frac{\varepsilon'_r}{\varepsilon_r} \approx 2^{(1-3\sigma_H/\overline{\sigma})} \qquad (2a)$$

where $\overline{\sigma}$ is the effective stress and σ_H is hydrostatic stress.

For the thread root region of the model bolt this equates approximately to:

$$\varepsilon'_r \approx 0.09\varepsilon \qquad (2b)$$

Table 2. *Summary of model bolt test results*

Bolt	Material	ε_o %	Hold Time h	σ_o MPa	$\sigma_{r[1]}$ MPa	$\sigma_{r[2]}$ MPa	$\sigma_{r[5]}$ MPa	N cycles	$\varepsilon_{p[N]}$ %	$\sigma_{r[N]}$ MPa	$\sigma_{TL2[N]}$ MPa	a_{final} observed mm	a_{crit} PD6493 mm
B13	CP D1055	0.15	144	235	161	171	211	75	0.94	213	405	[<5]	5.5
B15	CP D1055	0.20	144	310	197	232	264	20 21*	0.52	264	540	4	4.0
B16	CP D1055	0.15	1000	235	123	142	156	11 12*	0.56	163	315	8	7.5
B17	CP D1055	0.20	1000	310	142	144	133	5 6† 6†	0.55	133 (133)	405 405	5 12 12	5.5 12.5 5.5
B18	CP D1055	0.20	3000	235	100	114	123	8	0.60	119	355	[<6]	6.5
B21	CP D1055	0.20	3000	310	81	104	128	8	1.01	123	396	[<5]	5.5
B12	HPVV	0.15	144	235	174	192	216	75	0.80	217	400	[<6]	6.0
B14	HPVV	0.20	144	310	209	244	280	55	1.02	279	539	[<4]	4.0
B19	HPVV	0.15	1000	235	140	155	167	25	0.98	185	315	[<8]	8.0
B20	HPVV	0.20	1000	310	149	164	198	20	1.46	213	423	[<5]	5.5
B22	HPVV	0.15	3000	235	108	118	152	8	0.51	142	315	[<8]	8.0
B23	HPVV	0.20	3000	310	85	124	184	8	0.77	185	400	[<6]	6.0

ε_o: tightening strain in the shank; σ_o: notional initial stress in the shank at 550°C; N: cycles completed; $\sigma_{r[N]}$: relaxed stress at 550°C at the end of the Nth cycle; $\varepsilon_{p[N]}$: total accumulated plastic strain in the shank at the end of the Nth cycle; $\sigma_{TL2[N]}$: take-up stress during unloading at the end of the Nth cycle; *: denotes bolt failure at the end of the Nth cycle during unloading procedure; †: in this row the data represent the situation at the end of the 6th cycle, prior to unloading; ( ): value assumed from behaviour at end of 5th cycle; [ ]: value assumed from calculated critical crack sizes.

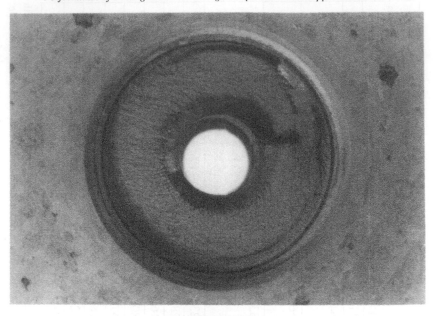

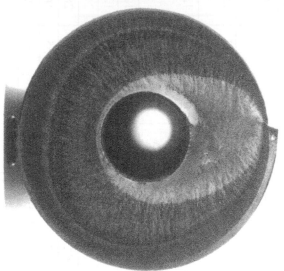

Fig. 7 *Fracture appearance of failed CP D1055 bolts after (a) 21 0.20%ε/144 h retightening cycles, and (b) 6 0.20%ε/1000 h retightening cycles.*

and hence a multiaxial rupture ductility of ≈1.3% is predicted for the high purity steel and this geometry. The accumullated plastic strain in the shank of only one HPVV bolt exceeded 1.3% (i.e. B20*). However, thread root surface inspection revealed no evidence of crack nucleation in any of the HPVV bolts.

---

*Identification numbers are given in Table 2.

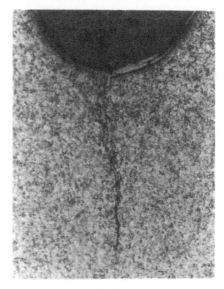

| 50μm |

| 0.5mm |

Fig. 8 Grain boundary cavitation at the root of the first engaged thread at the non fractured end of bolt B15.

Fig. 9 Microcracking in the second engaged thread at the fractured end of bolt B15.

The good experience with the high purity 1CrMoVTiB model bolt tests is mainly attributed to the high rupture ductility of the alloy.

4.2. Failure Diagnosis

CP D1055 bolt failures occurred after 21 cycles in the B15/0.20%ε/144 h test, after 12 cycles in the B16/0.15%ε/1000 h test, and after 6 cycles in the B17/0.20%ε/1000 h test. At the time of failure, the accumulated plastic strain in the shanks of each of the fractured bolts was around 0.5–0.6%. Equation (2b) predicts a multiaxial rupture ductility of ≈0.6% for this CP D1055 steel (i.e. $\varepsilon_r = 7\%$). Creep crack initiation can therefore be explained in terms of ductility exhaustion.

Fracture surface examination of the failed bolts reveals different extents of stable crack growth prior to final fracture. In the case of the B15/0.20%ε/144 h and the B16/0.15%ε/1000 h tested bolts, final crack depths[†] are 4 and 8 mm respectively (e.g. B15 in Fig. 7(a)). The final crack depth in the B17/0.20%ε/1000 h bolt is 12 mm (Fig. 7(b)). The observations are explained by the results of a fracture analysis of the failed bolts during their final unloading operations.

[†]The cracks tended to be part rather than fully circumferential. Nevertheless, a fully circumferential crack was assumed in the fracture analysis, using an 'equivalent' crack depth determined from the mean of 16 equally spaced measurements around the periphery.

Fracture calculations combined the results of LEFM and limit analyses using a PD6493 Level 2 failure diagram [7] (e.g. Fig.10). This defines the conditions for final fracture in terms of K_r and S_r, where K_r is the K/K_{IC} ratio and Sr is the σ_{net}/σ_f ratio. Room temperature K_{IC} values of 50 and 55MPa√m were used for the CP D1055 and HPVV steels [4], while the flow stress, σ_f, was taken to be $(Rp_{0.2}-Rm)/2$ using the information given in Table 1. A 550°C K_{IC} value of 70 MPa√m was taken for CP D1055 [4]. The critical defect sizes predicted to give unstable fracture during unloading are summarised in the last column of Table 2.

The B15/0.20%ε/144 h and the B16/0.15%ε/1000 h bolt failures occurred because their crack depths at the end of cycles 21 and 12 respectively exceeded the critical crack depth during the unloading procedure (cf. observed a_f and predicted a_{crit} in Table 2). The take-up stresses during final unloading are listed in Table 2.

The development of cracking was different in the B17/0.20%ε/1000 h bolt test (Fig.10). At the end of the 5th cycle, the crack depth was just less than the critical size during unloading. During the 6th 1000 h cycle, the crack extended in a stable manner from 5 to 12 mm (Fig. 7(b)). While this crack depth was still just safe at 550°C, it failed in a brittle manner during unloading at the end of the 6th cycle.

Since the plastic strain accumulated in the shanks of the three remaining CP D1055 bolts exceeded 0.6%, it is concluded that crack initiation had occurred but that in no case did the crack size exceed 5 mm.

The CP D1055 model bolt test results demonstrate the advantage of keeping the tightening strain to the lowest level necessary to maintain joint integrity for applications involving conventional purity 1CrMoVTiB bolts subject to multiple retightenings. In these circum-

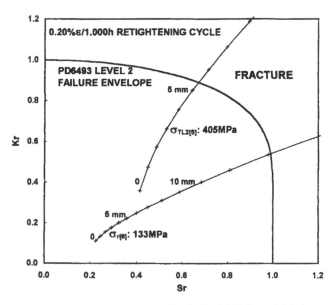

Fig. 10 *Failure assessment diagram for the B17/0.20%ε/1000 h model bolt test showing the variation of K_r, S_r for increasing crack sizes at (a) the take-up stress at the end of cycle 5, and (b) the relaxed stress at the end of cycle 6.*

stances, increasing the tightening strain from 0.15 to 0.20% leads to a reduction in bolt life (Figs. 4(b), 5(b)).

4.3. Stress Relaxation

Model bolt tests are also used as an alternative means of determining stress relaxation properties [8]. While this was not the objective of the present project, first cycle relaxed stress data are compared in Fig. 11 with existing uniaxial results for the CP D1055 and HPVV steels [4]. In view of the very different testing conditions, there is reasonable agreement between the two sets of relaxed stress results.

The relaxed strength at the end of each cycle initially increases with repeated retightening to a plateau level, before appearing to decrease once a crack has initiated (Table 2 and Figs. 4(a), 5(a), 6(a)). The increase in relaxation strength is due to strain hardening, the effect being clearly greater for the high purity steel when the retightening period is of a significant duration.

5. Conclusions

Full size bolt tests have been used to confirm the superior performance of high purity 1CrMoVTiB steel for critical large steam turbine bolting applications involving repeated re-tightening operations.

Tests at 550°C, involving re-tightening strains of 0.15 and 0.20% and retightening periods of 144, 1000 and 3000 h, indicate a substantial advantage of the high purity alloy over

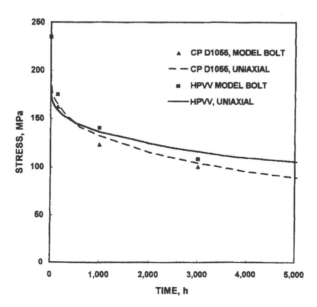

Fig. 11 *Comparison of first cycle model bolt relaxed stress levels with uniaxial stress relaxation data for the HPVV and CP D1055 steels.*

conventional 1CrMoVTiB steels. The superior performance of high purity 1CrMoVTiB is primarily due to its high creep rupture ductility and its greater resistance to stress relaxation following multiple reloading.

References

1. H. F. Tipler, *Conf. Proc. Residuals Additives & Materials Properties*, The Royal Society, London, 1978, May, p.235.

2. A. Strang and S. R. Holdsworth, *Conf. Proc. Rupture Ductility of Creep Resistant Steels*, The Institute of Metals, York, 1990, December, Paper 20, p.224.

3. B. L. King, *Conf. Proc. Residuals Additives & Materials Properties*, The Royal Society, London, 1978, May, p.235.

4. S. R. Holdsworth and A. Strang, unpublished results.

5. M. J. Manjoine, *Welding J. Res. Suppl.*, 1982, February, 50s.

6. D. J. Gooch, *Conf. Proc. Rupture Ductility of Creep Resistant Steels*, The Institute of Metals, York, 1990, December, Paper 27, p.302.

7. BSI PD6493:1991, 'Guidance on Methods for Assessing the Acceptability of Flaws in Fusion Welded Structures', 23.

8. K. H. Mayer and H. König, this proceedings, pp. 54–69.

7

Review of Current International Stress Relaxation Testing Standards

A. STRANG

GEC ALSTHOM Large Steam Turbines, Rugby, UK

1. Introduction

A primary requirement for the design of safe and reliable power plant is the establishment of accurate high quantity material property data. In the particular case of high temperature bolting for modern high performance steam turbine applications, improvements are required not only in the alloys used, but also in the accuracy and reliability with which their long term stress relaxation properties can be measured and predicted. In the past, the design of critical bolted joints, such as those used for HP and IP steam chests and main casings, was based on uniaxial stress relaxation data extending to 30 000 h duration. This ensured steam tight joints at least up to the normal turbine overhaul period of approximately 25 000 h. In modern power plant however, longer overhaul periods are demanded with the result that accurate and reliable uniaxial stress relaxation data extending to at least 50 000 h duration are now required [1].

The determination of higher accuracy long term uniaxial stress relaxation data requires improvements to be made in the currently used testing standards in order to minimise errors arising from test procedures, measurements and control of testing parameters [2–4]. In order to achieve this, the current British Standard for uniaxial stress relaxation testing [BS3500: Part 6: 1969 (1987)] is currently being reviewed by a working party of the BSI Technical Committee on the uniaxial testing of metals (ISM/NFE/4/1) consisting of representatives from manufacturing industry and alloy producers. The primary objective of this work is to establish a code of practice for stress relaxation testing which will lead to improvements in the measurement and accuracy of long term uniaxial stress relaxation data on high temperature bolting alloys. In addition, the code of practice will include procedures for the continuation of previously interrupted stress relaxation tests to longer durations. This paper attempts to identify those areas in the current British Standard where, in the author's view, improvements might be made thus leading to the possibility of producing a revised standard for uniaxial stress relaxation testing. In developing a revised standard, with an associated code of practice, it is hoped to realise the following benefits:

- a reduction in stress relaxation property measurement errors.
- the generation of more accurate, consistent and reliable long term data.

- an improvement in more accurately defining cast to cast variations in alloy properties with better definition of the 'true' material property scatterband.
- a reduction in interlaboratory testing scatter thus enabling improved validation of collaboratively generated data.
- the ability to restart and reliably continue previously interrupted tests to longer durations.
- exploitation of material properties to their full advantage in future high temperature bolting design applications.

In attempting to realise these benefits the revision of the current British Standard BS3500: Part 6: 1969 (1987) will take into consideration the provisions set out in other national standards for uniaxial stress relaxation testing as well as the recommendations of the European Collaborative Creep Committee (ECCC) for the validation of test data [5]. These are compared and contrasted below.

1.1. Current National Standards

Currently, national standards for stress relaxation testing exist in the UK, USA, Japan, France and Germany, although the latter is only in draft form (Table 1). Whilst the British, Japanese, French and German standards are exclusively concerned with uniaxial stress relaxation testing in tension the American Standard [ASTM E 328-86 (1991)] also covers stress relaxation testing in uniaxial compression, bending and torsion. Only stress relaxation testing under uniaxial tension will be considered in this paper. Although the French Standard, [NF A03716 (Nov. 1969)], primarily covers isothermal stress relaxation testing of reinforcement wires and bars for pre and post stressed structures at ambient temperatures, it could in principal be used as a guide for testing at higher temperatures.

1.2. Practicalities of Stress Relaxation Testing

Each of the standards referred to in Table 1 define, to varying degrees, the practical requirements for conducting stress relaxation tests and the controls which must be exercised to meet the requirements of the standard in terms of the technical specification of the test being conducted. The accuracy, reliability and repeatability of the test results are critically dependent on the following factors:

- Technical specification of testing conditions required in terms of (i) the initial loading condition (stress or strain), (ii) test temperature and (iii) test duration.
- Testpiece design, (which may be dependent on availability of material).
- Type and accuracy of extensometry available.
- Type of test machine available.
- Accurate control of strain and test temperature.

Close attention to these factors, which are themselves interdependent, is essential if accurate, reliable and repeatable long term stress relaxation test data are to be obtained. In this respect it is necessary to ensure that testing is conducted in the prescribed manner within the control limits of temperature, strain/load required and that all measuring and control equip-

Table 1. National standards for stress relaxation testing

National Standard	Country of Origin	Stress Relaxation Test Type					Comments
		Uniaxial Tension	Uniaxial Compression	Bending	Torsion		
BS3500: Part 6: 1969 (1987)	UK	√	–	–	–		
ASTM E328-86 (1991)	USA	√	√	√	√		
JIS Z 2276 (1975)	Japan	√	–	–	–		
Stahl-Eisen-Prufblatt 1973	Germany	√	–	–	–		Preliminary Draft
NF A03–716 (1969)	France	√	–	–	–		Ambient Temp.

ment used is calibrated in accordance with the requirements of the pertinent auxiliary standards. It is with the possible improvement of these aspects of BS3500: Part 6: 1969 (1987) that the BSI Stress Relaxation Working Group is primarily concerned.

According to Guest [3], the stress relaxation test is the most demanding elevated temperature test for which material standards exist, being more exacting and difficult to consistently perform than the highest sensitivity creep tests. Not only is the accuracy of the test dependent upon the precision of the extensometry and degree of temperature control, but also on the detailed setting up procedures as well as the skill and experience necessary to achieve consistent results. The latter is rarely specified and it is important factors such as these which may be able to be incorporated into a code of practice to be used in conjunction with a revised testing standard. Some of the main areas where reconsideration of the requirements of BS3500: Part 6: 1969 (1987) might lead to improvements in the accuracy of long term stress relaxation data are considered below.

2. Testing Equipment
2.1. Testing Machines

BS3500: Part 6: 1969 (1987) does not specify the type of testing machine to be used for conducting stress relaxation tests. The standard does however require the testing machine to be calibrated over the load range to be used in accordance with BS 1610, the standard covering methods for the load verification of testing machines. Furthermore, the machine used must

comply with a BS1610 Grade A classification which requires that the load must be known to within +1% of its value at the start as well as during the course of the test. The present testing standard does not specify verification for both increasing and decreasing loads but this would be the normal procedure now adopted when calibrating to the latest issue of BS1610. The only other points relating to test machine requirements implied in BS3500: Part 6: 1969 (1987) are:

- the testing machine used must enable the load to be applied smoothly without appreciable torque to the test piece, and
- reduction of the load can be achieved automatically or manually.

A comparison of the test machine requirements detailed in other national standards is shown in Table 2. From this it can be seen that testing machine types are never specified per se but some indication of their technical requirements in terms of precision, range and operational capabilities are given. Only in the case of the ASTM Standard E 328-86 (1991) is a calibration standard referred to, viz. ASTM E4-93. Thus any type of test machine meeting the required precision in load and enabling the same to be applied and reduced smoothly, either manually or automatically, is deemed to be suitable for conducting stress relaxation tests. In modern terms this encompasses servocontrolled hydraulic and electric machines as well as the more traditional gravity loaded machines normally used in the past [6]. Whilst the former types of test machine are eminently suitable for short term stress relaxation tests (< 1000 h) and or stress relaxation tests at strain levels above yield, lever loaded machines

Table 2. Comparison of national stress relaxation standards – testing machine requirements

National Standard	Test Machine Type	Calibration Standard	Load Precision	Other Requirements
BS3500: Part 6: 1969 (1987)	Not Specified	BS1610 – Part 3: 1990 currently used	Repeatability 1% Error < ±1%	Load to be able to be applied smoothly and axially without torque
ASTM E328-86 (1991)	Not Specified	ASTM E4–93 Under both increasing and decreasing load	< 1% throughout working range	Means of adjusting load continuously and automatically. Axiality to be < 15% as determined using ASTM E139-83 (1990)
JIS Z 2276 (1975)	Not Specified	Not Specified	±0.5% up to 20 tf ±1.0% 20 tf up to 100 tf ±2.0% > 100 tf	Load precision to be verified for both increasing and decreasing load. Means of adjusting load manually or automatically
Stahl-Eisen-Prufblatt 1973 (Preliminary Draft)	Not Specified	Not Specified	Not Specified	
NF A03-716 (1969)	Not Specified but to be reported	Not Specified but method to be reported	< 1% up to 100 tf < 2% and > 100 tf	Must only allow tensile stressing

such as the Dennison T45 type having an adjustable jockey weight are preferred in the UK for long term stress relaxation tests (Fig. 1). With appropriate control equipment, these machines are particularly suitable for long term stress relaxation testing operating either in manual or automatic mode.

In terms of a revised standard it is not envisaged that any specific test machine type would ever be specified. A revised standard would however need to recognise and permit different types of testing machine to be used for specific types of testing and qualify the test requirements accordingly.

2.2. Extensometers

As in the case of stress relaxation testing machines, particular types of extensometers are not specified in BS3500: Part 6: 1969 (1987) other than that they must be of the integrating type and be capable of measuring the strain on both sides of the testpiece. The standard does however require that the extensometers used be calibrated in accordance with, and for specified test conditions, meet the requirements of BS3846. This means that for stress relaxation tests conducted up to and including strain levels of 0.2% the extensometer used must have a minimum of a Grade B rating as defined in the calibration and grading standard BS3846. Thus a Grade B minimum rating requires the ability to detect changes of ≤ 0.15 µm (6 µ in.) and to measure to an accuracy of up to 0.511µm (20 µ in.). For strain levels of > 0.2% extensometers having Grade C and Grade D ratings can be used, depending on the requirements of the test. Furthermore, BS3500: Part 6: 1969 (1989) requires that for the entire

Fig. 1 5 ton Dennison stress relaxation testing machines.

duration of the test. the strain must be capable of being held to within ±1% of its initial specified value. Thus for a stress relaxation test being conducted at a strain level of 0.2% using a testpiece having a 100 mm gauge length, the strain control expected during the entire test, which may extend to 30 000 h or more, must be within ±0.002 mm, i.e. ±2 μm. Clearly, the use of high resolution extensometry coupled with continuous careful control of testing procedures and parameters are necessary to meet such stringent requirements.

Modern extensometers now tend to make use of transducer type measuring devices, particularly because of their suitability for use with automatic data logging systems. These are mainly of the linear variable differential transformer (LVDT) or super linear variable capacitance (SLVC) types. In the case of stress relaxation testing, the sensitivity, accuracy and long term stability of transducers and their associated electronics is of key importance [8]. In this respect some attempt to specify the maximum acceptable errors associated with their use in a revised standard is necessary. A comparison of the extensometry requirements and strain control limits specified in other national standards are shown in Table 3.

2.3. Temperature Control

Details of the test temperature tolerance requirements specified in BS3500: Part 6: 1969 (1987) are shown in Table 4 where they are compared with those stipulated in other national standards for stress relaxation testing. The tolerances quoted in the British Standard include allowable deviations of temperature variation and errors arising from all sources, including thermocouples, temperature measuring devices, cold junctions, furnace controllers, ambient temperature variations, etc. Apart from the French standard (NF A 03-716 Nov. 1969) which is only basically concerned with ambient temperature tests, the temperature tolerances specified in the other national standards are similar. Desvaux [4] has recently suggested that with modern improvements in measuring and control equipment some revision of the currently specified temperature tolerances may be possible. This would clearly be advantageous since it is generally held that temperature variations contribute largely to observed testing inaccuracies. By considering temperature variations and errors associated with thermocouples, measurement systems and drift, Desvaux claims that the total overall uncertainty in temperature measurement could be reduced to < ±2.2°C and that this would be largely independent of test temperature (Table 5). Adding to this the effects of temperature gradients along the testpiece gauge length (generally < ±1°C) Desvaux proposes the revised temperature tolerance limits for single specimen testing shown in Table 6. In comparison with the present British Standard, Desvaux's proposals only offer advantages for tests conducted at temperatures greater than 600°C. If however the gradient along the gauge length could be consistently controlled to < $1^{1}/_{2}$°C it may be possible to reduce the test temperature tolerances for tests conducted up to and including 600°C from ±3°C to ±$2^{1}/_{2}$°C or less. This would have significant advantages, since according to Guest [3], better temperature control could lead to the generally accepted data scatterband on stress being reduced, thus revealing the true extent of cast to cast variations in the materials being tested. Closer temperature control would also be greatly assisted by the use of noble metal thermocouples, such as Pt/Pt Rh, which are universally acknowledged to have good long term stability and are generally used in most laboratories conducting long term high sensitivity creep and stress relaxation testing. This requirement could be included to advantage in a revised stress relaxation test-

Table 3. Comparison of national stress relaxation standards – strain measurement and control

National Standard	Extensometer Type	Calibration	Extensometer Grade	Resolution	Required Strain Control	Gauge Length
BS3500: Part 6: 1969 (1987)	Averaging	BS3846: 1970 (1985) to Grades B, C and D	< 0.2% Strain – B 0.2% to 0.5% – C > 0.5% – D	< 0.15 μm < 0.30 μm < 0.75 μm	Within ±1% of initial value	Largest possible At least 5 times testpiece dia. > 63 mm min for high sensitivity test
ASTM E328-86 (1991)	Averaging	ASTM E83–93	Class B as indicated in E83–93	< 1 μm	±1 μm	As indicated in ASTM standards E8-93 & E13983 (1990) Normally in range 100–200 mm
JIS Z 2276 (1975)	Not Specified	Not Specified	Not Specified	< 1 μm	Within ±1.5% of initial value	Normally 100 mm but not less than 50 mm
Stahl-Eisen-Prufblatt 1973 (Preliminary Draft)	Rib Attachment Type	Not Specified	Not Specified	Not Specified	Not Specified	100 mm
NF A03-716 (1969)	Not Specified	Not Specified	Not Specified	Not Specified	Within 0.1% of initial value	> 100 mm

Table 4. Comparison of national stress relaxation testing standards – temperature measurement and control

National Standard	Test Temp Range °C	Temp Tolerance °C	Spread on Testpiece Gauge Length °C	No. of Thermocouples on Gauge Length	Thermocouple Type	Calibration Standard
BS3500: Part 6: 1969 (1987)	< 600 601 – 800 800 – 1000 > 1000	±3 ±4 ±6 By Agreement	Within Tolerance Range	< 50 mm – 2 > 50 mm – 3	Base and noble metal allowed	BS4937: Parts 2 & 4: 1973 (1981)
ASTM E328-86 (1991)	Not Specified	< ±3 or < ±0.5% whichever is greater	Within Tolerance Range	< 28 mm – 1 25 mm to 50 mm – 2 > 50 mm – 3	Base and noble metal allowed	Calibration — ASTM E220 Measurement — ASTM E139
JIS Z 2276 (1975)	35 – 600 601 – 800 801 – 1000 > 1000	±3 ±4 ±6 By Agreement	Within Tolerance Range	3 – position not specified	Not Specified	JIS C 1602
Stahl-Eisen-Prufblatt 1973 (Preliminary Draft)	< 600 601 – 800	±3 ±4	Within Tolerance Range	2 – outside the gauge length	Not Specified	Not Specified
NF A03-716 (1969)	20°C Unless otherwise specified	< 1	< 1	Not Specified	Not Specified	Not Specified

Table 5. *Summary of sources of temperature measurement errors in a practical creep laboratory (Desvaux Measurement of H.T. Mech. Props NPL Symposium, June 1981)*

Source	±°C	Cumulative ± °C
A. *Thermocouple:*		
Accuracy of standard	0.10	
Calibration procedure for operating couples	0.50	0.60
B. *Measurement System:*		
Compensating cable in lab maintained at ±	0.10	
3°C of nominal	< 0.10	
Electrical junctions	< 0.18	
Cold junction stability	0.13	0.51 digital option
(i) ADC (15 bit)		
or	0.20	
(ii) Potentiometer (average quality)	< 0.02	< 0.6 analogue option
Galvanometer		
Thermocouple drift in 10 000 h	±1°C	
Total overall temperature uncertainty from all sources	< ±2.2°C	

Table 6. *Test temperature tolerance standards*

Current		Proposed by Desvaux		Modified proposal	
BS3500: Part 6: 1969 (1987)					
< 600°C	± 3°C				
601–800°C	± 4°C	< 800°C	± 3°C	601–800°C	± 3°C
801–1000°C	± 6°C	801–1000°C	± 4°C	801–1000°C	± 4°C
> 1000°C	By Agreement	1001–1100°C	± 5°C	1001–1100°C	± 5°C
		1101–1200°C	± 7°C	110 –1200°C	± 7° C

ing standard particularly for those requiring high accuracy long term data or wishing to conduct more exacting tests.

Finally, an important aspect of a revised standard for stress relaxation testing should be the specification of a maximum temperature gradient along the testpiece gauge length of < 2°C, which for single specimen tests is well within modern furnace control capabilities. Excessive temperature gradients are known to be deleterions since they result in variations in the rate of creep damage accumulation along the gauge length of the testpiece and lead to strain levels which are locally greater than those measured by the extensometer [2, 8]. According to Branch [2] a 2°C temperature gradient can lead to an 8% error in the relaxed stress value. At present BS3500: Part 6: 1969 (1987) theoretically permits a temperature gradient along the testpiece gauge length of up to 6°C. In practice the recognised limits are probably < ±1°C and should be incorporated into any revision of BS3500: Part 6: 1969 (1987).

2.4. Testpiece Design and Manufacture

Careful attention to the design and manufacture of stress relaxation testpieces are important factors affecting the generation of accurate and reliable long term data, as is the means of attaching the extensometer. Ideally this should be fitted directly onto the testpiece gauge length, the dimensions of which must be appropriate for the degree of sensitivity and accuracy required. For the generation of long term, low strain stress relaxation data (≥ 30 000 h) circular cross section testpieces having gauge lengths of 100 mm or greater are generally preferred. Careful machining and close control of machining tolerances are also required, particularly with respect to parallelism of the gauge length and co-axiallity of the gripping ends to minimise bending errors during testing.

In terms of testpiece design BS3500 permits the use of testpieces with circular or rectangular cross sections with gauge lengths ranging from 20 to 56 mm. To enable small strains to be measured with a higher degree of accuracy, circular cross section testpieces having longer gauge lengths are permitted provided that the gauge length selected is not less than 5 times its diameter. Wherever possible the largest practicable size of testpiece should be used, preferably with a gauge length ≥ 100 mm. In all instances the parallel length must not vary in diameter by more than 0.03 mm. Furthermore, co-axiallity of the gauge length with the gripped ends must be within a maximum concentricity tolerance of ±0.03 mm. A comparison with the testpiece requirements specified in other national standards is shown in Table 7.

Clearly there is a close link between the requirements of the test conditions, the testpiece design and the extensometer to be used. On this basis a revised standard should reflect the capabilities and limitations of various testpiece/extensometer attachment locations in terms of the accuracy to be expected in conducting the test and if long term stress relaxation data are required, location of the extensometer on the gauge length should be mandatory.

2.5. Stress Relaxation Testing Procedures

Additional factors which can significantly effect the accuracy of stress relaxation test results are the procedures used to reach the initially specified testing conditions. The most important of these are:
   (i) the procedure used for applying the load;
   (ii) the loading rate; and
   (iii) the definition of the start of the test, i.e. $t = 0$.

Whilst the importance of these factors are generally recognised in qualitative terms in existing national standards (Table 8), the varying procedures used can lead to significant differences in results between laboratories, even when a common material is tested under the same specified testing conditions. For example, in a Round Robin stress relaxation testing programme conducted in nine UK and six German laboratories on a single cost of 1CrMoV bolting steel at 0.2% strain and 500°C, Hacon and Krause reported a scatter of approximately ±15% on the initial mean starting stress of 319 MPa and the mean relaxed stress after

Table 7. Comparison of national stress relaxation standards – test pieces

National Standard	Type	Test Piece Gauge Length	Test Piece Diameter Gauge length	Machining Tolerances	Concentricity	Extensometer Fixing
BS3500: Part 6: 1969 (1987)	Circular Section Rectangular Section	As long as possible — but not less than 5.65 √ area	To suit gauge length	< 0.03 mm	±0.03 mm	Not specified but ribs allowed by agreement
ASTME 328-86 (1991)	Circular Section	Generally 100 to 200 mm	Typically 10 to 12 mm	+0.025	< 0.01 mm	Gauge length preferred — shoulders allowed
JIS Z 2276 (1975)	Circular Section	> 100 mm	10 mm	< 0.04 mm	Not Specified	Not Specified
Stahl-Eisen-Prufblatt 1973 (Preliminary Draft)	Circular Section	100 mm	10 mm	+0.02 max.	Not Specified	Ribs
NF A03-716 (1969)	Circular Section	> 100 mm	No machining permitted	No machining permitted	Not Specified	Not Specified

Table 8. Comparison of national stress relaxation standards – test initiation procedures

National Standard	Prescribed Initial Loading Condition	Loading Procedure	Loading Rate	Additional Comments
BS3500: Part 6: 1969 (1987)	Specified Stress or Specified Strain	Smoothly and axially without shock	Not specifically specified — but to be recorded	Loading rate to be kept approximately constant from test to test. Loading stress/strain plot required
ASTM E328-86 (1991)	Specified Strain	Smoothly and rapidly without shock	Not Specified but to be recorded	
JIS Z 2276 (1975)	Specified Strain	Smoothly and rapidly without shock and vibration	Not Specified but to be recorded	Loading stress/strain plot required
Stahl-Eisen-Prufblatt 1973 (Preliminary Draft)	Specified Strain	Smoothly	Not Specified but to be recorded	Loading stress/strain plot required
NF A03-716 (1969)	Specified Stress	Without shock or sudden variation — continuous or in steps	Constant mean rate of $1/8$th of final load/min	Loading stress/strain plot — if loading is non-continuous a minimum of 8 steps are required

2000 h at 107 MPa [10]. Within the UK laboratories temperature control was within the requirements of BS3500 and in the German laboratories in accordance with DIN 50 118. Apart from this all other aspects of the testing technique were in accordance with the normal practices used in each individual laboratory. In this exercise the variations between repeat tests within individual laboratories was found to be small thus indicating that the major cause of the scatter in results was mainly due to differences in interlaboratory testing procedures. Results such as these provide strong evidence supporting the view that, wherever practically possible, quantitative specification of laboratory procedures which will enable testing errors associated with stress relaxation test initiation to be minimised, is desirable [3].

In terms of loading the stress relaxation test, all of the national standards require that the load should be applied smoothly without shock or vibration and that a load/extension plot be recorded for each test. With the exception of the French Standard [NF A03-716 (1969)] which requires the load to be applied at a constant mean rate of $F_o/8$ per min, where F_o is the final starting load, none of the other standards specify a loading rate. Both the British and German standards respectively recommend that a constant rate of loading and a constant loading time be used for each type of material tested in order to achieve consistency of results. Furthermore the American and German standards recognise that creep can take place during loading and advise that the initial loading conditions for the tests should be reached as quickly as possible to minimise loading errors due to the creep process.

The magnitude of these errors is to some extent dependent on the type of testing machine used, the load being able to be applied more rapidly in automatic servohydraulic or servoelectric machines than in manual jockey weight machines such as the Dennison T45. In the latter case, recent experiments conducted on Durehete 1055 at 550°C using loading times varying between approximately 3 and 10 min indicated that, for a starting strain of 0.15%, the error in initial load could be > 5% (Fig. 2). Assuming that stress relaxation curves 'parallel' each other for small differences in load, an error of 5% in an initial stress of say 230 MPa could result in an error of approximately 20% in the relaxed stress at 30 000 h. The loading error due to creep is both temperature and stress dependent, the largest errors occurring at the highest temperature and stresses. As is shown in Fig. 3 and Table 9 the error is clearly also loading rate dependant with smaller errors occurring for shorter loading times. These preliminary results clearly demonstrate the importance of standardising and controlling the initial loading rate in high temperature stress relaxation testing. The effects of varying loading rate are however also dependant on the creep strengths of the materials to be evaluated and consequently further studies will be necessary before loading rate limits can be quantified to the extent that they can be incorporated into a stress relaxation testing standard.

A final area where the effects of creep can result in errors in the stress relaxation data is found in the definition of the start of the test, i.e. when is t equal to 0. With the exception of the French standard (which is a specific case), the other national standards define $t = 0$ as the time when the planned starting stress (δ_0) or starting strain (ε_0) is reached. Provided that there are no delays in defining when δ_0 or ε_0 is reached there should be no significant additive effects of creep onto that which may have to be tolerated associated with the initial loading procedure used. Nevertheless a consistent procedure for defining $t = 0$ would be an advantage in minimising testing errors from this source.

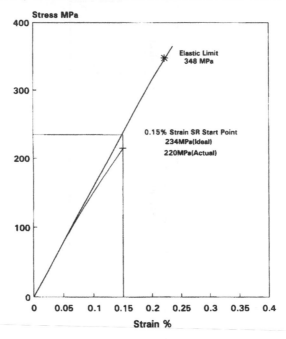

Fig. 2 *Starting conditions for SR test on Durehete 1055 at 0.15% strain and 550°C.*

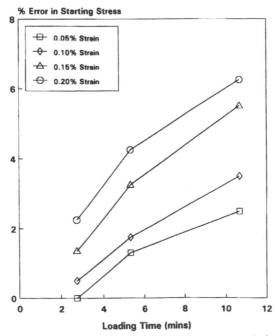

Fig. 3 *Effect of loading rate on the SR test starting stress at various strain levels for Durehete 1055 at 550°C.*

Table 9. Effect of loading time on stress relaxation test starting stress for Durehete 1055 at 550°C

| Loading Time | % Error in Loading Stress for Different Strain Levels at 550°C | | | |
	0.05% Strain	0.10% Strain	0.15% Strain	0.20% Strain
< 1 min	0	0	0	0
2 min 44 s	0	0.50	1.37	2.15
4 min 20 s	1.30	1.80	3.25	4.23
10 min 40 s	3.40	3.40	5.15	6.15

3. Final Observations

Based on the various points already discussed, it is clear that in order to reduce errors associated with stress relaxation testing, revision of the current British Standard BS3500: Part 6: 1969 (1987) is clearly merited. Specific attention to the following factors would be advantageous in improving the accuracy and reliability with which long term stress relaxation data could be determined. This is held to be an important objective in generating reliable long term stress relaxation data for the design of high temperature bolting for advanced power plant.

- Temperature Control — Reduction of temperature tolerance limits for tests conducted up to 600°C from +3°C to < $2^{1}/_{2}$°C. Specification of maximum temperature gradient in testpiece gauge length of < 2°C.

- Testpieces — Specification of standard testpieces with gauge lengths ≥ 100 mm. Specify that extensometer must be Class B minimum and fitted to testpiece gauge length.

- Testing Procedures — Specify loading rate conditions to minimise errors due to creep effects on loading. Clearly specify test start conditions on loading to δ_0 or ε_0, i.e. when $t = 0$ to minimise errors due to creep effects in loading.

- Test Machines — Specify calibration of testing machine for both increasing and decreasing loads. Specify limits on testing machine axiallity.

Finally it is important to ensure that calibration of thermocouples, transducers, extensometers and test machines are carried out to appropriate standards with full traceability and that the uncertainties in measurements are controlled in accordance with current CEN and ISO directives.

References

1. J. Bolton, 'Design Considerations for High temperature Bolting', this volume, pp. 1–14.

2. G. D. Branch, 'Design and Operation Requirements for High Temperature Mechanical Property Data', in *Measurement of High Temperature Mechanical Properties of Materials* (Eds M. S. Loveday, M. F. Day and B. F. Dyson), NPL, 1982, pp. 13–22.

3. J. C. Guest, 'Standards in Elevated Tensile and Uniaxial Creep Testing', *ibid.*, pp. 23–31.

4. M. P. E. Desvaux, 'Practical Realisation of Temperature Measurement Standards in High Temperature Mechanical Testing', *ibid.*, pp. 91–112.

5. J. Orr, 'Activities of ECCC Working Group on High Temperature Bolting Materials', this volume, pp. 115–124.

6. M. S. Loveday and B. King, 'Uniaxial Testing Apparatus and Testpieces', in *High Temperature Mechanical Properties of Materials* (Eds M. S. Loveday, M. F. Day and B. F. Dyson), NPL, 1982, pp. 128–157.

7. D. S. Wood and J. Wynn, 'Machine Design Requirements', *ibid.*, pp. 113–127.

8. M. F. Day and G. F. Harrison, 'Design and Calibration of Extensometers and Transducers', *ibid.*, pp. 225–240.

9. C. Downey St and J. H. M. Draper, *Proc. Inst. Mech. Eng.*, Vol. 178, Pt 3L, 1963-64, Paper 32.

10. J. Hacon and M. Krause, Relaxation Properties of a 1CrMoV Bolting Steel Under Uniaxial Tensile Load, *Proc. Conf. Properties of Creep Resistant Steels*, VDEL, Düsseldorf, 1972.

8

Activities of a European Creep Collaborative Committee Working Group on High Temperature Bolting Materials

J. ORR

British Steel Technical, Swinden Laboratories*, Rotherham, UK

ABSTRACT

The area of responsibility and activities of EC3 WG 3.4 with regard to the composition, heat treatment and properties of bolting materials used within European industries are described.

The working group comprises representatives of manufacturers, users and national committees within the major European countries where bolting materials are manufactured and used.

A wide range of materials from low alloy steels to nickel-based alloys, and the extent of property values for creep rupture and stress relaxation have been identified by the group. These data are available within the group on a mutual exchange basis and are used to provide comment and support to relevant European standards.

Interaction with CEN[†] and ECISS[†] committees on grade selection and composition control for bolting steels is described.

The future activities of the group in terms of data collection and assessment to agreed EC3 guide lines are defined.

1. Introduction

The European Creep Collaborative Committee ((ECCC) — see Glossary in Appendix 1), was formed in late 1992, to bring together the mutual interests of material producers, (mainly steelmakers), manufacturers, fabricators, utilities and research institutes, involved in high temperature materials, principally for the power generation industry.
The aims of the ECCC are:

(i) to exchange data for mutual benefit and thereby reduce some future test duplication;

(ii) to agree common rules and procedures for generation, collation and assessment of creep data (includes stress relaxation);

(iii) to interact with European Standards makers and so influence and improve the quality of standards relevant to industries related to power generations.

---

*See also Appendix 2.
[†]See Glossary Appendix 1.

The ECCC is an independent body but has already become recognised in Europe as a significant source of expertise with respect to creep properties of materials.

This paper describes the activities and objectives of one of the working groups of ECCC, that is involved with bolting materials. Its designation within the ECCC structure is WG 3.4.

2. The ECCC

The ECCC is structured so that working groups have well defined areas of responsibility, each reporting to a Management Committee on which most of the European Union and EFTA countries are represented. The working groups are bodies of experts and thus represent the areas of interest of individual countries. Whilst WG 1 and WG 2 have specific interests relating to data and technical publicity, the WGs 3.13.4 have specific interests defined by steel type and/or application (Fig. 1).

The ECCC has become recognised by COCOR as a relevant body from which comments on those draft European Standards which contain property values relating to stress rupture, creep and stress relaxation will be accepted and considered.

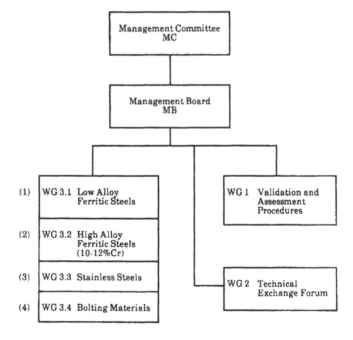

Chairman/Convener

(1)   Dr. W. Bendick - Mannesmann, Germany

(2)   J. Hald - Elsam/Elkraft, Denmark

(3)   Prof. R. Sandstrom - SIRM, Sweden

(4)   J. Orr - British Steel, UK

Fig. 1 *Structure of European Creep Collaborative Committee.*

The costs of organisation, co-ordination and management of the ECCC are funded by a Brite Euram II Concerted Action (BE5624) contract until December, 1996. The responsibilities for the cost of the WG 3 working groups and their agreed activities, apart from travel and subsistence, belong with the members of the groups. The Secretariat of ECCC and its working groups is held by ERA Technology Ltd.

3. ECCC-WG 3.4 Bolting Materials

The membership of WG 3.4 is drawn from France, Germany, Italy and UK (see Appendix 2), thus representing the principal manufacturers and users of bolting materials.
The objectives of WG 3.4 are to:

(i)   co-ordinate the collection and exchange of stress rupture and stress relaxation data for bolting materials used in Europe;

(ii) identify industry requirements related to such materials in terms of material type and data required;

(iii) interact with Committees CEN TC74 WG9 and ECISS TC22/23 SCI in relation to the standardisation of bolting materials; and

(iv) review the metallurgically important factors for testing and properties.

With regard to (i) above, ECCC-WG 1 will propose, by the end of 1994, methodologies to be used by the WGs 3.1–3.4, for data generation, collation and assessment, thus determining a high degree of commonality across Europe for testing and data assessment. This will ensure proper representation of data and an ability for fair comparison of property values.

The industry requirement and the interactions with the relevant CEN and ECISS committees activities are largely synonymous, through a degree of common membership on these and WG 3.4.

The metallurgical activity areas include type of testing methods, e.g. uniaxial and model bolts, factors influencing notch sensitivity and rupture ductility and the development and improvement of new and existing materials.

4. Bolting Materials in Europe
4.1. Standards Requirements

CEN and ECISS comrnittees are preparing application and material requirement standards respectively to support a European Pressure Vessel Code, which ought to be in place by the year 2000 initially on a voluntary basis.

CEN TC74 WG 9 is preparing a standard for bolting materials to be used with flanges, operating at cryogenic and elevated temperatures. Some material requirements have been defined by this Committee but this responsibility has been devolved to ECISS TC22/23 SCI. The ECISS committee has identified several steels and alloys commonly used for bolts and fasteners in the pressure vessel industries in Europe and has made proposals for the compositions and property levels required, the latter largely in terms of tensile and impact properties at ambient.

The material types, grades and their specified compositions identified by ECISS TC22/23 SCI for bolting applications at elevated temperatures are listed in Tables 1 and 2. (It should be noted that this committee is also responsible for grades to operate at cryogenic temperatures, but this is outside the scope of this paper.) Other grades were examined by ECISS TC22/23 SCI but not included — as indicated by the non-continuous sequence numbers.

4.2. Role of EC3 WG 3.4

EC3 WG 3.4 has been invited by ECISS TC22/23 SCI to examine the list of steel grades selected (Table 1), and to comment on:

(i)   temperature range of application;

(ii) range of available data appropriate to (i);

(iii) specific details relating to compositions and heat treatment.

4.2.1. Selection of Grades

The first response of EC3 WG 3.4 was to note, for example, that steels C35E, 25CrMo4 and 42CrMo4 were not usually used above ~450°C. Therefore, creep rupture and stress relaxation data, for which this committee is responsible, do not apply for these steels. Thus, further consideration of these steels in the context of EC3 WG 3.4 has not been pursued.

In addition to the steels listed in Table 1, EC3 WG 3.4 identified other grades of material which are used in Europe for bolts operating at elevated temperatures. These are 0.4% C, 1% Cr, 0.5 %Mo (BS 1506-631-850) and Nimonic 80A. The 0.4% C, 1% Cr, 0.5% Mo grade is similar to that of 42CrMo4 and is used, at least in UK designed turbines, within the 'creep' range of temperatures. Similarly, Nimonic 80 alloy is used sometimes in place of the higher strength low alloy steel grades in some turbine designs.

Furthermore, it has been indicated, without confirrnation at present, that the ECISS Committee wishes to extend the range of austenitic steels to include Types 304, 316, 316LN and 321.

Through links with the COST 501 Round III group, it has also been indicated that high strength austenitic steels, e.g. warm worked Esshete 1250 [1] may be required for future designs of steam turbines operating at higher temperatures than the maximum currently used, i.e. above 565°C. The composition ranges for the austenitic steels are included in Table 2 (p.120).

At present, these additional austenitic steels have not been included in the work programme of EC3 WG 3.4, but can be introduced into the work programme when their requirement is properly identified.

Table 1. *Bolting material grades in Europe (using TC22/23 designations and individual sequence numbers)*

A. Carbon Steels
1. C35E - relates to 1.1181.
B. Low Alloy Ferritic Steels
3. 25CrMo4 — relates to 1.7218 and 26CD4.
4. 42CrMo4 — relates to BS 1506-630 and ASTM A193 B7.
5. 40CrMoV4-6 — relates to 1.7711, BS 1506-670-860 and ASTM A193 B16.
6. 21CrMoV5-7 — relates to 1.7709 and 20CDV 5.07.
7. 20CrMoVTiB4-10 — identical to BS 1506 681-820.
C. High Alloy Ferritic Steels
10. X21CrMoNiV12-1 — relates to X22CrMoV12.1.
11. X12CrNiMoV12-3 — relates to turbine blade grade, e.g. DTD 5066.
12. X20CrMoNiNbVN11-1 — relates to X19CrMoVNb11.1.
D. Austenitic Steels
14. X8CrNiMoBNbl6-16 — relates to 1.4980.
15. X6NiCrTiMoVB25-15-2 — relates to 1.4986 and BS 1506 A286 531.

4.2.2. Available data

A survey of data collected by various organisations throughout Europe on most of the steel grades in Table 1 and for those additional steels identified by EC3 WG 3.4 is summarised in Table 3. This is a preliminary listing, since it does not yet include indications of the amounts of data available from notched specimen tests and embrittlement tendencies, these being recognised as important additional factors because of the requirement for acceptable ductility levels in bolt materials. Data on these factors are being examined currently by EC3 WG 3.4.

The data listed in Table 3 represent preliminary views of the data available from several sources. However, this does not necessarily represent that which will be available to all parties since mutual exchange has been agreed as a working principle within all EC3 WGs.

It is the responsibility of individual members of EC3 WG 3.4 to represent not just their own organisations but also more broadly based groups, where these exist within countries. Thus, for example, the German, Italian and UK representatives have been able to contribute inforrnation on behalf of national groups within their respective countries. This situation will be taken into account when data exchange begins. Examples of the range and amount of stress rupture data available from 560°C tests, for the steel 20CrMoVTiB4.10, is shown in Fig. 2.

Table 2
Bolting Materials - Specified Compositions

ECISS Steel Name	Trade Name	Number	Specified Elements wt.%																
			C	Si	Mn	P	S	Cr	Mo	Ni	Al	As	B	Cu	Nb	Sn	Ti	V	N
C35E	-	1.1181	0.32 0.39	≤0.40	0.50 0.80	≤0.035	≤0.035	≤0.40	≤0.10	≤0.40	-	-	-	-	-	-	-	-	-
25CrMo4	-	1.7218	0.22 0.29	≤0.40	0.60 0.90	≤0.035	≤0.035	0.90 1.20	0.15 0.30	-	-	-	-	-	-	-	-	-	-
42CrMo4	-	-	0.38 0.45	≤0.40	0.60 0.90	≤0.035	≤0.035	0.90 1.20	0.15 0.30	-	-	-	-	-	-	-	-	-	-
40CrMoV 4-6	A193 B16 (D 950)	1.7711	0.36 0.44	≤0.40	0.45 0.85	≤0.030	≤0.030	0.90 1.20	0.50 0.65	-	-	-	-	-	-	-	-	0.25 0.35	-
21CrMoV 5-7	-	1.7709	0.17 0.25	≤0.40	0.40 0.80	≤0.030	≤0.030	1.20 1.50	0.65 0.80	≤0.6	-	-	-	-	-	-	-	0.25 0.35	-
-	D 900	BS 1506 631-850	0.35 0.45	0.10 0.35	0.40 0.70	≤0.035	≤0.040	1.00 1.50	0.50 0.70	≤0.40	-	-	-	-	-	-	-	-	-
20CrMoVTiB 4-10	D 1055	BS 1506 681-820	0.17 0.23	≤0.40	0.35 0.75	≤0.020	≤0.020	0.90 1.20	0.90 1.10	≤0.20	0.015 0.08	≤0.020	0.001 0.010	≤0.20	-	≤0.020	0.07 0.15	0.60 0.80	-
X21CrMoNiV 12-1		1.4923	0.18 0.24	≤0.40	0.30 1.00	≤0.030	≤0.030	11.0 12.5	0.80 1.20	0.30 0.80	-	-	-	-	-	-	-	0.25 0.35	-
X12CrNiMoV 12-3		1.4938	0.08 0.15	≤0.35	0.50 0.90	≤0.030	≤0.025	11.0 12.5	1.50 2.00	2.00 3.00	-	-	-	-	-	-	-	0.25 0.40	0.020 0.040
X22CrMoNiNb V11.1		1.4913	0.18 0.25	≤0.50	0.30 1.00	≤0.025	≤0.015	10.0 12.0	0.50 1.00	0.30 1.00	-	-	≤0.005	-	0.25 0.55	-	-	0.10 0.30	0.05 0.10
X7CrNiMoBNb 16.16		1.4986	0.04 0.10	0.30 0.60	≤1.5	≤0.045	≤0.030	15.5 17.5	1.6 2.0	15.5 17.5	-	-	0.05 0.10	-	10 x 6 1.20	-	-	-	-
X6NiCrTiMo VB 25.15.2	A286	1.4980	0.03 0.08	≤1.0	1.0 2.0	≤0.025	≤0.015	13.5 16.0	1.0 1.5	24.0 27.0	≤0.35	-	0.003 0.010	-	-	-	1.90 2.30	0.10 0.50	-
NiCr20TiAl	Nimonic 80A	2.4952	≤0.10	≤1.00	≤1.00	≤0.030	≤0.015	18.0 21.0	-	Bal. ≥65	1.00 1.80	-	-	-	-	-	1.8 2.7	Co ≤2.00 Fe ≤3.00	-
X6CrNi 18.10	'304'	1.4948	0.04 0.08	≤1.0	≤2.0	≤0.035	≤0.015	17.0 19.0	-	8.0 11.0	-	-	-	-	-	-	-	-	-
X6CrNiMo 17.12.2	'316'	1.4919	0.04 0.08	≤1.0	≤2.0	≤0.035	≤0.015	16.5 18.5	2.0 2.5	10.0 13.0	-	-	-	-	-	-	-	-	-
X3CrNiMoBN 17.13.3	'316LN'	1.4910	≤0.04	≤0.75	≤2.0	≤0.035	≤0.015	16.0 18.0	2.0 3.0	12.0 14.0	-	-	-	-	-	-	-	-	0.10 0.18
X7CrNiTi 18.10	'321'	1.4941	0.04 0.08	≤1.0	≤2.0	≤0.035	≤0.015	17.0 19.0	-	9.0 12.0	-	-	-	-	-	-	5 x 6 0.80	-	-
	'1250'	BS 3059 215S15	0.06 0.15	0.20 1.00	5.50 7.00	≤0.040	≤0.030	14.0 16.0	0.80 1.20	9.0 11.0	-	-	0.003 0.009	-	0.75 1.25	-	-	0.15 0.40	-

Table 3. *Summary of data available in Europe on bolting materials*

Steel Grade	No. of Casts Tested	Stress Rupture Max. Duration, h	Stress Relaxation Max. Duration, h
21CrMoV5.7/4.7	28	50 000	40 000
21CrMoV5.11	25	120 000	40 000
40CrMoV4.7	7	90 000	10 000
D900 (1CrMo)	28	90 000	30 000
D950 (1CrMoV)	22	70 000	30 000
D1055 (1CrMoVTiB)	64	100 000	40 000
X19CrMoNbN11.1	27	130 000	40 000
X20CrMoV12.1 (600 MPa)	62	105 000	–
X22CrMoV12.1 (>700 MPa)	21	120 000	40 000
A286	2	20 000	–
Nimonic 80	6	50 000	40 000

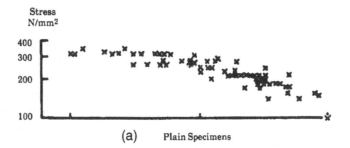

(a)   Plain Specimens

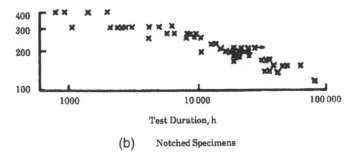

Test Duration, h

(b)   Notched Specimens

Fig. 2 *Stress rupture data sets at 550°C for steel X20CrMoVTiB 4.10.*

Such data compiled by EC3 WG 3.4 will be assessed using standardised methodologies developed by EC3 WG1 [2], to determine property values which can be used either directly by turbine manufacturers, as at present from material data sheets, or via appropriate (European) standards where they will be included as part of the material characteristics.

4.2.3. Detail responses

Following the agreement with COCOR that EC3 can participate in discussions regarding relevant European standards, an opportunity has been taken to react to a draft document of ECISS TC22/23 SCI on bolting materials. Similarly, the ECISS Committee has raised detailed questions about composition ranges for certain elements in two of the steels listed in Table 1. Thus, a working relationship has been established.

One example of the former situation is that the ECISS Committee, having taken an overall view of specified compositions ranges, redesignated a steel commonly known as X19CrMoVNbN11.1, as in DIN 17240, as X20CrMoNiNbVN11.1 (see Table 1). Since this steel is often referred to as 'X19' and another related but different steel type referred to as 'X20', this change was considered to be confusing for manufacturers and users. Therefore, EC3 WG 3.4 has responded requesting small changes in carbon content range, e.g. 0.17–0.22% rather than 0.17–0.23%, to allow the standard designation of 'X19' to be retained.

This may seem to be a trivial point, but it is an example of a considered unified industry response to an early stage of development of a European standard.

5. Conclusions

The activities of ECCC (EC3) WG 3.4 'High Temperature Bolting Materials Working Group' within the contexts of the ECCC and interactions with European standards bodies have been described.

The working group consists of representatives from individual industries principally in France, Germany, Italy and UK. Some of the members also represent national interest groups so that in total a wide sector of European industry which manufactures and uses bolting materials, principally in power generation, is represented.

The purposes of the group are to collect and assess data and to define appropriate values to assist in the design of bolted joints and to understand relevant factors determining those properties. The data collected will be reviewed at regular intervals and made available on a mutual exchange basis.

Interactions have already occurred with relevant European Standards Committees, through mutual membership and in fulfilment of the agreement reached between ECCC and COCOR. Comments from EC3 WG 3.4 have been passed to ECISS TC22/23 regarding grade selection and composition control of bolting steels.

EC3 WG 3.4 covers a wider range of materials than envisaged in the ECISS standard with the inclusion of nickel base alloy.

6. Acknowledgements

The author thanks the members of EC3 WG 3.4 for assistance in compiling this paper and support of EC3 Management Committee, Dr. M. J. May, Manager, Swinden Laboratories and Product Technology and Dr. R. Baker, Director Research & Development, British Steel for permission to publish this paper. The financial assistance from Brite Euram Co-ordinated Action Contract BE 5524 is also acknowledged.

References

1. J. Orr, H. Everson and G. Parkin, 'Warm Worked Esshete 1250 — A High Strength Bolting Steel', this volume, pp. 203–219.
2. European Creep Collaborative Committee Working Group 1 (Convener: Dr. S. R. Holdsworth), GEC Alsthom, Rugby.

Appendix 1

Glossary of Abbreviations

CEN	Comité Européen de Normalisation (European Committee for Standardisation)
COCOR	Co-ordinating Committee on the Nomenclature for Iron and Steel Standards
ECCC (EC3)	European Creep Collaborative Committee
ECISS	European Committee for Iron and Steel Standardisation
EFTA	European Free Trade Association
Technical Committee	Sub-Committee Responsible to TC.

Appendix 2
Membership of ECCC WG 3.4

Technology Ltd (Secretary)
h Steel
ricité de France
Steels
Kraftwerke AG
Energie
ns AG
h Steel (Convener)

Technology Ltd (for Project 2021).

SESSION 3

Ferritic Bolting Steels

Chairman: DV Thornton
GEC ALSTHOM Large Steam Turbines, Rugby, UK

9

ESB Approach to the Assessment of High Temperature Ferritic Steel

J. H. BULLOCH and A. BISSELL

Electricity Supply board, Power Generation, Dublin 2, Ireland

ABSTRACT

The present paper describes the findings to date of an ongoing investigation aimed at understanding the factor which control the toughness degradation characteristics of a series of CrMoV steel turbine bolts during service. Essentially it has been demonstrated that the primary factors which dictate the degree of toughness loss through reverse temper embrittlement were microstructural grain size, d, bulk phosphorus, %P and accumulated plastic strain, %ε . Indeed a unique relationship was shown to exist between grain size and %ε phosphorus for a fixed value of %ε such that

$$d \times (\%P) = \text{constant}.$$

When the LHS was greater than the value of the constant embrittlement occurred whereas when the value was lower no toughness degradation was evident. From a creep damage assessment to was found that RTE and widespread cavitation, representing a situation which could seriously affect integrity considerations, were prevalent in the majority of the bolts at accumulated strain levels of around 0.5%.

1. Introduction

In the vast majority of instances important steels are ordered to various specifications, viz.

(i) chemical composition which covers both purpose alloy additions (Cr, Mo, V) and residual tramp elements (P, S, Sn), and

(ii) basic mechanical properties.

However, as a result of steelmaking and subsequent fabrication procedures cast to cast variations in bulk chemical composition and microstructure are commonplace. Indeed numerous mechanical properties (yield strength, impact toughness, ductility, etc.) are directly controlled by the strength of grain boundary locations which are essentially dependent upon the extent or concentration of certain metalloid impurity elements, e.g. P, Sb, Sn [1–3].

Woodvine [4], as early as over four decades ago, indicated that the segregation of such tramp elements to prior austenite grain boundaries in low alloy steels caused a significant reduction in Charpy impact strength. This phenomenon is commonly known as reverse temper embrittlement, RTE, and usually occurred upon slow cooling or isothermal holding at temperatures within the range 300–600°C. Since certain critical steam raising plant components, such as a turbine, are subjected to prolonged service within this temperature range it is expected that toughness degradation will occur in turbine rotors casings and bolts. At such temperatures creep processes are also prevalent and in combination they can, in certain instances, precipitate catastrophic brittle fracture of turbine components.

During a routine maintenance overhaul of a 250 MW steam turbine the removal of the intermediate pressure inner cylinder casing revealed the presence of a number of cracked and through thickness section fractured flange studs. Indeed it was observed that 50% of the studs (totalling 16) were completely fracture, 25% contained sizeable cracks, while the remaining 25% contained sizeable cracks, while the remaining 25% were defect free [5]. These particular studs (i) were fabricated from a 1CrMoV Steel, (ii) were around 140 mm in diameter and 700 mm long, (iii) were subjected to temperatures around 490°C for some 50 000 h and (iv) had seen around 600 and 300 hot and cold starts respectively.

Much evidence of significant creep damage was observed on the fracture surfaces, viz. primary and secondary macrocracking as well as isolated and linked cavities (incipient microcracks). As well as irreversible creep damage the studs were in the reverse temper embrittled, RTE, condition through the diffusion of phosphorus solute atoms to prior austenite grain boundary locations.

As a direct result of this experience the ESB has conducted a lengthy assessment study on RTE and creep damage of a series of CrMoV steel turbine bolts and studs. The present paper describes this study and the incorporation of such findings into bolt inspection guidelines particular to the ESB.

2. RTF and Creep Assessment Procedures

Two sets of IP and HP turbine bolts fabricated from a 1.3% Cr, 0.8% Mo, 0.35V steel, from two sister 120 MW units, designated units 1 and 2, were (i) subjected to a magnetic particle inspection aimed at detecting any surface cracking in the bolt thread locations and (ii) given length measurement to assess the accumulated strain during the 122 600 h at temperatures approaching 490°C. Essentially no cracks were detected in any of the one hundred and eighty two bolts and measurable accumulated strains were only observed in a few IP bolts. These strains varied from around 0.1% minimum to a maximum of 0.58%. A selection of bolts which exhibited strains towards the maximum were sectioned, and after conventional metallographic preparation, subjected to a scanning electron microscope assessment in an effort to determine if any creep damage had occurred during service.

A small metallographic section was removed from the keyway location. From each section microstructural hardness, microstructural type, average grain size diameter and full bulk chemical analysis were obtained. The bolt embrittlement assessment was based on two criteria; (a) full size Charpy V notch impact specimens and (b) Auger Electron Spectroscopy (AES). Two Charpy criteria were utilised, viz. the fracture appearance transition temperature (FATT) and the room temperature Charpy energy (RTCE). In the present AES study, phosphorus was identified as the grain boundary segregant species.

3. Results

From the various Charpy and AES data it was established that the bolts could be simply classed as:

(a) embrittled — FATT = + 85–125°C, RTCE = 5–35 J and % phosphorus monolayer > 0.1,

(b) partially embrittled — FATT = + 15–40°C, RTCE = 60–80 J and % phosphorus monolayer = 0.5, and

(c) non-embrittled — FATT < 0°C, RTCE > 100 J, zero phosphorus monolayers.

At this figure only around 30% of the fracture surface exhibited significant grain boundary phosphorus segregation. The corresponding AES analysis indicated that the embrittled bolts exhibited phosphorus segregation levels of mainly between 0.2 to 0.3 monolayers.

The relationships between hardness level and average grain size for Unit 1 and Unit 2 are given in Figs. 1 and 2 respectively. These figures illustrate that hardness is not a reliable factor in signalling embrittlement. Figures 3 and 4 show the relationship between grain size and a compositional factor of so-called 'J factor' for Units 1 and 2 respectively, where $J = (Mn + Si)(P + Sn) \times 10^4$.

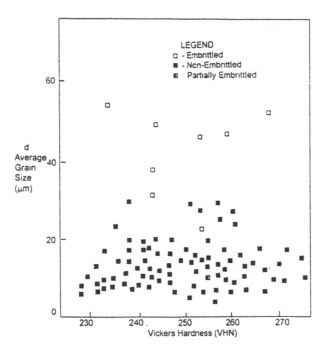

Fig. 1 *Relationship between average grain size and the microscopic hardness value for all IP and HP bolts from Unit 1.*

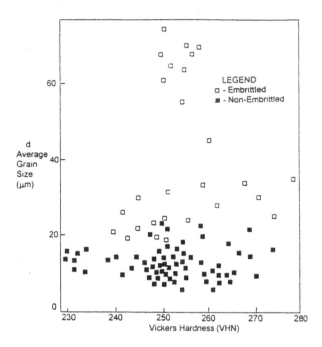

Fig. 2 *Relationship between average grain size and the microscopic hardness value for all IP and HP bolts from Unit 2.*

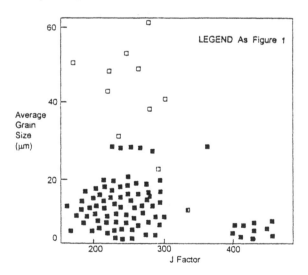

Fig. 3 *Relationship between grain size and the composition-based J factor for all the IP and HP bolts from Unit 1.*

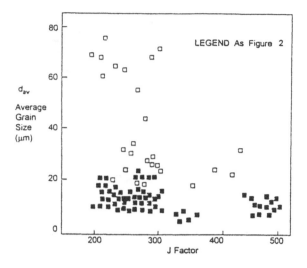

Fig. 4 *Relationship between grain size and the composition-based J factor for all the IP and HP bolts from Unit 2.*

From these figures it is seen that there is no clear indication whether *J* dictates the occurrence of embrittlement during service. Interestingly through, from Figs. 1–4, it appears that significant effects of grain size on embrittlement are operative inasmuch that most embrittlement bolts occurred at grain sizes greater than 20–30 μm.

The average grain size values of each individual HP and IP bolt from Units 1 and 2 are plotted as a function of bulk phosphorus level and are shown in Figs. 5 and 6 respectively. Upon inspection of these figures it is evident that two distinct regimes are evident, viz. an embrittled regime and a non-embrittled regime, which are separated by a critical interface or line which can be described by the expression

$$d = 0.28 \, (\% \, P) \quad \text{Unit 1} \tag{1}$$

$$d = 0.18 \, (\% \, P) \quad \text{Unit 2} \tag{2}$$

and clearly the embrittled condition can more readily occur in Unit 2, i.e. embrittlement was seen at lower grain size – % phosphorus combinations.

From the bolt length measurements it was found that the average accumulated strain in Unit 2, 0.36%, was more than double that observed in Unit 1, 0.17% and it is suggested that the differences in embrittlement observed between Unit 1 and 2 are the result of powerful effects of accumulated strain on RTE. By assuming that the strain effect is linear in nature Fig. 7 (p.134) can be constructed, and it is evident that at strain levels of around 0.5% the majority of the CrMoV bolts will have become embrittled under the prevailing conditions of the present study; at a 0.01% P level only those bolts with very fine grain sizes (below 10 μm) will not undergo embrittlement.

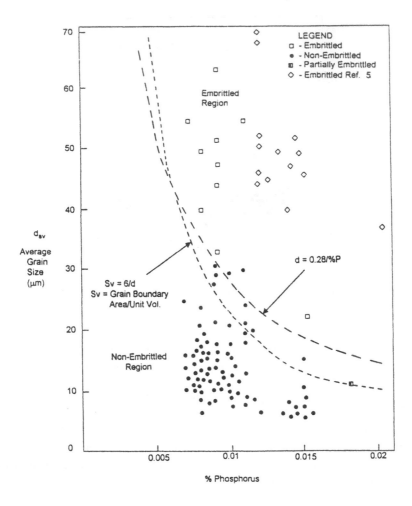

Fig. 5 *Relationship between grain size and phosphorus level for all the IP and HP bolts from Unit 1.*

The influence of accumulated strain, ε, on RTE susceptibility is portrayed graphically in Fig. 8, from which it is evident that increased accumulated strain induced the occurrence of embrittlement at finer grain sizes. Again, two distinct embrittled and non-embrittled regimes are evident and the interface line between them can be expressed as

$$d = 36\text{--}48 \ (\%\varepsilon) \qquad\qquad (3)$$

where d is in μm.

The findings of the creep damage assessment are presented in a plot of grain size against % accumulated strain in Fig. 9 (p.135). From such a figure certain important information regarding RTE and creep damage are evident. Indeed the figure is comprised of three discrete regions:

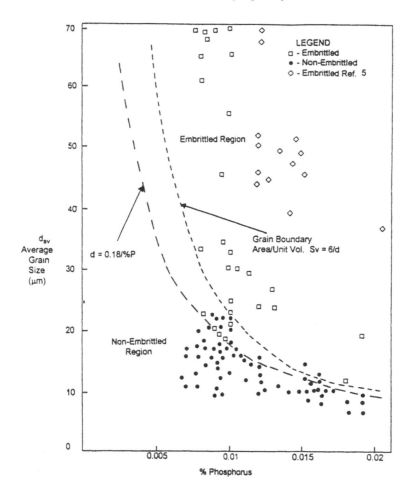

Fig. 6 *Relationship between grain size and phosphorus level for all the IP and HP bolts from Unit 2.*

(i)   Non-embrittled zero creep-damaged region,

(ii) an embrittlement zero creep–damaged region, and

(iii) an embrittled creep-damage region and demonstrated the powerful effects that grain size can have on RTE and, to a certain extent, creep damage.

Only the bolts within the embrittled creep damage region would cause safety concerns and for bolts with grain sizes above around 10 μm (the majority of the bolts in the present study) this would be the case at accumulated strains of around 0.5% or greater.

The extent of cavitation or creep damage in this highest strained bolt, where the average

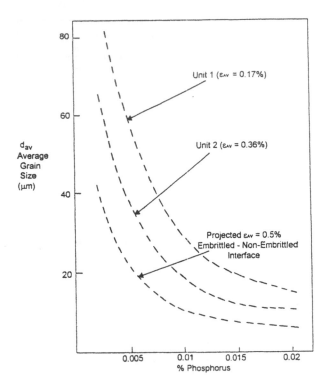

Fig. 7 Influence of accumulated average strain on the critical EM-NON-EMB interface.

void size was 1.1 μm, is shown in Fig. 10. Note that cavitation is widespread and in some instances grain boundary cavity linkage is about to ensue. Effectively the message obtained from Fig. 9 is that, for the present study, i.e. CrMoV steel bolts exposed to temperature of 490°C for some 122 000 h, at accumulated strain levels of about 0.5%, the vast majority of the bolts will (i) have suffered RTE and (ii) contain widespread grain boundary creep damage in the vicinity of the first engaged thread: such a situation would seriously affect integrity considerations.

During a recent study examining creep cavity growth in a commercial CrMoV rotor steel [6] of very similar composition to the present bolts it was observed that the relationship between cavity size, *l*, and creep strain, ε, during secondary creep could be described as

$$l = 0.63\varepsilon \qquad (4)$$

where *l* is in μm.

From the measured cavity sizes in the four bolts which exhibited creep damage and using eqn (3) it was observed that the accumulated strain at the thread root could be estimated by multiplying the strain value generated from bolt length measurements by a factor of 2.7, i.e. the local strain at the thread root was of the order of 1.5%. A strain based continuum damage assessment suggested that crack initiation could occur in as little as 20 000 h. From the

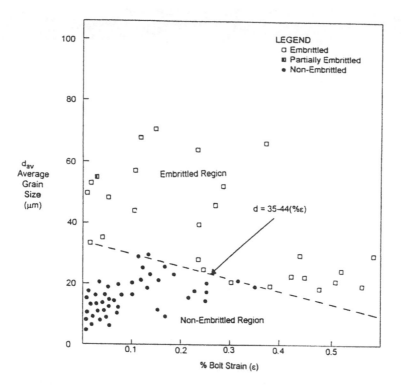

Fig. 8 *Relationship between grain size and accumulated bolt strain and the embrittlement characteristics of selected IP and HP bolts from Units 1 and 2 which exhibited measurable strain levels.*

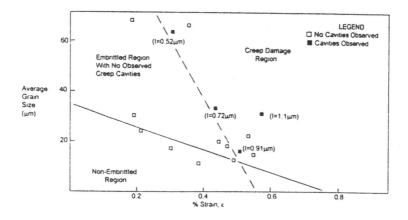

Fig. 9 *Reverse temper embrittlement and creep damage regions in selected CrMoV bolts from the present study. The average cavity lengths are given in parentheses.*

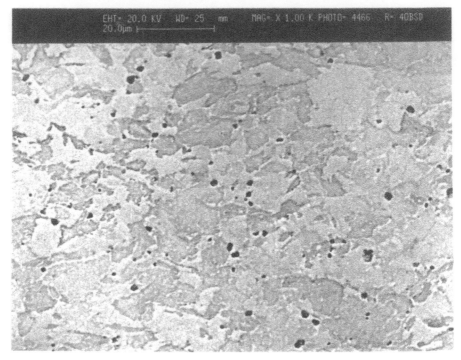

Fig. 10 *Grain boundary cavities at the thread root of a Unit 2 IP bolt with an average accumulated strain of 0.58% and a grain size of 30 mm. The average cavity length was assessed at 1.1 mm.*

present study replacement could be based on bolts which are identified as embrittled and exhibit average strains of around 0.5%. However, Fig. 9 highlights the possibility that significant creep damage can occur at lower average strains as the grain size increases. Hence for the present this figure is used for selection of bolts for assessment. It is well known that strain will not accumulate evenly when there is a temperature gradient along the length of the fastener. ESB's assessment procedure therefore also includes hardness measurements along the length of the bolt shank. Damage levels would be expected to be higher for a given average strain under such conditions and it is expected that the trend lines in Fig. 9 may shift to the left. Work on this aspect is ongoing.

At present, ESB's approach to managing high temperature fastener integrity involves the measurement of the following on bolts cleared by conventional NDT:

• average strain
• grain size
• phosphorus content
• hardness.

In time it is hoped to be able to optimise bolt replacement relative to metallurgical/operational conditions, i.e. different replacement strains for specific grain size ranges, temperature gradients, etc.

4. Concluding Remarks

It has been shown that the primary factors affecting RTE of CrMoV bolts during service at elevated temperatures were grain size, bulk phosphorus level and the level of accumulated strain. For a given value of strain the interface between the embrittled and the non-embrittled regions could be simply described as

$$d \times (\% \text{ P}) = C \tag{5}$$

where the value of the constant C decreased with increasing accumulated service strain ($\%\varepsilon$).

Currently ESB are using such relationships in the selection process of CrMoV steel bar stock materials from which new turbine bolts will be fabricated. Essentially, prior to ordering, the bar stock material is analysed for grain size and % phosphorus and if the combination reside within the embrittled region (see Figs. 5 or 6), the bar stock is rejected. Only bar stock materials which fall within the non-embrittled region and meet conventional mechanical testing requirements will be classed as acceptable.

Interestingly, recent work on other turbine bolts fabricated froln other steels, viz. a CrMo and a CrMo3/4V, have demonstrated that this simple relationship could also be applied to differentiate between bolts which had embrittled during service and those which did not.

References

1. D. L. Newhouse, Temper embrittlement study of NiCrMoV rotor steels: Part I — effects of residual elements, ASTM, STP 499, American Soc. Testing Materials, Philadelphia, PA, 1972, pp. 3–36.
2. J. Olefjord, Temper Embrittlement, *Int. Metals Rev.*, 1978, **4**, 149–163.
3. R. Viswanathan and T. P. Sherlock, Long Time Isothermal Thermal Embrittlement in NiCrMoV steels, *Met. Trans.*, 1972, **3A**, 459–467.
4. B. C. Woodvine, Some Aspects of Temper Embrittlement, *JISI*, 1953, **183**, 229.
5. J. J. Hickey and J. H. Bulloch, *Proc. 5th Int. Conf. on Creep of Materials*, Lake Buena Vista, Florida, May 1992, pp. 321–328.
6. K. Ijima, M. Sukekawa, Y. Fukui and R. Kaneko, *ibid.*, pp. 281–286.

10

The Continuing Development of Durehete 1055 and Other High Integrity Bolting Steels

J. ORR, H. EVERSON*, D. BURTON and J. BEARDWOOD*

British Steel Technical, Swinden Laboratories, Rotherham, UK
*UES Steels, Stocksbridge, UK

ABSTRACT

This paper describes the continuing development of Durehete 1055 in terms of steelmaking and compositional control, leading to improved creep rupture ductility values and notched rupture strength which give increased confidence to the manufacturers and users of such high temperature bolting material. Data are presented showing how Durehete 1055 may be used up to 570°C, such as in the steam conditions likely to occur in future high performance plant, when other more highly alloyed steels might otherwise be used.

The benefits to notched rupture strength and ductility derived from the reduction in residual element levels are described and quantified.

The paper also includes experience and reference data for other high integrity bolting materials produced by UES Steels, Stocksbridge and tested in the same manner as Durehete 1055.

1. Introduction

UES Steels produces a wide range of alloy ferritic/martensitic steels for bolting applications for service at elevated temperatures ranging in compositions from 1CrMo to 12CrMoV. Several of the steels are specified in BS1506 [1] and will appear in a future European Standard which is currently being drafted by an ECISS committee [2]. The latter also includes austenitic bolt steel grades. Another paper in this conference discusses warm worked Esshete 1250 as a candidate austenitic bolting steel [3]. The austenitic steels tend to be less compatible with the conventional casing and flange materials and become most useful at temperatures ≥ 600°C [3]. Such austenitic steels lie outside the scope of this paper which deals with the more commonly used 'ferritic' steels.

Durehete 1055 was designed originally to operate in the UK at the normal steam conditions of 565°C [4], which is higher than the majority of European and USA steam temperatures (typically 538°C), and is used in UK-designed turbines installed throughout the world.

Although mature in its basic design concept, Durehete 1055 has been the subject of continuous improvement in specifications to reflect developments in steel processing and increased expectations from users in terms of higher operating temperatures and rupture ductility [5].

Greater integration of the European power generation industry, and a move towards higher operating temperatures outside the UK have led to increased interest in Durehete 1055 from American and European manufacturers. Conversely, interest from British manufacturers in

German and French steels such as 21 CrMoV5.7 and X19 CrMoVNbN 11.1 [6] has arisen, and work on these latter steels is now established in the UK from both a manufacturing and testing point of view.

This paper concentrates on the continuing development of Durehete 1055, but reference is made to the work in hand on continental grades, where metallurgical and manufacturing expertise gained in the production and development of D1055 and similar alloys is being applied.

2. Development of Durehete 1055

A range of bolting steels produced by UES Steels is given in Table 1.

Durehete 1055 was first developed in the 1960s from Durehete 1050 [4], and has continued to be developed into a high strength, ductile material of the 1990s. During these 30 years, the basic alloy content of 0.2%C, 1%Cr, 1%Mo, 0.7%V has remained relatively unchanged. However, there have been adjustments to the composition ranges, for both major and minor elements, and changes to the heat treatments, which have been reflected in the specifications drawn up for this alloy steel since Durehete 1055 was first specified formally in 1969 (Table 2).

2.1. Composition Developments

Narrower composition ranges have been adopted, since about 1982, for carbon, sulphur, phosphorus, chromium, molybdenum, vanadium and titanium, reflecting the more precise steelmaking control available which enables a greater compositional consistency to be achieved.

Surveys of data and experimental work for Durehete 1055 and other low alloy steels have shown, however, the beneficial effect of reducing the residual elements, particularly on rupture ductility [5, 7, 8]. It is in this area where the most significant compositional changes have occurred for Durehete 1055, which is reflected in the specification details for the residual elements — phosphorus, arsenic, tin and antimony, in particular (see Table 2). A combined effect of these grain boundary segregant elements in CrMo/CrMoV type alloys is usually recognised and summarised by empirically based equations. These include J and X factors developed for specific applications such as forgings and weldments [9, 10]. The factor most commonly used for Durehete 1055 is the R value [7] defined as

$$R = P + 2.43(As) + 3.57(Sn) + 8.16(Sb) + 0.13(Cu)$$

where ( ) = wt% of each particular element.

The R values for casts of Durehete 1055 made in recent years by UES Steels are illustrated in Fig. 1. Thus, since the previous reporting of such data in 1987 [5], there has been a further decrease in the average value and also a greater consistency in the R values achieved up to the most recent production in Spring 1994 (Fig. 1). A revised specification for this steel recognises this improvement by restricting the R value to no greater than 0.14 (Table 2).

Samples of material from various casts representing a wide range of R values are being

Table 1. A selection of bolt steel grades produced by UES Steels

UES Grade Name	Nominal Composition, %											Related Grades and Specifications	Service Temperature Range °C
	C	Si	Mn	Cr	Mo	Ni	B	Nb	Ti	V	N		
Durahete 900	0.40	0.25	0.55	1.25	0.7	–	–	–	–	–	–	BS1506 631–850 (40CrMo5.7)	Up to 480
Durehete 950	0.40	0.25	0.55	1.1	0.55	–	–	–	–	0.25	–	BS1506 671–850 ASTM A193 B16 (40 CrMoV 4-6)	Up to 510
Durehete 1055	0.20	0.25	0.50	1.00	1.00	–	0.003	–	0.10	0.70	–	BS1506 681–820 (20 CrMoVTiB 4-10)	500–570
21 CrMoV 5-7	0.20	0.25	0.6	1.25	0.70	–	–	–	–	0.30	–	DIN 17240 W1.7709 NFA 35.558 20 CDV 5.07	Up to 540
Jethete X19	0.20	0.30	0.60	10.8	0.80	0.60	–	0.40	–	0.20	0.075	DIN 17240 X19 CrMoVNbN 11.1 NFA 35.578 Z21 CDNbV 11 NFA 35.558 Z20 CDNNbV 11	Up to 580
Jethete X20	0.20	0.30	0.60	11.8	1.0	0.60	–	–	–	0.03	–	DIN 17240 X20 CrMoV 12.1 NFA 35.578 Z21 CDV12	Up to 580

Table 2. Development of Durehete 1055 specifications

Specification	Year	C	Si	Mn	P	S	Cr	Mo	Ni	V	B	Ti	Al	As	Cu	Sb	Sn	R	Harden	Temper	R_m	$R_{p0.2}$
CEGB 02596 Issue 1	1969	0.15 0.25	0.10 0.35	0.35 0.75	- 0.04	- 0.04	0.90 1.30	0.85 1.10	-	0.60 0.80	0.001 0.010	0.05 0.20							970-1010	660-730	850-1050	≥680
02596 GDCD-2 Issue 2	1982	0.17 0.23	0.10 0.35	0.35 0.75	- 0.02	- 0.02	0.90 1.20	0.90 1.10	- 0.20	0.60 0.80	0.001 0.010	0.07- 0.15	- 0.08	- 0.020	- 0.20	-	0.020		970-990(2)	680-720	820-1000	≥660
BS1506 681.820	1986 (1990)	0.17 0.23	0.10 0.35	0.35 0.75	- 0.02	- 0.02	0.90 1.20	0.90 1.10	- 0.20	0.60 0.80	0.001 0.010	0.07 0.15	- 0.08	- 0.020	- 0.20	-	0.020		970-990(2)	680-720	820-1000	≥660
Current D1055	1994	0.17 0.23	0.10 0.35	0.35 0.75	- 0.02	- 0.02	0.90 1.20	0.90 1.10	- 0.20	0.60 0.80	0.001 0.010	0.07 0.15	- 0.08	- 0.020	- 0.20	*	0.020	(1)R≤ 0.14	970-990(2)	680-720	820-970	≥660

Composition, Wt %

Heat Treatment Temperatures °C

Room Temperature Tensile Properties N/mm^{-2}

*Value to be reported to customer

(1) $R = R + 2.43$ As $+ 3.57$ Sn $+ 8.16$ Sb $+ 0.13$ Cu

(2) Also a pre-hardening treatment at 660–700°C

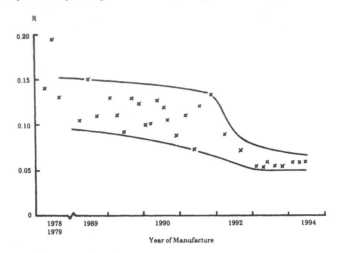

Fig. 1 R *values of Durehete 1055 casts made by UES.*

tested at British Steel Technical, Swinden Laboratories, on behalf of UES Steels, to determine stress rupture and relaxation properties.

2.2. Heat Treatment Developments

The strength of Durehete 1055 is based around a stoichiometrically balanced composition
with respect to V_4C_3 and is therefore controlled primarily by dissolution and reprecipitation
of those carbides. The hardening temperature has therefore been around or just below 1000°C
from the outset, and this has led to a generally consistent level of specified strength values
for Durehete 1055. However, in the 1970s it was considered, in the light of bolt failures in
service for which high hardness and low impact toughness values were implicated, along
with other factors related to bolt design and service temperatures [11], that the strength of
Durehete 1055 may have been too high. Thus, the specified tensile strength values have
been lowered progressively from 850–1050 through to 820–1000 to the values accepted in
1994 of 820–970 N^{-1} mm^{-2}, through reduction of the maximum temperature for hardening
and narrower ranges for the tempering temperature, and the inclusion of a pre-hardening
treatment (see Table 2).

These changes in heat treatment have had no significant influence on the strength (rupture and relaxation) of Durehete 1055, as determined for steels made before and since 1982
[5] (see Fig. 2).

The other property of relevance to bolt performance and ultimately service life is rupture
ductility. There have been significant changes in the rupture ductility of Durehete 1055 since
about the mid-1980s. It is considered that these differences are more related to composition
changes rather than to those of heat treatment, though it is recognised that, all other things
being equal, general strength/ductility relationships hold for Durehete 1055 as for other
creep resisting materials.

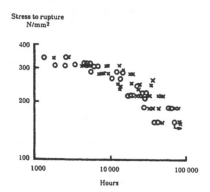

Fig. 2 *Rupture strength at 550°C of Durehete 1055.*

3. Low Residual Durehete 1055

The effects of reducing the metalloid residual element and copper contents in Durehete 1055 have been described in general terms thus far. In this section more specific and detailed data and relationships are presented, which have led to the introduction of a draft revised specification for Durehete 1055 (Table 2).

3.1. Stress Rupture Strength

The plain and notched stress rupture strengths of several Durehete 1055 air melted casts are presented in Fig. 3. The results for plain specimens all lie within a very narrow scatter band, which indicates no significant relationship between R value and rupture strength, Figs. 3(a) and (b). The reference lines in Figs. 3(a) and (b) represent the mean values determined from data for Durehete 1055 casts made before 1979, i.e. prior to the introduction of CEGB 02596 GDCD-2 Issue 2, Table 2 [12]. However, for notched specimens the results are more scattered and a contribution from the residual element content (R) is evident. Cognisance should be taken of the different number of data points for plain and notched samples. However, the two sets of data can be rationalised by interpolation within each data set and then comparing the derived results. This interpolation leads to stress values representing individual cast characteristics for 10 000, 20 000 and 30 000 h at 550°C and 10 000 h at 575°C.

These interpolated values were used to calculate a ratio representing the relative notch and plain strengths for the same test duration/temperature for each cast. A reasonable correlation between notched to plain strength ratio (HPSR) and R values is shown in Fig. 4, with a beneficial effect of reducing the residual element content very apparent.

The data in Fig. 4 thus confirm, for a greater number of casts and to longer durations than the indications given previously [5], that ductility and notched strength of Durehete 1055 are significantly influenced by residual element content. However, for the longer term, the aim is for R values of ≤ 0.1, so that the notched to plain strength ratio (NPSR) becomes equal to, or greater than unity.

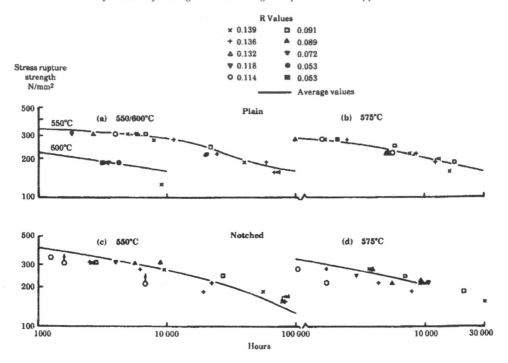

Fig. 3 *Durehete 1055 — stress rupture strengths.*

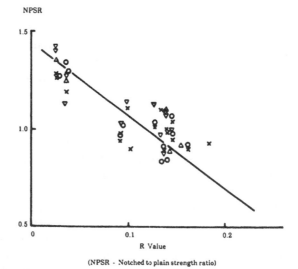

(NPSR - Notched to plain strength ratio)

Fig. 4 *NPSR values for Durehete 1055.*

The low R values in casts made in 1993–94, shown in Fig. 2, indicate the ability of the steelmaker to meet this specification and give confidence to the bolt user that excellent properties will ensue. Stress rupture tests are in progress for these casts but to date only short term plain specimens have completed testing (Figs. 3(a) and (b)). However, the matching notched specimens already show longer test durations than broken plain specimens at 575 and 600°C, albeit for test durations below 5000 h, thus confirming that notched strengthening is present in these materials.

3.2. Stress Relaxation Strength

The stress relaxation strengths (for 0.15% strain at 550°C) of several Durehete 1055 casts with R values in the range 0.07–0.14, are independent of R value and lie close to the mean stress values established before 1982 (Fig. 5). Stress relaxation tests are planned for the most recently manufactured material to confirm that the strength of Durehete 1055 is not significantly affected by reducing the residual element content.

4. Continental Grades

As previously mentioned, increasing requirements from customers for continental grades such as 21 CrMoV5.7 and X19 CrMoVNbN 11.1 have created a demand from established customers. UES Steels has therefore commenced the manufacture of these grades, and, in line with a policy of product verification, a long term elevated temperature test programme was commenced on these two specifications.

The manufacturing route used is the same as that for Durehete 1055, and expertise gained over many years by UES Steels in the manufacture of steels for high temperature use has been applied.

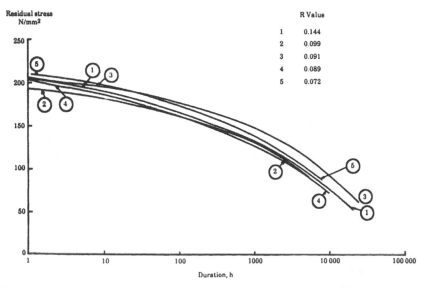

Fig. 5 Durehete 1055 stress relaxation strengths at 550°C.

4.1. 21CrMoV 5.7 Steel

A typical composition for steel made by UES Steels to the related German and French specifications is listed in Table 1, with typical R values of 0.07 having been achieved for the casts made to date.

Stress rupture testing at 450–550°C, with current durations to ~7000 h, on samples from 55–190 mm dia. bars, hardened from 920–930°C and tempered at ~700°C, show that the plain strengths of this cast are above the mean values given in French and German specifications for this steel type [6]. Furthermore notched specimens have higher strengths than equivalently tested plain specimens, indicating a notch strengthening condition to be present.

Testing is continuing to confirm these initial trends, indicating that a satisfactory composition and heat treatment ranges have been selected for this steel.

4.2. X19 CrMoVNbN 11.1 Steel

This grade is similar to the 11% CrMoVNbN Jethete M160 and FV448 steels, which have been produced for many years in the UK. However, the composition of X19 appears to be better balanced with regard to achieving optimum creep strength and notched properties. Following discussions with users of the steel and a study of specifications and technical literature, a hardening temperature of 1150°C, with tempering in the range 680–700°C, was chosen. Stress rupture tests on this material showed results conforming with data produced by a German high temperature data working group. The results obtained to date at 550, 575 and 600°C are given in Fig. 6(a). It should be noted that, particularly over the longer durations, the average values given in Fig. 5 do not correspond with those in DIN 17240 (1976). This is because test data to 100 000 h are now available, allowing long term values to be derived from data without extrapolation [13]. The preliminary results at 575 and 600°C show fairly low ductility values and relatively low notched rupture strengths (Fig. 6(b)). However, at 575 and 600°C the longer term results indicate the notched strength begins to approach that of the plain specimens after 10–20 000 h.

An examination of the data shows that the material hardened from 1150°C may have been 'too strong', thus accounting, at least in part, for the low ductility and notched strength values (see Fig. 6 and Ref. [13]). Therefore. it was considered that a lower austenitising temperature, resulting in less Nb(CN) in solution, could be tolerated. Laboratory experiments confirmed that lowering of the hardening temperatures from 1150 to 1120°C gave satisfactory tensile properties. Consequently 1120°C was adopted as the preferred hardening temperature. Material from 95 mm dia. bars, heat treated in this manner, is on test at 500–600°C. The short term results, available to date, show that although the plain strength is reduced it remains within the scatter band (Fig. 6(a)), and the notched strength at 550 and 575°C is above the scatter band of equivalent results for samples hardened from 1150°C. A direct comparison of results for plain and notched specimens emphasises the relative improvement in notched strength achieved by reducing the hardening temperature (see Fig. 7). The ductility values for completed tests at 550°C–324 N^{-1} mm^{-2}, i.e. 17 and 53% (% Z) for steel hardened from 1150 and 1120°C respectively is further confirmation.

In terms of the NPSR value, this is greater than unity for ~1000 h at 575°C for the material hardened from 1120°C, whereas for samples hardened from 1150°C the NPSR value is below 0.8 (Fig. 6). although in the case of the latter material the NPSR value increases with

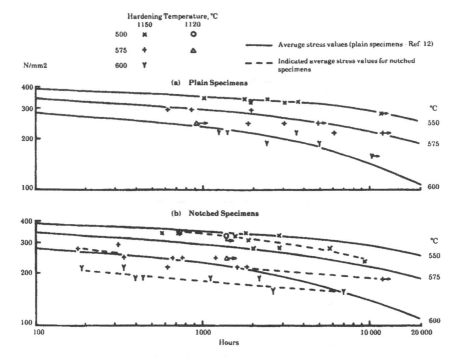

Fig. 6 *Jethete X19 — stress rupture strengths.*

increasing test duration, it will not attain unity until well beyond 10 000 h at 575 and 600°C (see Fig. 6).

Several further stress rupture tests are in progress for materials hardened from both 1150 and 1120°C (+ tempered) to confirm the trends indicated by the results in Figs. 6 and 7, and to establish long term test results for Jethete X19. Stress relaxation tests are also included in the test programme.

5. Conclusions

The continuous improvements of hardened and tempered alloy steel grades made by UES for bolting have been described. Durehete 1055, developed first in the 1960s, has undergone several refinements in composition and heat treatment over the intervening period, which have led to a high strength ductile steel capable of being used up to 570°C. The heat treatment refinements have led to a planned reduction in tensile strength and the inclusion of a pre-hardening treatment, neither of which has had any significant effect on the stress rupture strength at 500–600°C. More significant benefits, particularly on rupture ductility and notched strength, have been achieved by the progressive decrease in residual element content, firstly by the introduction of specified upper limits for particular elements and then by a specification limiting the cumulative addition of the same elements. Data are presented in the paper showing that the notched to plain strength ratio (NPSR) for Durehete 1055 are related linearly to a factor (*R*) describing the cumulative residual element content. The residual ele-

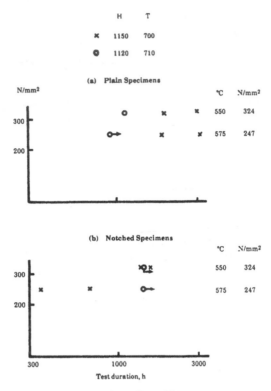

Fig. 7 Jethete X19 — stress rupture strengths; effect of heat treatment.

ment content which gives higher than average notched strength and NPSR values above unity is recognised in the 1994 draft specification. An aim is to achieve R values ≤ 0.1.

Other alloy grades made more recently by UES Steels are the 21 CrMoV 5.7 (1% CrMoV) and X19 CrMoVNbN 11.1(11% CrMoVNbN) steels. Stress rupture tests at appropriate temperatures, in the range 450–600°C, demonstrate that the strengths of these steels are in accord with those in international data sets. Preliminary results for the latter indicate that steel hardened from the traditionally accepted 1150°C resulted in low notched strength, and rupture ductility values at 550–600°C. A significant improvement in these properties has been achieved by hardening from 1120°C. This is indicated by an increase in the NPSR value to a value greater than unity for 1120°C hardened material for material tested for 1000 h at 575 and 600°C, compared with an NPSR value of ~0.8 for 1150°C hardened material. Longer duration test data at 575 and 600°C indicate that the NPSR value on the 1150°C hardened material will probably increase, but not attain unity until well beyond 10 000 h.

Stress rupture and stress relaxation tests are in progress on all the grades described in this paper, to establish long term strength values for design and to confirm the metallurgical trends indicated by the data obtained to date.

6. Acknowledgements

The authors would like to thank Dr I. G. Davies, Technical Director of UES Steels, Dr M. J. May, Manager, Swinden Laboratories and Product Technology and Dr R. Baker, Director, R&D, British Steel plc for permission to publish this paper.

References

1. BS1506:1990, 'Specification for carbon, low alloy, and stainless steel bars and billets for bolting material to be used in pressure retaining applications', BSI, London.

2. pr EN 10 fel Draft EN Standard for steel bars and rods for fasteners for use at elevated and low temperatures, ECISS TC22 document — ECISS/TC22/N183, April 1993.

3. J. Orr, H. Everson and G. Parkin, 'Warm worked Esshete 1250: A high strength bolting steel', this volume, pp. 203–219.

4. P. G. Stone and J. D. Murray, 'Creep ductility of Cr–Mo–V steels', *J. Iron Steel Inst.*, 1965, **203**,1094.

5. H. Everson, J. Orr and D. Dulieu, Low alloy ferritic bolting steels for steam turbine applications: The evolution of the Durehete steels, in *Int. Conf. on Advances in Material Technology for Fossil Power Plant*, Chicago, 1–3 September 1987, Paper 21.

6. DIN 17240, July 1976, High temperature and extra high temperature materials for nuts and bolts, Quality Specifications, FES im DIN.

7. B. L. King, Intergranular embrittlement in CrMoV Steels: An assessment of the effects of residual impurity elements on high temperature ductility and crack growth, *Phil. Trans. Roy. Soc., Lond.*, 1980, **A295**, 235.

8. N. G. Needham and J. Orr, The effect of residuals on the elevated temperature properties of some creep resistant steels, *Phil. Trans. Roy. Soc., Lond.*, 1980, **A295**, 279.

9. S. Beech, Rolls-Royce Associates, IRD, Private communication.

10. R. Bruscato, Temper embrittlement and creep embrittlement of 2 1/4CrMo shielded metal arc weld deposits, *Welding Res. Suppl.*, 148s, April 1970.

11. S. T. Kimmins and D. J. Gooch, Austenite memory effect in 1Cr1Mo0.75V(Ti,B) Steel, *Metal Sci.*, 1983, **17**, 519.

12. Durehete 1055 Brochure, UES Steels, Stocksbridge Engineering Steels, 6/MM, 5 June 1989.

13. H. Konig, Langzeitverhalten des hochwarrnfesten schrauben und turbinenschaufelstahles X19 CrMoVNbN 11-1 in zeitstandversuchen bis 600°C, in *Langzeitverhalten warmfester stahle und hochtemperaturwerkstoffe,* Meeting, 26 November 1993, Dusseldorf. Organised by VDEh.

11

Operational Characteristics of 10–12% CrMoV Bolt Steels for Steam Turbines

K. H. MAYER and H. KÖNIG

MAN Energie, Nürnberg, Germany

ABSTRACT

For 30–40 years 11–12% chromium bolt steels of X22CrMoV 12 1 and X19CrMoVNbN 11 1 have been used successfully in steam turbines in the temperature range of about 450–550°C, where these steels are characterised by higher relaxation strength than the 1% CrMoV steel 21CrMoV 5 7. Subject to optimum adjustment of the alloying, trace and deoxidation elements, the Nb and N-bearing steel X19CrMoVNbN 11 1 is superior to the steel X22CrMoV 12 1. In spite of a distinctly higher creep strength, the improved 9–10% CrMoV steels developed in the 1980s with optimum adjustment of all alloying elements and variations of the molybdenum, tungsten and boron contents surprisingly do not feature a higher relaxation strength than the conventional steel X19CrMoVNbN 11 1. Consequently, the 9–12% CrMoV steels are only suitable as turbine bolt steels up to a temperature of max. 550–560°C.

1. Introduction

The intensive efforts currently being made in the direction of 'Improved Fossil-fired Steam Power Plants' with steam admission temperatures of *ca.* 600 °C have recently focused renewed interest on highly heatresistant ferritic steels of the type 9–10% Cr [1–4]. Figure 1 provides an overview of research programmes carried out on an international scale in Japan, USA and Europe with a view to developing advanced power plants and of the contracts and planning resulting from this development work.

In the light of the expectation that the newly developed 9–12% CrMoV steels will also offer higher relaxation strengths for bolting applications up to operating temperatures of *ca.* 600 °C, it is useful to summarise the experience accumulated over the past 30–40 years with similar steels at steam admission temperatures up to 565°C and to compare the properties of these steels with the properties of the newly developed 9–10% Cr steels.

The major factors which govern the life span of heatresistant bolt materials are shown schematically in Fig. 2. A primary requirement is a low loss of preload due to creep and relaxation in order to warrant a sufficiently high residual clamping force or sealing force for the pressure-containing components during the overhaul intervals of 30 000 to 80 000 h service.

Past cases of damage suffered by heat-resistant bolts have clearly revealed that the ductility and toughness of the bolt materials are also major factors for the life span of a bolt. On the one hand, the material must feature adequate ductility under conditions of creep stress in order to eliminate the formation of creep cracks in the first few stress-bearing threads. On

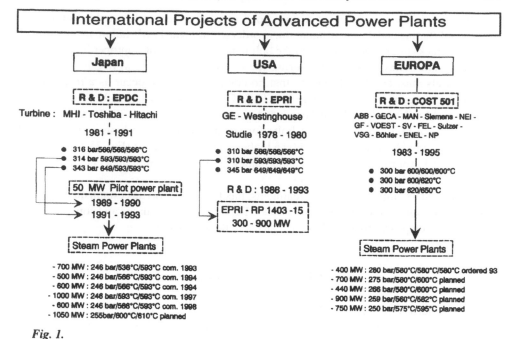

Fig. 1.

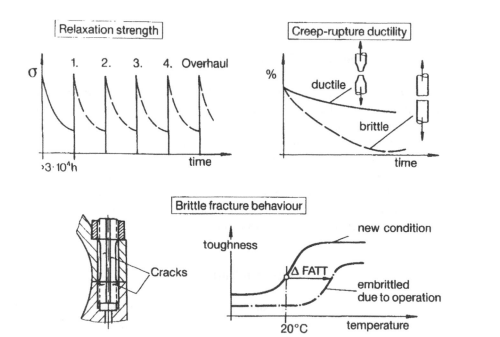

Fig. 2 *Operational requirements for bolt materials (schematic).*

the other hand, the material must feature good toughness also at low temperatures, i.e. during cold starting and during installation, in order to eliminate brittle failure in the event of any creep cracks or cracks occurring in the bolt holes on account of improper heating while preloading and relieving the bolts during the overhauls. In other words, heat-resistant bolt materials must be of such quality that no embrittlement can occur as a result of the stresses arising in operation. The 11–12% Cr DIN steels (5) X22CrMoV 12 1 and X19CrMoVNbN 11 1 have been used successfully for over 30–40 years in the manufacture of highly heat-resistant bolts for steam turbines. The higher alloyed steel X19CrMoVNbN 11 1 is characterised by a higher relaxation strength (Fig. 3). On the other hand, it has a lower rupture toughness and a somewhat lower ductility under creep stressing compared to the X22CrMoV 12 1 steel.

The chemical compositions, heat treatments, mechanical properties established in short-time and long-time tests of the abovementioned DIN steels are discussed and compared with the newly developed 9–10% CrMoVNbN steels in the following.

2. Chemical Composition

Table 1 provides an overview of the characteristic compositions of the DIN and newly developed steels. The newly developed 9–10% CrMoV steels as a rule have 0.06% Nb, 0.06% N and some have B and W additionally.

Moreover, the alloying constituents are matched up to a great extent so that optimum utilisation of their effectiveness is possible. The basic structure is 100% martensitic and, consequently, free of Delta ferrite.

3. Mechanical Properties at Room Temperature

Information on typical 0.2 limits and toughness values of the DIN steels are given in Fig. 4. In addition, this Figure includes actual values of test melts of the 9–10% Cr steels which have been newly developed under the COST 501-2 programme and the applicable heat treatment data. While the 0.2 limit is roughly comparable, the X19CrMoVNbN 11 1 DIN steel has the lowest impact energy in (Charpy-V-notch specimen). In the case of the X22CrMoV 12 1 DIN steel and the newly developed steels, the impact energy is at least twice as high. The lower toughness of the X19CrMoVNbN 11 1 steels, however, has not proved to be a disadvantage in long years of operation because it is not susceptible to creep crack formation in the first few stress-bearing threads of the bolts [6].

4. Long-term Embrittlement

As a matter of experience, the two DIN steels X22CrMoV 12 1 and the X19CrMoVNbN 11 1 will have roughly the same transition temperature of the impact energy (FATT50) as in the initial condition even after long periods of operation, i.e. they are not susceptible to long-term embrittlement as are some of the low alloy steels. Figure 5 gives an overview of FATT50

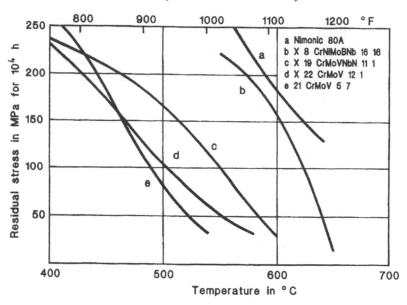

Fig. 3 *Residual stress bolt material for 10 000 h to DIN 17 240 for an initial strain of 0.20%.*

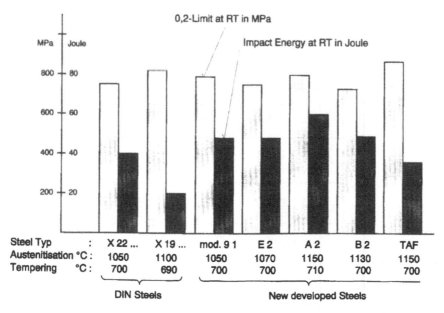

Fig. 4 *0.2-limit and impact energy at RT of DIN 11–12% Cr bolting steels and newly developed 9–10% Cr ferritic steels.*

values which were established on turbine bolts of the traditionally used bolt steels removed after up to 150 000 h service at bolt temperatures of 350–530°C [7]. In the case of X19CrMoVNbN 11 1 steel, the transition temperature FATT50 in the virgin condition ranges

Table 1. *Chemical composition of 9–12% CrMoV (NbNB) steels*

Steel Typ	Chem. Composition in wt%											
	C	Si	Mn	Cr	Mo	W	Ni	V	Nb	N	B	Al
1. X 22 CrMoV 12 1	0.22	0.20	0.50	12	1	–	0.55	0.30	–	–	–	?
2. X 19 CrMoVNbN 11 1	0.19	0.30	0.30	10.5	0.7	–	0.45	0.18	0.45	0.05	0.0015	0.015
3. X 10 CrMoVNbN 9 1 (P 91)	0.10	0.15	0.40	9	1	–	0.20	0.20	0.08	0.05	–	0.007
4. X 12 CrMoWVNbN 10 1 1 (Rotor Steel E)	0.12	0.04	0.50	10	1	1	0.80	0.20	0.06	0.05	–	0.005
5. X 5 CrMoVNbN 9 1 (N alloyed Steel)	0.05	0.33	0.60	9	1	–	0.10	0.20	0.06	0.10	–	–
6. X 18 CrMoVNbB 9 1 (Rotor Steel B 2)	0.18	0.12	0.06	9	1.5	–	0.10	0.25	0.06	0.01	0.01	0.01
7. X 20 CrMoVNbNB 10 1 (TAF Steel)	0.20	0.33	0.90	10	1.5	–	0.02	0.25	0.18	0.02	0.03	0.004

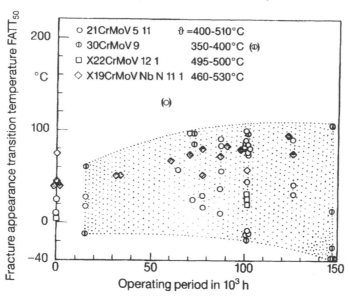

Fig. 5 *Toughness of bolt material as a function of operating period.*

from 40–80°C. After long-time service, values were found to vary between 50 and 90°C. The body of information available on the X22CrMoV 12 1 steel is less because this steel has been used less on account of its lower relaxation strength. In the case of the few bolts tested after about 100 000 h service in the temperature region of about 500°C, FATT50 values were recorded between 20 and 30°C. Published data gathered by competitors on bolts made of X22CrMoV 12 1 steel are summarised in [8]. There is no tendency towards embrittlement evident from these published values either. Figure 6 affords an overview of the long-term behaviour of the newly developed 9–10% Cr steels as determined under the COST 501-2 programme in exposure tests at 480, 600 and 650°C up to 10 000 h for the X12CrMoWVNbN 10 11 and X18CrMoVNbB 9 1 steels. At 480°C, only a minor loss in toughness was found to have occurred. At 600°C, the change in toughness was most pronounced with a rise of the FATT50 of *ca.* 22–25 °C. However, this temperature would be outside the future range of application for heat-resistant ferritic 9–10% Cr bolts on account of the marked relaxation occurring at this temperature (see discussion below).

5. Creep Ductility

Ductility under conditions of creep is of great importance in the case of heat-resistant bolt materials in view of the high peak stresses arising in the region of the first few stress-bearing threads. With a low creep ductility, there is no possibility for a stress redistribution by plastification and, consquently, there is the risk of early creep crack formation at the bottom of the thread. Both in the case of the X22CrMoV 12 1 and in the case of the newly developed 9–10% Cr steels, a high ductility was invariably recorded in the creep test on smooth and notched specimens. In contrast to this, the X19CrMoV NbN 11 1 steel tends to have lower

.splay a notch-weakening behaviour after heat treatment at
ε (*ca.* 1150°C) and higher Al contents. Figure 7 summarises
/ at 600°C [9]. A reduction of area above 20% was recorded
ıd austenitisation temperatures of 1100–1130°C.

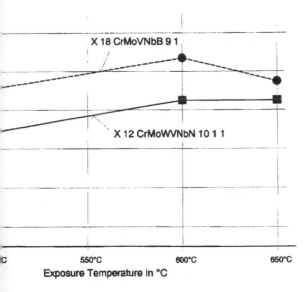

of newly developed 9–10% Cr steels in the virgin condition and 600 and 650°C.

Test temperature 600°C

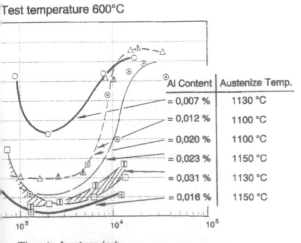

Al Content	Austenize Temp.
= 0,007 %	1130 °C
= 0,012 %	1100 °C
= 0,020 %	1100 °C
= 0,023 %	1150 °C
= 0,031 %	1130 °C
= 0,016 %	1150 °C

Time to fracture in h

The favourable service properties obtained with the X19CrMoVNbN 11 1 bolt steel in the past were invariably related to these limited Al contents and austenitisation in the range of 1100–1130°C.

6. Relaxation Strength

The higher relaxation strength of the X19CrMoVNbN 11 1 steel referred to above in comparison with X22CrMoV 12 1 will be obtained only if the C content is at least 0.17% and the N content at least 0.05%. Apart from that, it is necessary for an austenitisation temperature of at least 1100°C to exist for a sufficient amount of niobium carbides to go into solution. Figure 8 supplies information on corresponding results which were established in the 1960s when heats from various steel makers [10] were tested. The good long-time service experience recorded by the authors over the past 30 years are based on the variant characterised by a C content of 0.19%, an N content of 0.06% and austenitisation at 1100–1130°C. Recent investigations have shown that an Ni content of less than 0.50% is important for a high relaxation strength to be obtained.

These requirements are also complied with by the test results compiled in Fig. 9 of long-term relaxation tests on bolted joint models using bolts and nuts made of X19CrMoVNbN 11 1 and a flange of X22CrMoV 12 1 steel. The elastic strain [11] determined by means of electric strain gauges applied to the bolt shank [11] is plotted against the exposure time at temperatures of 425, 480, 540 and 600°C. The initial strain was 0.18% in each case. At 425°C, a slight negative creep effect was witnessed over a sustained period. At 600°C the relaxation strength over a sustained period was too low, as expected. For the

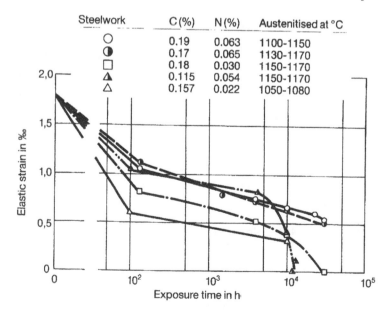

Fig. 8 *Influence of C, N and heat treatment on the relaxation stress of steel X19CrMoVNbN 11 1 at 540°C.*

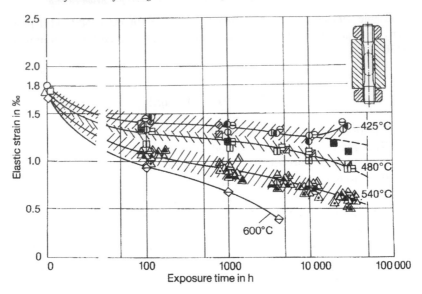

Fig. 9 *Relaxation strength of X19CrMoVNbN 11 1 bolted joints.*

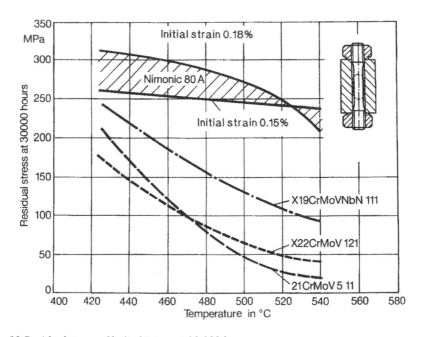

Fig. 10 *Residual stress of bolted joints at 30 000 h.*

traditional range of application of this bolt steel, Fig. 10 compares the bolted joints of the X19CrMoVNbN 11 1 material combination with the alternative material combinations in respect of the residual stresses at 30 000 h [6].

Bolts and nuts	Flange
Bolts and nuts	**Flange**
Nimonic 80A	21CrMoV 5 11
X22CrMoV 12 1	X22CrMoV 12 1
21CrMoV5 11	21CrMoV 5 11

n comparison with the other ferritic steels X22CrMoV 12 1 and 21 CrMoV 5 11 there is a distinct superiority, especially at temperatures above 450°C. Compared with the nickel-based alloy Nimonic 80A the residual stress of the X19CrMoVNbN 11 1 combination reveals a 50% lower value in the higher temperature range [12].

Figures 11–13 provide information on the relaxation strength up to 10 000 h established under the COST 501-2 programme for bolted joints at 540, 570 and 600°C [13]. The flanges of the bolted joint models in each case consist of the G-X10CrMoVNbN 9 1 cast steel (chemical composition according to P91) and the nuts of the bolt material. The initial strain or preload was invariably 0.20% for the ferritic bolt steel joints. In the case of Nimonic 80A bolted joints, initial strains of 0.20 and 0.25% were adopted. Due to the higher thermal expansion coefficient of the Ni alloy compared to the ferritic flange steel, an initial strain of *ca.* 0.10% or respectively, 0.15% was obtained with these values at the relaxation test temperature.

Surprisingly, the results of the relaxation tests revealed that the newly developed ferritic 9–10% Cr steels do not have any higher relaxation strength than the traditionally used X19CrMoVNbN 11 1 bolt steel although they possess a distinctly higher creep rupture strength at temperatures above 540° C. A general overview of all residual stresses, and, respectively, relaxation strength values established after 10 000 h is given in Fig. 14 (p.161). It is evident from this that it is possible with the bolted joints using Nimonic 80A bolts and nuts to obtain

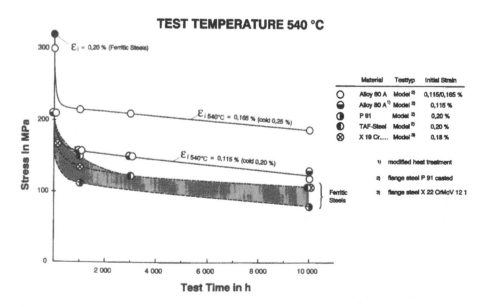

Fig. 11 *10 000 h relaxation strength of steam turbine bolting material at 540°C/model tests.*

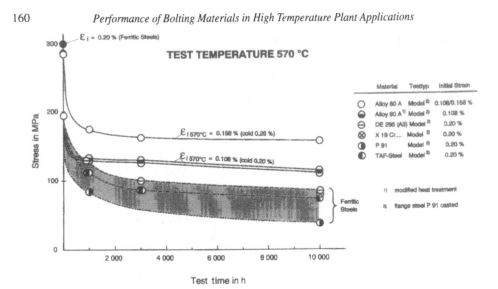

Fig. 12 *10 000 h relaxation strength of steam turbine bolting material at 570°C/model tests.*

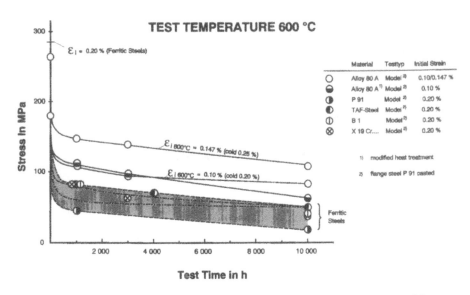

Fig. 13 *10 000 h relaxation strength of steam turbine bolting material at 600°C/model tests.*

distinctly higher relaxation strength values over the entire temperature range covered by the tests, especially where a higher initial strain of 0.25% is applied.

7. Summary and Conclusion

The 10–12% CrMoV DIN bolt steels X22 CrMoV 12 1 and X19CrMoVNbN 11 1 have been most successfully used during the last 30–40 years for casing, valve and pipe flange bolts in

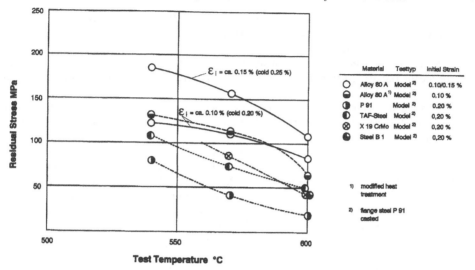

Fig. 14 *10 000 h relaxation strength of steam turbine bolt materials/model tests.*

a temperature range from *ca.* 450–550°C. The higher alloyed X19CrMoVNbN 11 1 steel is distinctly superior to the X22CrMoV 12 1 steel in terms of relaxation strength. Its susceptibility to a notch-weakening behaviour can be kept within acceptable limits if the Al content is limited to a maximum of 0.020% and, in addition, a maximum austenitisation temperature of 1130°C is specified.

The tests made with the newly developed 9–10% Cr steels have revealed a distinctly higher ductility and toughness of these steels compared to the DIN steel X19CrMoVNbN 11 1. Surprisingly, however, their relaxation strength is not higher although the newly developed steels have a distinctly higher creep rupture strength in the temperature range from about 550–600°C. In accordance with this result, the range of application of the ferritic 9–12% Cr steels for highly heat-resistant bolts in steam turbines is limited to about 550°C maximum. Higher bolt temperatures call for the use of Ni base alloys, such as Nimonic 80 A in order to warrant adequate relaxation strengths.

References

1. *Third EPRI Int. Conf. on Improved Coal-fired Power Plants*, 2–5 April, 1991, San Francisco, USA.
2. Y. Nakabayashi *et al.*, 'Development of Advanced Steam Plants in Japan', *COST 501 Conference on Materials for Power Engineering Components*, 13–14 October 1992, Jülich.
3. C. Berger, K. H. Mayer, R. B. Scarlin, W. Engelke, T. C. Franc and L. Busse, 'Turbinenkonstruktionen mit den neuen Stahlen fur hohe Dampftemperaturen', Int. *VGB Conf. on Fossil-fired Power Plants with Advanced Design Parameters*, 16–18 June 1993, Kolding, Denmark.
4. C. Berger, R. B. Scarlin, K. H. Mayer, D. V. Thornton and S. M. Beach, 'Steam Turbine Materials: High Temperature Forgings', *5th Int. Conf. Materials for Advanced Power Engineering*, Liège, Belgium, 3–6 October, 1994.
5. DIN 17 240, Issue 1976, 'Creep resisting materials for bolts and screws for service at elevated and at high temperatures; quality specifications'.

6. K. H. Mayer and H. König, 'Creep, Relaxation and Toughness Properties of the Bolt and Blade Steel X19CrMoVNbN 11 1', *Int. Conf. on Advances in Materials Technology for Fossil Power Plants*, 1–3 September, 1987, Chicago, IL, USA

7. K. H. Mayer and H. König, 'The Determination of the Residual Life Expectancy of Heat-resistant Bolts for Steam Turbines', *EPRI Conf. on Fossil Plant Life Extension and Assessment*, June 1986, Washington, DC, USA.

8. K. H. Mayer and K. H. Keienburg, 'Operating Experiences and Life Span of Heat-resistant Bolted Joints in Steam Turbines of Fossil-fired Power Stations', *Int. Conf. on Engineering Aspects of Creep*, University of Sheffield, 15–19 September, 1980.

9. H. König, Langzeitverhalten des hochwarmfesten Schrauben- und Turbinenschaufelstahles X19CrMoVNbN 11 1 in Zeitstandversuchen bis 600°C, 16. Vortragsveranstaltung "Langzeitverhalten warmfester Stähle und Hochtemperaturwerkstoffe" 26 November, 1993, Düsseldorf.

10. A. Erker and K. H. Mayer, 'Relaxations- und Sprodbruchverhalten von warmfesten Schraubenverbindungen', VGB Kraftwerkstechnik, Vol. 2, February 1973, pp. 121–131, *Int. Conf. on Creep of Heat-resistant Steels*, Düsseldorf, 3–5 May, 1972.

11. K. H. Mayer and H. König, 'Testing of the Relaxation Strength of Bolted Joints', this volume, pp.54–69.

12. K. H. Mayer and H. König, 'Operational Characteristics of Highly Creep-resistant Nimonic 80A Bolts for Steam Turbines', *Int. Conf. of Advances in Material Technology for Fossil Power Plants*, 1–3 September 1987, Chicago, IL, USA.

13. K. H. Mayer, 'Critical Components for Turbines with High Steam Parameters', Final COST 501-2 Report of MAN Energie, 3 June 1993.

12

Embrittlement of 12CrMoV Bolting Steel

J. D. PARKER and B. WILSHIRE

Department of Materials Engineering, University of Wales, Swansea, UK

ABSTRACT

The integrity of creep resistant fasteners is of critical importance to maintaining steam-tight joints in turbine casings, valve covers, etc. Stress relaxation during high-temperature service necessitates periodic retightening and useful bolt life is a function of the number of tightenings plus the accumulated operating hours. Examination of material prior to service indicates that significant microstructural variations occur and these differences result in scatter of creep strength and other mechanical properties. Ageing during operation can then further reduce creep and impact properties, resulting in cracking of 12CrMoV fasteners.

1. Introduction

Components in steam turbine applications are required to operate for long periods under stress at elevated temperatures. Thus, the materials selected for this type of service must be resistant to oxidation and have the appropriate creep strength. In particular, the fasteners used in casings and valve covers are loaded to required levels during maintenance. Since stress relaxation will take place during high temperature service, the ability of a specific joint to remain free from leaks will depend on the level of initial loading and the creep behaviour of the material. In general, both creep deformation and fracture properties are important since resistance to deformation will control the rate of relaxation and sufficient ductility must be available to prevent crack initiation and growth.

In the majority of cases, the materials used for high temperature fasteners are based on precipitation hardened 1CrMoV and 12CrMoV steels. Service problems were encountered with early 1CrMoV alloys which had the necessary creep strength but were found to exhibit low ductility [1]. Modification of the alloy composition, involving additions of boron and titanium, largely alleviated these problems by increasing the creep strength of grain boundary regions preventing localised strain concentrations. Despite these modifications, in-service problems in 1CrMoV steels have been reported [2, 3], with cracking developing primarily at the first engaged thread. A significant amount of work has been undertaken investigating these problems. In general, it has been suggested that failure occurs because of in-service embrittlement which leads to

(i)   a shift of the impact transition curve to higher temperatures

(ii) a reduction in room temperature impact energy, and

(iii) lowering of the upper shelf fracture energy.

Detailed research has indicated that the in-service embrittlement in 1CrMoV steels is primarily the result of diffusion of impurity elements to prior austenite grain boundaries. In particular, elements such as P, As, Sn are believed to be particularly deleterious to performance [4]. Operating experience with 12CrMoV fasteners has also identified that cracking develops during elevated temperature service. The location of defect development is similar to that found with lower chromium steels. However, the details controlling damage have not been established. The present paper evaluates the factors affecting cracking in these alloys, compares the critical issues involved in component performance for 1CrMoV and 12CrMoV steels and discusses approaches for condition assessment of high temperature fasteners.

2. Microstructure and Properties

A range of 12CrMoV steel fasteners has been considered. All components were approximately 89 mm in diameter and 350 mm in length. The compositions were within the appropriate specification and the manufacturing procedure involved normalising at 1035°C, followed by tempering at 650°C [5]. Four new fasteners were examined, with studies also performed on eight fasteners withdrawn from operation after about 65 000 h service at 550°C, each of which had been subject to 10 tightening operations. Cracking had developed in four of the service-exposed fasteners.

Despite the fact that these components were fabricated in accordance with the same specification, the operating performance was markedly different. Thus, satisfactory behaviour was noted for some situations, with components operating under nominally the same conditions found to have developed cracks. Since service life is related to creep and fracture behaviour, laboratory of examples of the ex-service samples was undertaken to establish these properties. Moreover, similar tests were carried out on the new fasteners to allow evaluation of the as-fabricated properties. Detailed metallographic studies were also performed to determine the range of structures present after manufacture and also any microstructural changes taking place during service.

2.1. Microstructure

Microstructural characterisation was performed using standard metallographic techniques with sections taken in the body of each fastener as well as in the threaded region. Evaluation by optical and electron microscopy was undertaken for each specimen, with these results supported by Vickers hardness testing.

An optical micrograph of a typical as-fabricated sample is shown in Fig. 1(a). As expected, cooling from the normalising temperature had resulted in the formation of martensite, with the subsequent heat treatment leading to carbide precipitation around the lath boundaries. Thus, the martensitic nature of the original microstructure was clearly evident even from electron microscopy of carbon extraction replicas (Fig. 1(b)). However, the microstructure of one of the new fasteners was significantly different, with carbide precipitates apparent even by optical microscopy (Fig. 2). Furthermore, no evidence of martensite lath boundaries was detected. In both specimens, the microstructures were uniform throughout the components and these observations were supported by the results of the hardness

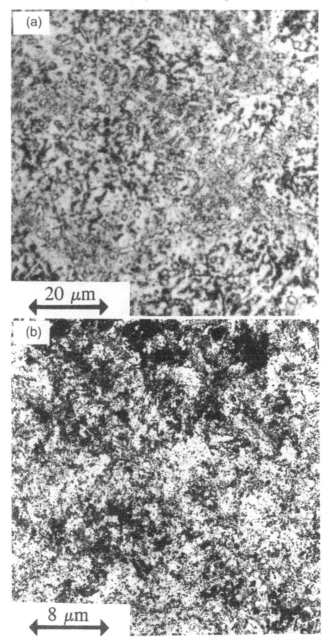

Fig. 1 *Typical microstructure of an as-fabricated fastener with a martensitic structure revealed (a) by optical microscopy and (b) by transmission electron microscopy of carbon extraction replicas.*

testing. Within the body of each component and in the threaded regions, average hardness values of 346 and 325 HV were found for the materials with a martensitic structure and coarse-carbide structure respectively.

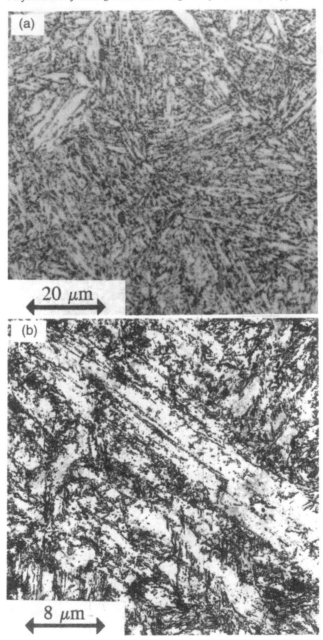

Fig. 2 *Microstructure of a new fastener containing networks of coarse carbides revealed (a) by optical microscopy and (b) by transmission electron microscopy of carbon extraction replicas.*

Examination of the ex-service components revealed that the microstructures and hardness values were in two distinct groups. Thus, of the eight components examined, four were found to have developed cracks at the first engaged thread. The maximum length of these defects was about 3 mm. Microstructural examination indicated that coarse carbides were present in each of these cases, with no evidence of martensite lath boundaries detected. In this group of samples, the hardness values were in the range 267–280 HV. No significant variation in microstructure or hardness was identified within a particular fastener. However, it was noted that the cracking and microdamage present had developed in association with the carbide precipitates. In contrast, all of the fasteners which had shown satisfactory in-service performance exhibited tempered martensitic microstructures similar to those shown in Fig. 1, i.e. while some coarsening of the carbides had apparently occurred during elevated temperature service, the precipitates still delineated the original martensite lath boundaries. Despite the tempering which had taken place during service, in these cases, the measured hardness was greater than 295 HV.

2.2. Fracture Behaviour

The fracture behaviour was examined by consideration of the results of Charpy impact tests undertaken in the temperature range of–20 up to 250°C. A comprehensive set of data was evaluated for the new materials to establish the full extent of the change in fracture energy with temperature and to examine whether the values obtained varied with the location of specimen removal. Data obtained for the material with a coarse-carbide structure (Fig. 3), indicate a 50% fracture appearance transition temperature (FATT) of approx. 100°C, with upper and lower self energies of 78 and 15 J respectively. For the martensitic material, room temperature values were higher, with an average fracture energy of *ca*. 26 J. These results are in good agreement with other data reported for new components of this type [2].

Evaluation of the performance of the ex-service fasteners revealed that components with a martensitic structure exhibited fracture behaviour similar to that measured for the new components, with service exposure reducing the room temperature energies to within the range 14–18 J. Data for components with a coarse carbide structure are included in Fig. 3. These results also reveal a decrease in room temperature fracture energy and suggest an increase in the 50% FATT, i.e. the components which had developed in-service cracks exhibited the lowest fracture energy.

2.3. Creep Behaviour

The results of creep tests on new samples [5] are compared with published data [6] for similar steels in Fig. 4. As shown, the behaviour of the new material was microstructure dependent. Thus, the material with a typical tempered martensitic microstructure exhibited a creep strength in agreement with published results. However, the material with the coarse precipitates was significantly weaker [7], i.e. for a given applied stress, the deformation rate was approximately a factor of ten greater than expected. The deformation rates noted in tests on ex-service samples were also found to be greater than the typical values. In this case, the performance of fasteners with a martensitic structure approached that for new components found to contain coarse carbides. However tests on studs which had developed cracks during operation revealed that the strength was markedly reduced, i.e. for these components, the strength was about three orders of magnitude less than that expected for typical material (Fig. 4).

rgy (J)

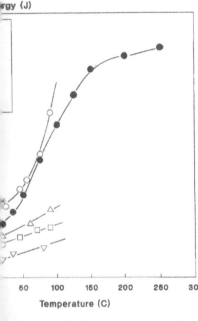

Temperature (C)

t Energy with test temperature. (A): data for new material with
ew material with a martensitic structure; and (F, G and H):
service. Other data for new material with a martensitic struc-

juctility for 12CrMoV fastener materials is shown in Fig. 5
e scatter in results is between *ca.* 10–20% strain. However,
elation to in-service performance. Firstly, in common with
steels, there is a trend which suggests that rupture ductility
]. Thus, for typical values of sustained stress in operating
less than 10% may be encountered. The limited test results
coarse carbides [7] then fall near the lower bound of the
aps of greater concern, is the fact that in these tests the strain
d 5% for stresses of 400 and 500 MPa respectively. These
tiary creep may commence at creep strains of less than 1%

tors Affecting Performance

microstructure and performance of 12CrMoV steels demon-
was observed even with samples fabricated to the same speci-
ration resulted in degradation in properties. However, the
iponents were not simply a function of exposure to elevated

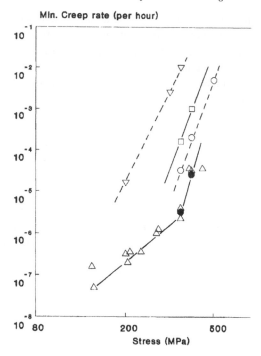

Fig. 4 *Variation of minimum creep rate with applied stress for a range of new material with a martensitic structure (• ∆) and with a coarse carbide structure (○), compared with results for ex-service samples with martensitic (square) or coarse carbide structures (∇).*

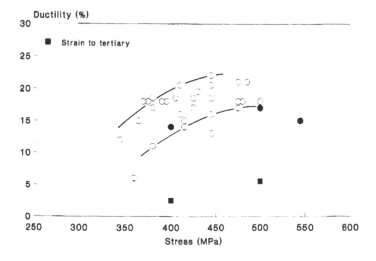

Fig. 5 *Creep ductility of 12CrMoV steel used in fastener applications. New component with a coarse carbide structure is designated •.*

procedures used for retightening of fasteners following a maintenance period will markedly affect performance. However, in the present situation, the procedures adopted were similar in all cases, so that this is not considered to be a factor with the current study. Furthermore, the fact that different behaviour was identified for both new and ex-service components is in agreement with previous work which evaluated the properties of some 65 ex-service bolts. This examination revealed that no systematic relationship between service time and impact strength could be established [2]. Indeed, this study also identified that bolts which had brittle behaviour were present very early in life.

In the present case then, while creep strength and impact properties were degraded by operation, the variability observed cannot simply be a function of time in service. Instead, the variation in structure noted for as-manufactured components appears to be the critical factor in establishing performance. Thus, the existence of coarse precipitates in new components resulted in reduced short term performance and would also contribute to rapid ageing under service conditions. Furthermore, the coarse precipitates present would be expected to reduce creep strength since they will not be of an optimal size and distribution to impair dislocation movement. However, simply reducing creep strength should not result necessarily in failure in fasteners. In these components, the initial level of stress is developed during tightening and the rate of subsequent stress relaxation is related to creep deformation. Thus, a low strength material would develop creep strain more rapidly than expected and so would sustain a lower stress level than predicted. In extreme situations, this low stress may be such that steam leaks would occur. However, in the present case, the low strength is the result of the formation of a network of coarse carbides which can also lead to an increased susceptibility to the nucleation of creep voids [9], ie both creep and fracture resistance appear to be reduced.

Reductions in ductility can result from the development of carbides on grain boundaries. Moreover, the presence of coarse carbides resulted in a reduction in both room temperature impact strength and an apparent increase in the 50% FATT in all fasteners which contained cracks (Fig. 3). However, despite the fact that, in the worst case room temperature Charpy values of 6 J were measured, rapid brittle fracture had not taken place. Thus, while the reductions in hardness and fracture resistance provide an indication of potential problems, the levels of damage developed were the result of a microstructure susceptible to low-strain creep cavity nucleation. On this basis the conditions established are such that rapid creep deformation will be accompanied by cavity nucleation and growth, with the effect that creep cracks are initiated.

This type of behaviour pattern may even be similar to that occurring with 1CrMoV steels. A comprehensive review of the performance of these bolts has recently been published [3]. In this study, it was noted that bolts which exhibited embrittlement had significant carbides on prior austenite grain boundaries and tended to have developed greater permanent strain in service. These observations were interpreted as evidence that strain accumulation had accelerated the embrittlement process. However, it could be that bolts in which brittle behaviour was observed also had a lower strength. Thus, for 12CrMoV steels and potentially for 1CrMoV alloys, the presence of brittle behaviour in fasteners appears to be due to the presence of extensive networks of coarse carbides on prior austenite grain boundaries. Indeed, the current examination shows that these carbide networks are not simply a function of ageing during service, since deleterious microstructures were also present following fabrication.

Current assessment practices generally utilise factors such as hardness and tensile strength as acceptance criteria for new fasteners. Clearly, hardness does not provide an unambiguous measure of future performance for 12CrMoV steels and previous work has also shown that hardness is not a good indicator of performance in 1CrMoV steels [3].

4. In-Service Assessment

Evaluation of in-service performance requires that critical component specific properties are measured the anticipated operating conditions are known and an appropriate failure criterion has been established.

In the case of fasteners, experience suggests that the level of tightening strain adopted will have a major effect on performance. Thus, control of the methods of tightening must ensure that appropriate limits are not exceeded. Since rapid catastrophic fracture of fasteners should obviously be avoided, a number of different criteria have been suggested as representing failure. In general, these involve detection of macroscopic defects or measurement of dimensions to establish the pattern of permanent strain accumulation with time. In most situations non-destructive testing using ultrasonics or methods of surface crack detection can be applied to identify significant defects. Defective components will normally be replaced since detailed structural integrity assessment is not justified, i.e. components containing cracks are considered to have failed.

Evaluation of dimensional measurements has shown that significant increases in length were noted for some fasteners whereas the strains for other components operating at nominally the same conditions were markedly less (Fig. 6). Consideration of these data suggests [2] that

(i)   provided appropriate baseline data are taken, reliable changes with time can be assessed;

(ii) strain increases with time for all bolts but the rate of increase varies, and

(iii) a strain limit of 1% may be proposed as indicating effective failure.

Approaches utilising crack development or strain accumulation may thus be applied as indicators of performance. However, in many situations the detailed baseline measurements have not been taken to allow strain estimates to be made with the appropriate resolution. Moreover, even though NDE techniques can be applied, these approaches will only be successful in identifying cracks of a significant length. Since these methods can only be adopted during maintenance periods, metallurgical information is necessary to ensure that specific components will not fail in the operating periods between outages.

Evaluation of 12CrMoV fasteners indicates that no significant differences in the structure and properties exist between different regions of specific component. This is in agreement with observation on 1CrMoV steel [3] where detailed evaluation of material from the keyway at the end of a large number of bolts was found to be representative of bulk. Removal of material from non-critical keyway locations should not modify subsequent in-

sample removal is effectively non-destructive. The present
graphic characterisation can be useful in identifying micro-
ely to be susceptible to crack initiation. Furthermore, even
ounts of material are available in most cases, this should be
pecific material properties to be measured using miniature
ures have now been established which allow specimen of
eliably measure mechanical properties, fracture toughness,
12]. Typical specimens before and after testing are shown in
ces can be prepared from small amounts of material, includ-

nt elongation (%)

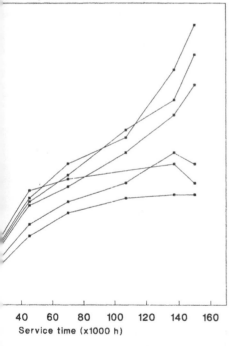

40   60   80   100   120   140   160
Service time (x1000 h)

s a function of operating time for a range of 12CrMoV bolts

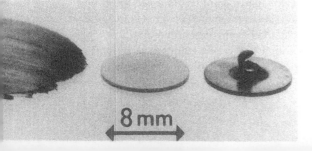

8 mm

ing the button shaped samples removed by the SSam system [13]. These approaches should prove useful in quantification of specific properties. However, at the present time insufficient metallurgical knowledge has been developed to allow a critical level of any particular property to be specified. Further work linking materials behaviour to in-service performance is therefore required before detailed life assessments can be made.

5. Concluding Remarks

The microstructure and performance of 12CrMoV steel used for the manufacture of high temperature fasteners indicates that significant differences can occur in as-fabricated components. It appears that this variation can arise, even for compositions within the allowable range, due to the failure of the normalising heat treatment to fully dissolve the carbides present. Under these circumstances, networks of coarse precipitates exist prior to service resulting in further property degradation. Microstructures of this type exhibit relatively low creep strength and an enhanced susceptibility to form creep voids on further ageing during service.

Fasteners which develop in-service cracks were invariably found to have microstructures containing networks of coarse carbides. Laboratory testing revealed that these components exhibited low creep strength and brittle fracture behaviour compared to anticipated properties. Defects therefore appear to develop as a consequence of the low levels of creep strength which result in enhanced deformation in a microstructure with a low threshold for cavity nucleation. The critical defect size is then dependent on fracture toughness. Assessment of service performance can be based on metallographic examination in combination with inspection to identify defects or the level of permanent strain. However, direct measurement of component specific properties can be carried out in an effectively non-destructive manner using miniature disc test procedures. If relationships between specific properties and remaining life can be established, these approaches should give increased confidence in decisions on reinspection intervals or replacement.

References

1. G. D. Branch, J. H. M. Draper, H. W. Hodges, J. B. Marriott, M. C. Murphy, A. T. Smith and L. H. Toft, *Int. Conf. on Creep and Fatigue in Elevated Temperature Applications*, Inst. Mech. Engrs, Philadelphia 1973 and Sheffield, 1974, pp.192.1.
2. M. De Witte and J. Stubbe, 'Failure Mechanisms and practical approaches for the following up of high temperature turbine bolts', Laborelec Report, 1975.
3. J. H. Bulloch and J. J. Hickey, 'Some observations concerning reverse temper embrittlement and creep damage in a series of CrMoV steel turbine bolts after 120 000 h of service', *Materials at High Temperatures*, 1993, **12**, (1), 13–24.
4. B. L. King, 'Intergranular embrittlement in CrMoV steels: an assessment of the effects of residual impurity elements on high temperature ductility and crack growth', *Phil. Trans. R. Soc. Lond.*, 1980, **A295**, 235–251.
5. J. D. Parker, 'The Behaviour of High Temperature Fasteners in Electricity Generating Plant', *Proc. 2nd Int. Conf. on Creep and Fracture of Engineering Materials and Structures*, (B. Wilshire and D. R. J. Owen, Eds), 1984, Pineridge Press, pp.723–738.

6. G. V. Smith, 'Evaluation of the Elevated Temperature Tensile and Creep Rupture Properties of 12 to 27% Chromium Steels', ASTM Data Series, DS59, 1980.

7. J. Kruszynska, 'Evaluation of 12% Cr Steel for High Temperature Services', MSc Thesis, University of Waterloo, 1983.

8. R. W. Evans and B. Wilshire, *Creep of Metals and Alloys*, 1985, The Institute of Metals, 314p.

9. D. J. Gooch, 'Creep fracture of 12CrMoV steel', *Met. Sci.*, 1982, **16**, 79–89.

10. M. De Witte, 'Power plant life estimation and extension: The Belgian experience from the users' point of view', *Int. J. Pres. Ves. and Piping*, 1989, **39**, 41–55.

13

Consideration of the Role of Nb, Al and Trace Elements in Creep Resistance and Embrittlement Susceptibility of 9–12% Cr Steel

V. FOLDYNA and Z. KUBON

Vitkovice Research Institute, Pohranicni 31, CZ-706 02 Ostrava, Czech Republic

ABSTRACT

The creep resistance of advanced chromium steels can be significantly increased due to precipitation of very small particles of vanadium nitride (VN). The volume fraction of VN is controlled by content of nitrogen which is available for the formation of VN–N_{AV}. The solubility and precipitation of VN, Nb(C,N) and AlN in austenite and ferrite was analysed using relevant solubility products. The dependence of NAV on the Nb and Al contents was determined. Based on results obtained over many heats of chromium modified steels, the calculated values of nitrogen in solid solution — N_{ss} — were used to assess the creep rupture strength of chromium bolting steel (mean considered chemical composition in wt%: 0.18C; 11Cr; 0.8Mo; 0.3V; 0.4Nb; 0.06N; 0.02Al). Increasing Nb content from 0.1 to 0.5% leads to decreasing creep rupture strength over a period of 100 000 h at 600°C of *ca.* 30%. Lowering Al content from 0.03 to 0.003 wt% produces higher creep rupture strength of *ca.* 15%.

1. Introduction

The creep properties of chromium modified steels are governed by the chemical composition of these steels and their microstructure. Given a certain chemical composition of the steel, its microstructure depends on the heat treatment.

The precipitation strengthening (PS) of 9–12% Cr steels is effected predominantly by $M_{23}C_6$ carbides. With decreasing interparticle spacing (IPS) of $M_{23}C_6$ the proof stress at room temperature ($R_{p\,0.2}$) increases while the creep rate decreases. Therefore, a close relationship can be assumed between $R_{p\,0.2}$ and the creep resistance of these steels. In vanadium-containing steels the effective IPS of secondary phases decreases during creep due to precipitation of vanadium nitrides. The volume fraction of VN is controlled by the content of nitrogen available for the formation of VN–N_{AV}. The creep rate is then controlled by the mean IPS of $M_{23}C_6$ and VN particles. Very fine VN particles play the most important role in PS of newly developed steels, with N content up to 0.08 wt%. In niobium-bearing steels, the effect of Nb(C,N) particles should be taken into account, too [1–16].

The strengthening of solid solution is caused primarily by the presence of Mo and/or W in solid solution. When assessing the creep properties we must take into account the Mo and/or W in solid solution during time of creep exposure. Precipitation of Laves phase and coarsening of $M_{23}C_6$ carbides are significant degradation processes occuring in the course of creep. Such structural changes indicate that there is no point to increase the Mo content in modified 9–12% Cr steels over 1 wt% [5, 9, 13, 16, 19, 22].

The aim of the present paper is to explain the role of Nb and Al in creep resistance and the role of Al and trace elements in embrittlement susceptibility of chromium steels. The attention is concentrated to the bolting steel X19CrMoVNbN 11.1 with the following target chemical composition (in wt%): 0.18C; 11.0Cr; 0.8Mo; 0.3V; 0.45Nb, 0.06N and 0.02Al.

2. Procedures

Analysis of equilibrium composition was based on solubility products of VN, AlN and Nb(C,N) in austenite and VN and AlN in ferrite. Solubility products used in steels with higher contents of alloying elements must be modified with respect to interaction parameters between alloying elements (e.g. chromium) and interstitials (C,N) or precipitate forming elements (V, Nb, Al, etc.). In the present analysis the first-order interaction parameters between chromium and carbon e_C^{Cr} and chromium and nitrogen e_N^{Cr} were exploited. The forms of equations used for normalising conditions were as follows:

$$\log([V]\cdot[N]) = 5.20 - 10\,500/T - e_N^{Cr} \cdot [Cr] \qquad (1)$$

based on Tatsuro and Hiroo equation [23];

$$\log([Al]\cdot[N]) = 1.03 - 6770/T - e_N^{Cr} \cdot [Cr] \qquad (2)$$

according to Leslie [25] and

$$(\log([Nb] \cdot [C + 12/14 \cdot N]) = 2.\,26 - 6770/T - 1/2\,(e_C^{Cr} + e_N^{Cr}) \cdot [Cr] \qquad (3)$$

published by Irvine, Pickering and Gladman [20].

In the last formula, the equivalent carbon content (C + 12/14 N) was used and the ratio of interstitials in Nb(C,N) was assumed to be 1:1 [21]. The interaction parameters e_C^{Cr} and e_N^{Cr} were used in accordance with Kunze [26] in the form:

$$e_C^{Cr} = -180/T + 0.09$$

$$e_N^{Cr} = -\,145.\,8/T - 0.\,056 + 0.017 \cdot \log\,(T)$$

No data are available to the authors' knowledge on the solubility product of Nb(C,N) in ferrite. As all the calculations of creep rupture strength are based on available nitrogen or nitrogen in solid solution, the only solubility products of VN and AlN were considered. The interaction parameter e_N^{Cr} in ferrite, published by Hertzmann and Jarl [27, 28] seemed to be too low.

Consequently, the solubility products of AlN and VN were deduced with respect to the experimental results gained on 9 and 12% chromium steels with vanadium content up to 0.35 wt% and nitrogen content up to 0.022 wt%, respectively. Vanadium nitride was never detected after tempering at 750°C, although precipitation of VN was observed during creep at 600°C [1, 5, 9]. As the temperature dependence of VN precipitation was unknown, only the absolute member was changed in equations for solubility products of VN [29] and AlN [30]. The adjusted solubility products were used in the following form:

$$\log([V] \cdot [N]) = 5.54 - 7830/T \tag{6}$$

and

$$\log([Al] \cdot [N]) = 5.80 - 10\,062/T \tag{7}$$

The creep rupture strength in 100 000 h at 600°C was estimated by means of the following equation:

$$\text{Rm T}/100\,000\,\text{h}/600°\text{C} = a + b \bullet R_{p\,0.2} + c \cdot N_{ss} \tag{8}$$

where a, b and c are temperature dependent regression quantities, $R_{p\,0.2}$ is the proof stress at room temperature and N_{ss} represents the amount of N in solid solution, which is not bound as AlN and/or TiN and/or Nb(C,N). The product $c\,N_{ss}$ determinates the contribution of VN to the creep rupture strength [16,19]. In this work, N_{ss} in ferrite at 600°C was calculated in the following way:

$$N_{ss}/600°\text{C} = N_{ss}/1100°\text{C} - \text{NAlN}/600°\text{C} \tag{9}$$

where N_{ss} 1100°C represents the nitrogen content in solid solution after solution treatment at 1100°C and NAlN 600°C means the nitrogen content which can be bound at 600°C as AlN. Titanium was not considered. It is clear that N_{ss} is practically equal to N_{AV} as the only small amount of N which remains in solid solution in equilibrium with all nitride forming elements is not considered. The effect of Nb (from 0.1 to 0.5 wt%) and Al (0.003 and 0.03 wt%) on the creep rupture strength was analysed using calculated values of N_{ss} and eqn (8). Correctness of some assumptions concerning thermodynamical values or embrittlement susceptibility was verified by means of transmission microscopy (thin foils and carbon extraction replicas).

3. Results

In Fig. 1 the solubility curves for VN, AlN and Nb(C,N) at 1100°C are shown for steel containing 11 wt% Cr. The solubility product of VN is significantly greater than these for Nb(C,N) and AlN. It can be expected that at 1100°C all VN and AlN will be dissolved, while Nb(C,N) will be only partially dissolved even at the austenising temperature 1150°C. On the assumption that the amounts of C, N and Al in the steel X19CrMoVNbN 11.1 are as follows (in wt%): 0.18C, 0.06N and 0.03Al, we have analysed how the concentrations of C, N and Nb in the solid solution at 1100°C depend on the total Nb content in the steel. The results are shown in Fig. 2. With increasing Nb content (up to 0.5 wt%) the N content in the solid solution after normalising at 1100°C decreases to one half of that in the steel containing only

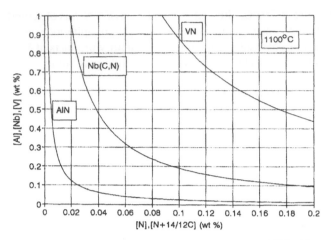

Fig. 1 *Solubility curves for VN, Nb(C,N) and AlN in austenite at 1100°C with respect to chromium content in the steel (11 wt%).*

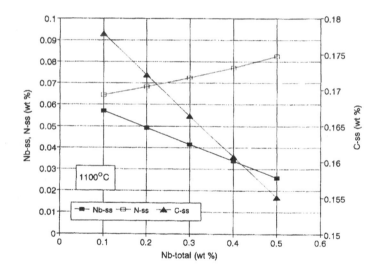

Fig. 2 *Dependence of solid solution concentrations of Nb, N and C in austenite on the total niobium content in X19CrMoVNbN 11.1 steel.*

0.1 wt% Nb. The drops of C content in the solid solution was also observed after mentioned solution treatment.

The dissolution of precipitates in austenite during normalising is only the prerequisite for obtaining high C and N contents in solid solution and good properties of chromium modified steels. Dissolved V and N at solution annealing promote the precipitation of VN during tempering or creep exposure. Besides of VN, precipitation of AlN in ferrite can be expected. The solubility curves of VN and AlN in ferrite at 700 and 600°C are shown in Fig. 3. The amount of vanadium which can be bound as VN at 700°C decreases with increasing Nb and/ or Al content in the steel (Fig. 4). Due to lower solubility product of VN at 600°C, it can be assumed that higher vanadium content can be bound as VN during creep at this temperature (Fig. 5 (a, b)). Practically all Al content in the steel can be bound as AlN during creep at 600°C (compare Fig. 5 (a) and Fig. 5 (b)). In these figures it is also shown, how the N_{ss} at 600°C depends on the Nb and Al contents in the steel.

Figure 6 indicates how the creep rupture strength in 100 000 h at 600°C depends on the Nb and Al content. The proof stress at room temperature of the mentioned bolting steel was considered to be 780 MPa. Given identical Al content, increasing Nb content from 0.1 to 0.5 wt% leads to decreasing creep rupture strength of *ca.* 30%. Lowering Al content from 0.03 to 0.003 wt% produces higher nitrogen content in solid solution (N_{ss}) and the creep rupture strength increases of about 15% for all mentioned niobium contents (see Table 1).

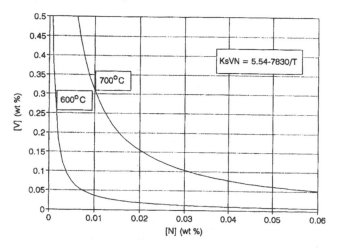

Fig. 3 (a) *Solubility curves for VN in ferrite at 700 and 600°C with respect to chromium content in the steel (11 wt%).*

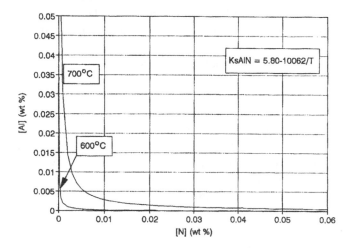

Fig. 3 (b) *Solubility curves for AlN in ferrite at 700 and 600°C with respect to chromium content in the steel (11 wt%).*

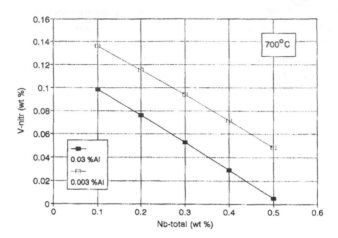

Fig. 4 *Differences in vanadium bound as VN at 700°C between steel with 'high' and 'low' aluminium contents.*

4. Discussion

Orr and Di Gianfrancesco have analysed the solubility data for VN in austenite at 1050 and 1100°C in the steel 91 (the target composition in wt%: C 0.1; Cr 9.0; Mo 1.0; Nb 0.1; V 0.2 and Al 0.5), assuming that all niobium is combined as Nb(C,N) and 50% of the interstitials in Nb(C,N) is nitrogen, reducing that N for combination with vanadium [17].

They used the solubility products of VN valid for C–Mn steel [20]. But the presence of 9% chromium in steel increases the solubility product of vanadium nitride at 1050°C approximately nine times and thus raises the amount of V and N in solid solution significantly [21, 31]. This assumption was verified by transmission electron microscopy analyses which did not reveal any precipitates of VN in normalised P91 steel [7, 32, 35], although according to solubility product of VN without respect to Cr content, large amount of VN should be expected in this steel.

In accordance with [17], the available nitrogen N_{AV} can be calculated using relation (10):

$$N_{AV} = N_t - 1/2 \cdot 0.15 \cdot Nb \qquad (10)$$

here N_t and Nb represent total nitrogen and niobium contents in the steel.

Recently we have analysed how the nitrogen content in solid solution depends on the Nb (up to 0.1 wt%) and Al (up to 0.05 wt%) contents in the 9% CrMoVNbN steel [33,35]. Providing that Nb content in the steel is constant, the amount of N in solid solution after normalising at 1050°C decreases with increasing Al content in the steel due to precipitation of AlN. The amount of undissolved Nb(C,N) depends on the Nb and N contents in the steel, but does not depend on the Al content. In the steel with low Nb content (up to about 0.05 wt%), practically all Nb(C,N) can be dissolved after normalising at 1050°C [35]. In spite of that, the N content in solid solution after normalising can not be considered as nitrogen which can be bound as VN during tempering and/or creep exposure. It was shown that

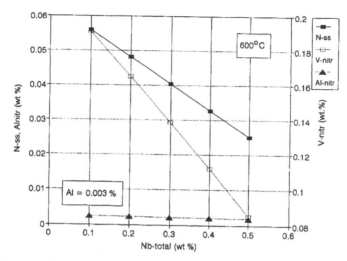

Fig. 5 (a) *Amounts of 'available' nitrogen, V and Al in nitrides as a function of total niobium content in steel X19CrMoVNbN 11.1 with 0.003% Al.*

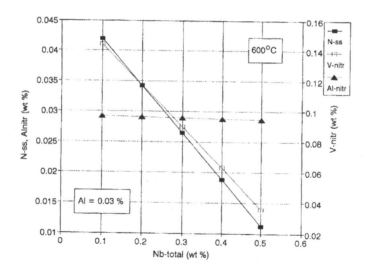

Fig. 5 (b) *Amounts of 'available' nitrogen, V and Al in nitrides as a function of total niobium content in steel X19CrMoVNbN 11.1 with 0.03% Al.*

besides of VN, precipitation of AlN in ferrite should be expected at least. The precipitation of AlN was observed in the well known steel X20CrMoV 12.1 [2, 9]. In the heat containing 0.048 wt% Al and 0.0172 wt% N (Al:N ratio = 2.79), the size of observed AlN particles varied from 0.25 to 1.6 μm. As a result of scavenging effect of aluminium on N in solid solution, no discernible signs of precipitation fine VN particles were observed. At the same time, this heat displayed about ten times higher creep rates when compared with other investigated heats of this steel with aluminium contents ranging from 0.003 to 0.013 wt% and Al:N ratios from 0.25 to 0.71:1. Moreover, the creep rupture strength in 100 000 h of this

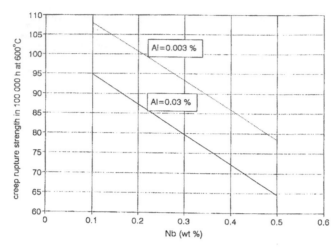

Fig. 6 *Dependence of creep rupture strength in 100 000 h at 600°C on niobium and aluminium contents in steel X19CrMoVNbN 11.1.*

Table 1. *Calculated creep rupture strength in 100 000 h at 600°C in dependence on aluminium and niobium contents of X19CrMoVNbN 11.1 steel*

Basic composition in wt% 0.18°C; 11.0Cr; 0.7Mo; 0.5Ni; 0.3V; 0.06N	Creep rupture strength in 100 000 h at 600°C				
	Nb in wt%				
Al in wt%	0.1	0.2	0.3	0.4	0.5
0.003	107.9	100.7	93.3	85.9	78.4
0.03	94.8	87.3	79.7	72.2	64.6

heat was lower, by as much as 30%, than those for the corresponding heats with lesser Al contents and lower (under-stoichiometric) Al:N ratios. The large AlN particles observed in the heat containing high Al content were about a thousand times bigger than VN particles found in all other heats. Obviously, coarse AlN particles can not contribute to any reduction of the mean inter-particle spacing of the secondary phases, and thereby to improve the creep resistance of the steel [2, 9]. Furthermore, the precipitation of large AlN particles especially on grain boundaries impairs the creep rupture ductility [36].

Recently it has been shown [34] how Al content in the steel deteriorates the creep rupture ductility at 600°C of the steel X19CrMoVNbN 11.1 (see Fig. 7). The reduction of area decreases with increasing Al content in the steel. In conformity with [2, 9], it can be assumed that precipitation of AlN on grain boundaries impairs the creep ductility. The only exception of the rule represents the heat containing 0.016 wt% Al. With respect to relatively low Al content, the observed creep ductility is very low. It is evident that in this case other embrittlement mechanisms play decisive role in ductility drop. Two main mechanisms are possible in accordance with [37, 38]:

— embrittlement due to segregation of impurities on grain boundaries;
— long term isothermal embrittlement due to precipitation of intermetallic compounds such as Laves phase in service.

With respect to relatively low Mo content in the steel X19 CrMoVNbN 11.1, the precipitation of Laves phase is not very probable, but the possibility of temper embrittlement can not be eliminated, especially when no impurity analysis of mentioned heat is available.

The temper embrittlement of 9–12% chromium steels, which is caused by segregation of P, As, Sb and Sn in grain boundaries, is accelerated by Mn and Si. The sussceptibility of steel to the temper embrittlement is, therefore, supressed by reducing the contents of mentioned elements [37, 38].

The reliability of the used creep rupture strength estimation depends on the correct assessment of the nitrogen content in the solid solution (N_{ss}) and of the regression quantities a, b and c in eqn (8), respectively. The assessment of N_{ss} is based on the authenticity of used solubility products. In fact, the scatter of published solubility products is wide [24]. We chose or adjusted solubility products to be in accordance with available microstructural analyses gained on modified chromium steels. The estimation of quantities a, b and c in eqn (8) is based on statistical analyses of creep rupture strength of modified chromium steel. For that purpose 33 heat with different chemical composition as 9%Cr–1%Mo, 9%Cr–1%Mo(V), 12%Cr–1%Mo–0.3%V, 12%Cr–1%Mo–1%W–0.3%V and 9%Cr–1%Mo–0.2%V–0.05%Nb–0.05%N were used [1, 2, 5, 7–9, 39–43]. All materials were investigated in normalised or quenched and tempered conditions (N + T or Q + T). Long-term creep rupture data were required to enable reliable assessment of creep rupture strength in 100 000 h [19]. Very good correlation between the calculated and measured creep rupture strength (r^2 = 0.791;

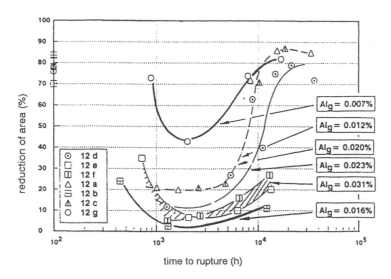

Fig. 7 *Ductility drop in X19CrMoVNbN 11.1 steel during creep as a result of increased aluminium content [34].*

r = correlation coefficient) can be comprehended as a confirmation that used quantities a, b and c in eqn (8) are right (see Fig. 8). Moreover, in this way it is confirmed that proof stress at room temperature and nitrogen content in the solid solution are very significant factors controlling the creep resistance of the modified chromium steels [16, 19]. Consequently, we are entitled to predict creep rupture strength of modified chromium steels using eqn (8). Results of such a prediction are shown in Table 1 for steel X19CrMoVNbN 11.1 with niobium content varying from 0.1 to 0.5% and for two levels of aluminium content (0.003 and 0.03%). The predicted effect of Nb content in studied chromium steels is in good agreement with Park and Fujita [44]. They have shown that presence of excessive amounts of undissolved NbC leads to coarsening of MX precipitates, and a marked drop in long-term creep rupture strength. A good correspondence was also observed between creep rupture strength in 100 000 h at 600°C calculated by eqn (8) and experimentally established [34] in case of 11%Cr–0.7%Mo–0.3%V–0.06%N steel with Nb and Al contents 0.5 and 0.03%, respectively (compare Fig. 6 and Fig. 9). Calculated creep rupture strength for 0.1% Nb and 0.003% Al is well above 100 MPa, the target value for the advanced grades of Cr steels (see Table 1).

5. Conclusions

The following conclusions concerning creep properties of chromium modified steels can be drawn from the matters discussed above:

1. The decisive role in precipitation strengthening of the steels discussed play small and dispersed particles of vanadium nitride.

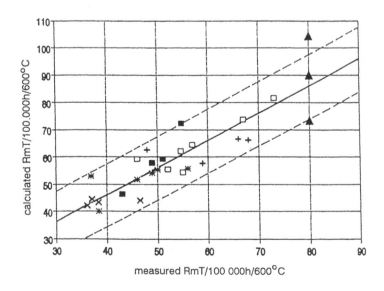

Fig. 8 *Correlation between experimental and calculated creep rupture strength of various types of chromium modified steels [35].*

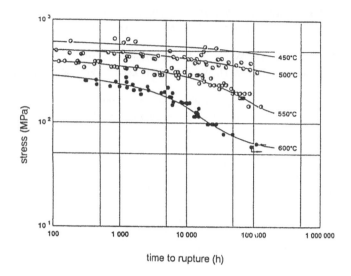

Fig. 9 *Time–temperature dependence of creep rupture strength of X19CrMoVNbN 11.1 steel [34].*

2. The effect of chromium on the solubility of VN in austenite and ferrite is very pronounced, thus enabling dissolution of VN during austenising and following precipitation of dispersed particles during tempering and/or creep exposure.

3. In studied chromium steels, the N content in solid solution after normalising depends on the Nb and/or Al content in the steel. Therefore, the available nitrogen content (N_{AV}) must be calculated with respect to all nitride forming elements in the steel.

4. The creep rupture strength of studied steels can be assessed by using eqn (8) provided the proof stress and nitrogen content in solid solution are known.

5. With increasing niobium and aluminium contents in 19CrMoVNbN 11.1 steel creep rupture strength decreases due to reducing nitrogen content free for VN precipitation.

6. Low creep ductility can be improved by reducing aluminium, manganese, silicon, phosphorus, arsenic and antimony contents in the steel.

References

1. V. Foldyna, A. Jakobova, R. Riman and A. Gemperle, *Creep and Fracture of Engineering Materials and Structures* (Eds B. Wilshire and D. R. J. Owen), Pineridge Press, Swansea, UK, 1984, p. 685.
2. V. Foldyna, A. Jakobova, A. Gemperle and R. Riman, *Creep and Fracture of Engineering Materials and Structures* (Eds B. Wilshire and R.W. Evans), The Institute of Metals, London, UK, 1990, p. 507.
3. P. Battaini, D. Angelo, G. Marino and J. Hald, *ibid.*, p. 1039.

4. V. Foldyna and J. Purmensky, *Czech. J. Phys.*, 1989, **B39**, 1133.

5. V. Foldyna, *Creep of Low Alloy and Modified Chromium Steels*, Technicke aktuality, Vitkovice, No. 1,1988 [in Czech].

6. E. Schnabel, P. Schwaab and H. Weber, *Stahl und Eisen*, 1987, **107**, 691.

7. F. Bruhl, K. Haarmann, F. Kalwa, H. Weber and M. Zschau, *VGB-Kraftwerkstechnik*, 1989, **69**, 1214.

8. F. Bruhl, H. Cerjak, P. Schwaab and H. Weber, *Steel Res.*, 1991, **62**, 75.

9. V. Foldyna, A. Jakobova, R. Riman and A.Gemperle, *ibid.*, p. 453.

10. A. Jakobova and V. Foldyna, *8th Int. Symp. on Creep-Resistant Metallic Materials*, Zlin, Czechoslovakia, September 1991, p. 167.

11. V. Foldyna, A. Jakobova and V. Kupka, *ibid.*, p. 186.

12. J. Hald, *ibid.*, p. 203.

13. V. Foldyna, A. Jakobova and V. Kupka, *6th Czechoslovak–Japan Joint Symp.*, The Society for New Materials and Technologies, Prague; The Iron and Steel Institute of Japan, Tokyo; TANGER, Ltd., Ostrava. Frydek-Mistek, Czechoslovakia, October 1991, p.84.

14. K. Tokuno, K. Hamada and T. Takeda, *JOM*, 1992, **44**, 25.

15. K. Hamada *et al.*, in *Proc. 7th JIM Int. Symp. on Aspects of High Temperature Deformation and Fracture in Crystalline Materials*, Nagoya, Japan, July 1993, p. 333.

16. A. Jakobova and V. Foldyna, *ibid.*, p. 317.

17. J. Orr and A. Di Gianfrancesco, in *Proc. ECSC Information Day on The Manufacture and Properties of Steel 91 for Power Plant and Process Industries*, VDEh, Dusseldorf, November 1992, paper 2.4.

18. J. Orr, D. Burton and C. Rasche, *ibid.*, paper 2.3.

19. V. Foldyna, A. Jakobova and V. Kupka, in *Proc. 15. Vortragsveranstaltung — Langzeitverhalten warmfester Stahle und Hochtemperaturwerkstoffe*, VDEh Dusseldorf, November 1992, p. 55.

20. K. J. Irvine, F. B. Pickering and T. Gladman, *JISI*, 1967, **205**, 161.

21. R. C. Sharma, V. K. Lakshmanan and J. S. Kirkaldy, *Met. Trans. A*, 1984, **15A**, 545.

22. F. Abe and S. Nakazawa, *Met. Trans. A*, 1992, **23A**, 3025.

23. K. Tatsuro and O. Hiroo, *Sumitomo Search*, 1973, No. 10, p. 78; cited in [24].

24. M. Gottwald, Possibilities of Predicting the Occurrence of Interstitial and Intermetallic Phases in Steels, *Hutnicke aktuality VUHZ*, 1983, **24**, (1) [in Czech].

25. W.C. Leslie *et al.*, *Trans. ASM*, 1954, **46**, 1470.

26. J. Kunze, Nitrogen and Carbon in Iron and Steel — Thermodynamics, *Physical Research*, Vol. 16, Akademie-Verlag Berlin, Berlin, 1990.

27. M. Jarl, *Scand. J. Metall.*, 1978, **7**, 93.

28. S. Hertzman and M. Jarl, *Met. Trans.*, 1987, **18A**, 1745.

29. R. W. Fountain and J. Chipman, *Trans. Met. Soc. AIME*, 1958, **212**, 737.

30. ? Iwayama, Nihon Kinzoku Gakkai Koen Gayo, ISI Japan, Tokyo 1977, p. 340.

31. K. Narita, in *Proc. 67th Conf. JIM*, ISI Japan, 1971, p. 162.

32. K. Tokuno *et al.*, *Scripta Metallurgica et Materialia*, 1991, **25**, 871.

33. V. Foldyna and Z. Kubon, in *Proc. 16. Vortragsveranstaltung — Langzeitverhalten warmfester Stahle und Hochtemperaturwerkstoffe*, VDEh Dusseldorf, November 1993, p. 10.

34. K. Konig, *ibid.*, p. 45.

35. V. Foldyna, A. Jakobova, V. Vodarek and Z. Kubon, in *Materials for Advanced Power Engineering 1994*, C.R.M., Liège, Belgium, October 1994 (in press).

36. A. Jakobova, unpublished results.

37. I. Kitagawa, K. Morinaka and A. Fujita, *Proc 11th Int. Forgemasters Meeting*, Terni/Spoleto, Italy, June, 1991, paper IX.6.

38. K. Sato *et al.*, *Proc. Symp. on Clean Materials Technology*, ASM Materials Week, Chicago, IL, USA, November 1992, p. 145.

39. NRIM Creep Data Sheet, No. 10A, Tokyo, 1979.
40. Ergebnisse deutscher Zeitstandversuche langer Dauer, Verlag Stahleisen GmbH, Dusseldorf, 1969.
41. P. Mumme and G. Belka, *Neue Hutte*, 1964, **9**, 736.
42. G. Belka, G. Munch, E. Preissler, V. Foldyna, V. Brazdil and J. Koucky, *Neue Hutte*, 1978, **24**, 424.
43. Data package for T91/P91 produced by Vallourec Industries, March, 1990.
44. I. M. Park and T. Fujita, *Trans. ISIJ*, 1982, **22**, 830.

14

Relaxation Behaviour of Flange Joints in the Creep Range

K. MAILE, H. PURPER, G. HÄNSEL and H. KÖNIG*

Staatliche Materialprüfungsanstalt (MPA) Universitat Stuttgart, Germany
MAN Energie, Nürnberg, Germany

1. Introduction

The electricity generating industry uses bolts in steam power plant in the many joints which need to be seperated for maintenance or repairs. Wherever used, they have to maintain a steam tight joint throughout all operating periods in the life of the plant. At the design stage not only the initial preload of the bolts but also additional operating loads must be taken into account. In high temperature application also the relaxation behaviour of the materials, which is a function of time, stress and temperature must be considered. This stress relaxation could lead to a severe loss in initial preload and can therefore seriously influence tightness of the joints. The two main requirements of a bolt material within the creep range are:

- high yield strength at service temperature. But it also must have a sufficient toughness in order to avoid the danger of brittle fracture.
- high stress relaxation resistance to ensure that the remaining preload is above its lower limit during the whole service interval.

A major problem is the inability to consider accurately the real relaxation behaviour of the bolt joints. Although values for the remaining stress in bolts are given in national standards, i.e. the german standard DIN 17 240, current design standards for pipe flanges does not take into account the loss of initial preload due to relaxation or plastic deformation.

Current experimental methods to determine the stress relaxation behaviour vary widely all over Europe. They include:

- the direct measurement of the relaxed stress at constant strain in automatic servo hydraulic (or servo-electric) machines. Due to the high costs this investigation is mostly used only for short test times.
- long-term tests using creep machines.
- long term tests using model bolt assemblies.

In relaxation tests using smooth specimen either in automatic servo-hydraulic or creepmachines the influence of temperature, test time and initial preload on the relaxation curves could be determined. As the model bolt assemblies are designed to simulate the practical situation the influence of the combination of materials, the area of the cross-sections (stiffness of the components) of turbine casing flanges and the contact surfaces could be obtained additionally. The real plastic deformation of the pipe flange joints which causes a superimposed bending stress in the bolts and the interaction between flange, bolts and gasket could also only be checked by experimental investigations on model flanges.

To extend the service intervals of flange joints in plant, data describing the stress relaxation properties up to 100 000 h are necessary. The data currently available (e.g. in DIN 17 240) are based on results oftests up to 10 000 h and are extrapolated up to 30 000 h. To simulate the stress situation of turbine casing different ferritic and austenitic bolt models were tested up to a test time of more than 40 000 h in a research project. In a second project three different types of pressurised model pipe flanges are tested. Results of both projects are described within this paper.

2. Testing technique and test materials
2.1. Bolt Joint Models

The different types of the used models to simulate the relaxation behaviour of bolted turbine casing flanges are shown in Fig. 1. In the cylindrical shaft of each bolt are hardness indentations to measure the remaining elongation of the model after testing. The type 1 model is primarily used to determine the relaxation behaviour of the bolt material. Therefore all parts of the model are made from the same material. The results obtained from these tests should be similar to those gained in relaxation machines.

The type 2 and 3 models are used to examine a possible component like combination of two different materials. The bolts are manufactured from a 12Cr-steel (X22CrMoV 12 1) and the flanges from a 1.25 Cr-cast-steel (GS-17 CrMoV 5 11). One nut is also made from the cast material in order to simulate the screw fixing in a casing.

In Table 1 the different test materials used for the bolts are given. For each material a

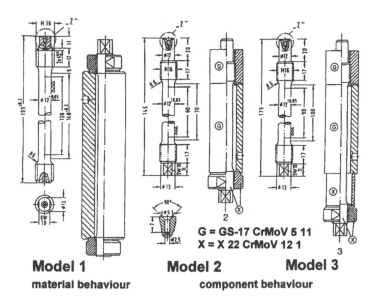

Model 1
material behaviour

Model 2

Model 3

component behaviour

G = GS-17 CrMoV 5 11
X = X 22 CrMoV 12 1

Fig. 1 *Types of the tested bolted joint models.*

Table 1. Materials used for bolted joint models

State of Delivery (initial state)	Service loaded with/without annealing
21 CrMoV 5 7	21 CrMoV 5 11
X22 CrMoV 12 1	X22 CrMoV 12 1
X19 CrMoVNbN 11 1	
X8 CrNiMoBNb 16 16	

single heat was used. For the 1Cr and the 12Cr steel also service loaded material with and without additional annealing was used.

The bolt is tightened in a device at room temperature. Special attention has to be paid that the flange is centered around the bolt and no additional torsion moment acts in the bolt. During tightening the initial elastic elongation of the bolt is determined with the help of a dial gage which measures the distance between the end faces of the bolt. The initial strain in the bolt can be calculated using a reference length. Each joint is pre-stressed three times to the initial strain in order to eliminate settling effects on the contact areas [1].

The assembled models are heated up in the furnace to test temperature with < 100°C/h. The deviations from the controlled test temperature correspond to the requirements of DIN 50 118 (creep test). After reaching the target test time the model is cooled down with < 100 °C h^{-1}. At room temperature the nut made from the bolt material (Fig. 1) is cut down to the root of the thread at two opposite sides using a milling cutter. After destroying the nut the model could be opened and measured. Therefore one model must be used for each test point.

The relaxation behaviour of the model bolts can be described on the basis of the remaining stress σ_2 due to unload. For the type 1 model it can be calculated from

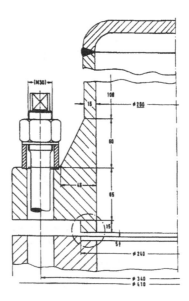

Fig. 2 *Dimensions of the pipe flange joints.*

$$\sigma_2 = \varepsilon_2 \cdot E_T = \Delta \, l'_B/l_B) \cdot E_T$$

ε_2 = elastic strain of the bolt at RT at the end of the test, E_T = average static Youngs-Modulus of the bolt at test temperature and $\Delta l'_B$ = elastic elongation of the bolt at the end of the test. This is determined by measuring the difference of the total length of the bolt before and after opening of the assembly. Therefore influences on the total length (i.e. oxidation) are of no interest, and l_B = reference length.

In the type 2 and 3 models the specific Youngs-Modules and the thermal expansion behaviour of the different materials must be taken into account as well as the influence of the stiffness [2]. The remaining stress of the bolt at test temperature results from:

$$\sigma_{2BT} = (a \cdot b \cdot \varepsilon_2 + a \cdot T)) E_{BT}$$

where E_{BT} = Youngs Modulus of the bolt at test temperature T, ε_T = thermal expansion.

$$a = \left(\frac{C_F}{C_F + C_B} \right)_T \quad \text{at test temperature} \quad T$$

$$b = \left(\frac{C_F + C_B}{C_F} \right)_T \quad \text{at troom temperature} \quad RT$$

$$C_F = \text{stiffness constant of the flange;} \quad C_B = \frac{A_B \cdot E_B}{l_B}$$

$$C_B = \text{stiffness constant of the bolt;} \quad C_B = \frac{A_B \cdot E_B}{l_B}$$

where A_F, A_B = cross-section flange/bolt, and l_F, l_B = length flange/bolt.

2.2 Pipe Flange Joint

In Fig. 2 (shown opposite) the main dimensions of the models are shown. All bolts, nuts and sleeves are manufactured according to DIN 2510. The dimensions ofthe flange were chosen to get results (plastic deformation of the flange and relaxation of the bolts) within a reasonable test duration. The test parameters and used materials are given in Table 2.

The test procedure of the pipe model flanges is quite similar to that of the bolted joint models. The bolts are tightened crosswise at room temperature. At the same time an on-line measurement of the strain on 8 of the 16 bolts is performed. Again the elastic strain of all bolts is determined by measuring the distance between the end faces of the bolts. The flanges are soaked at 550°C resp. 530°C in a furnace for a given time span. After testing the elastic

Table 2. Test materials and parameters for pipe flange joint tests

No.	Flange material	Bolt material	Sleeve material	Gasket	Internal pressure [bar]	Test temperature [°C]	Initial strain at RT [%]
1	GS-17	21 CrMo	21 CrMo	–	55	530	0.15
2	GS 17	X 19	X 19	–	55	550	0.15
3	GX-22	X 22	X 22	Ring joint	55	550	0.15

GS-17 = GX-17 CrMoV 5 11   X 19 = X 19 CrMoV 12 1

GX-22 = GX-22 CrMoV 12 1   X 22 = X 22 CrMoV 12 1

                                             21 CrMoV = 21 CrMoV 5 7

remaining strains of the bolts is measured. In addition the plastic elongation of all bolts is determined by means of hardness indentations.

During the tests on-line strain and displacement measurements are perfomed. The distance between the two flanges is measured on three different points as well as the strain of three sleeves and the deformation (in axial and circumferential direction) of the flange using capacitve strain gages. As the sleeves are stiff compared to the bolts, it is assumed that the sleeve deformation is elastic during the whole test. Knowing the elastic strain of the sleeves the remaining stress of the bolts could be easily calculated. Also the temperature on the flanges is measured on different points. Figure 3 shows one instrumented pipe flange model.

3. Test Results
3.1. Bolted Joint Models

This paper summarises onlythe results for the 12 Cr-steel (X 20 CrMoV 12 1).

In Fig. 4 the results for the type 1 model tests (MPA) are compared with values given in the german standard DIN 17 240 and obtained from an other research program (KEG) [4]. For temperatures up to 550°C the coincidence is remarkably good. At 600 °C the remaining stress given in the DIN-standard is higher than that obtained from the MPA tests. As longer the test time is, as higher are the deviations. For long times the difference in stress is larger than 20%. A significant decrease of the remaining stress at 600°C for test times near 20 kh could be stated. The S-shape of the relaxation curve is quite similar to that of creep rupture curves. In all experiments performed, this behaviour was found for all materials at high temperatures and long test times.

Comparing the results of bolted joint models model bolt tests with those of relaxation tests with smooth specimen some aspects have to be considered:

- As for each test point a single model was used the possible scatter band of experimental results have to be taken into account.
- Compared to the smooth specimen the stress distribution in the jointed bolt model is more complex. Additionally settling effects and a local plasticity of the contact surfaces between bolt, nut and sleeve must be considered.

Fig. 3 *Instrumented pipe flange model.*

In Fig. 5 the results of the bolted joint model tests are compared with relaxation tests performed in a servo-hydraulic machine. The coincidence between the remaining stress measured at both types of specimen is quite good.

Figure 6 shows the results for the different types of models. If the different thermal expansion coefficients were taken into account, only slight deviations can be observed. For the type 3 model tests performed at 550°C the remaining stress at room temperature was nearly zero. This means that the remaining stress at test temperature is identical or lower than the value of the stress due to the different thermal expansion cofficients.

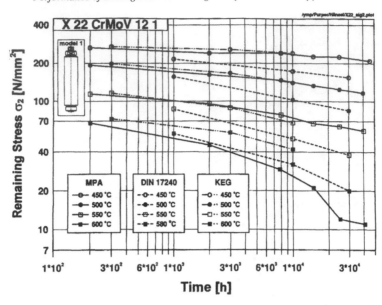

Fig. 4 *Results of the jointed bolt model (Type 1).*

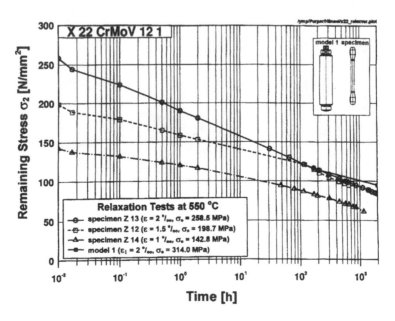

Fig. 5 *Comparison of results of bolted joint models tests and smooth specimen relaxation tests.*

3.2. Pipe Flange Joint

As the experiments are not finished yet, no final results can be given. In Fig. 7 the online measured strains at the sleeves and the flange for joint No. 1 are given for the first 3000 h. Also the measured distance between the two Mange casts is depicted in this figure.

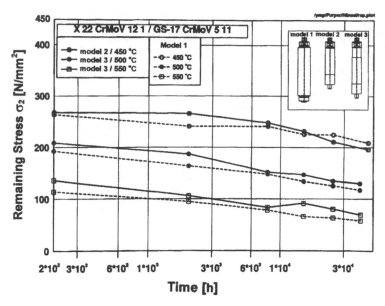

Fig. 6 *Comparison of the results of the different bolted jont models.*

After a test duration of 5 000 h, the experiments were interrupted. The joints were dismantled at room temperature. The elastic remaining strain of all bolts was determined by measuring the distance between the end faces of the bolts before and after the joint was opened. The plastic deformation of the bolts was determined by measuring the distances between the hardness identations on the cylindrical shaft of the bolt. Also the plastic deflection of the flange plate

was measured at several points using a dial gauge. In Fig. 8 the results for joints No. 1 and 2 are shown together with the elastic remaining strain of the bolts at the beginning of the test.

3.3. Finite Element Analyses

For Finite Elemente Analyses in the creep range a constitutive model of the creep behaviour must be derived. This model is based on three parameters:

- a uniaxial creep law describing the experimental uniaxial creep curves,
- a so called 'flow rule' for multiaxial conditions,
- a hardening rule for prescribing the specific manner in which the formulation applies to variable stress conditions.

Particular emphasis has to be put on the determination of the creep law. It must take into account, that the creep strain is a function of time, temperature and stress. For the numerical analyses of the relaxation behaviour it is necessary to describe the primary creep stage as accurate as possible, tertiary creep could be neglected. A very good fit of the measured data at an early stage of creep can be reached if two strain hardening terms are used as it is in the so called Garofalo/Blackburn equation.

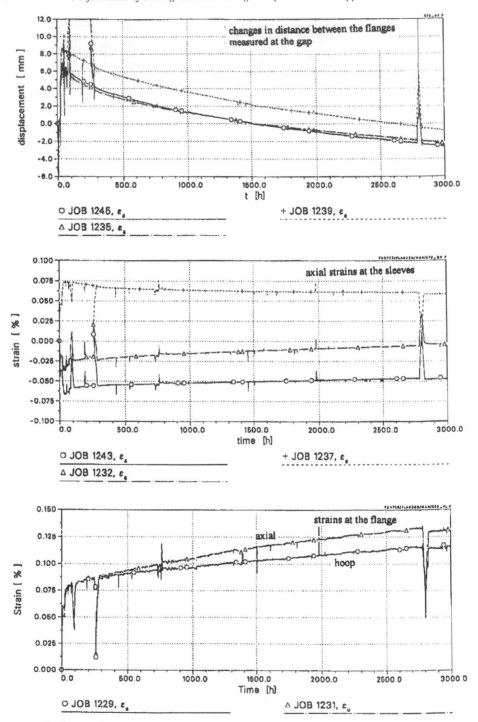

Fig. 7 *Online measured strains for Pipe Flange Joint No. 1.*

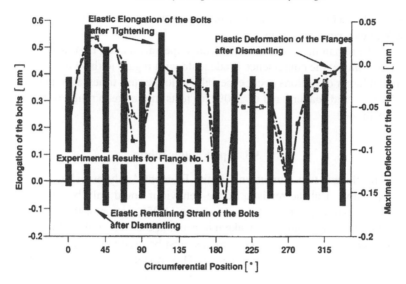

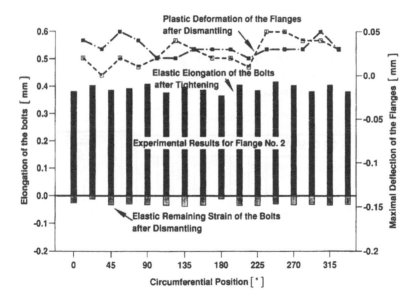

Fig. 8 *Experimental results of Pipe Flange Joints No. 1 and 2.*

3.4. Bolted Joint Models

For the numerical analyses of the bolted joint models it is sufficient to use axisymmetric elements if settling effects in the thread are not considered. These effects were studied qualitively in other calculations [3].

In the analyses of the type 3 models the different thermal expansion coefficients must be considered. The time-dependent material behaviour of the 12Cr-steel (X 22 CrMoV 12 1) was described by a Garofalo creep law whereas for the cast material (GS-17 CrMoV 5 11) a Graham-Walles creep law was used.

In Fig. 9 the remaining stress obtained from experiment and FE-Analyses are compared. For the type 1 model the coincidence is adequate with regard to the whole test time. For the type 3 model larger deviations occur especially at longer test times. But they are still within the usually assumed scatter band of ±20% for tests in the creep range if different material heats are examined.

3.5. Pipe Flange Joints

Also for pipe flanges two dimensional Finite Element Analyses using axisymmetric elements are possible. But the realistic flange behaviour could only be calculated using 3D models. Due to symmetry it is sufficient to use a 11.25° section of the flange if reasonable boundary conditions are chosen. To take into account the wear and settling effects between the different surfaces special contact elements must be used in the model. Figure 10 shows the Finite Element Model of the pipe flanges. This model could still be simplified by using a beam Element for the bolt and if the nut is ommtted. In the FE-Analyses the different load cases:

- tigthening of the bolts,
- heating of the assembly,
  - change of material properties (e.g. Youngs-Modulus)
  - different thermal expansion of the materials
- internal pressure must be considered.

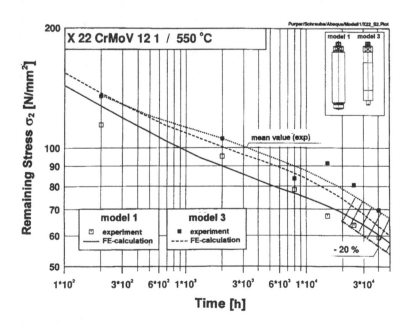

Fig. 9 *Comparison of Finite-Element and experimental results.*

Up to now not all creep curves for the specific materials are available. Therefore only linearelastic material behaviour can be considered. Figure 11 shows the von Mises equivalent stress ofthe pipe flange after the bolts were tightened.

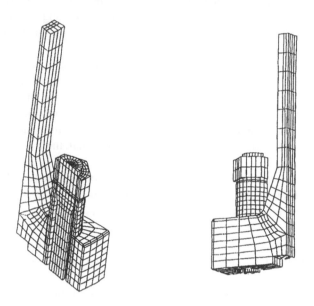

Fig. 10 *Finite Element Model for pipe flange models.*

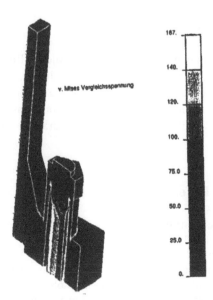

Fig. 11 *Von Mises equivalent stress of the flange model after the bolts were tightened.*

4. Summary

For the design and supervising of pipe flange connections used in high temperature plant application detailed knowledge about their relaxation and deformation behaviour is required. Due to the interaction between the deformation occuring in the flange, bolts and gasket and the different time dependent material behaviour of the components, the determination of the stresses on a flange connection is very complex.

Usually the relaxation behaviour is determined by direct measurement of the relaxed stress of a smooth specimen in a automatic servo hydraulic machine under constant strain loading conditions. However these tests are very cost-intensive and therefore focussed on relative short times. Long term investigations could be done at reasonable prices using bolted joint models. Another advantage of bolted joint models is that the influence of the thread, the flange dimension and the different material behaviour of the nut, the flange and the bolt, can be taken into account.

For different ferritic and austenitic steels bolted joint models were tested up to a total test time of more than 40 000 h. The results obtained from the model bolt tests were compared to those available from relaxation tests and Finite-Element-Analyses.

A suitable coincidence between both, experimental and numerical results could be observed. Obviously the analytical formulae to calculate the remaining stress can be used to describe the relaxation behaviour of the model bolts. Bolted joint model tests can be classified as a reliable and proven test technique.

To obtain a simplified calculation methodology and design rules for pipe Idange joints, three different types of pressurised model pipe flange joints are tested. In the tests the deformation of the flanges and the remaining stress on the bolts is measured on-line by capacitive high-temperature strain gages in order to verify the inelastic FE-calculations.

5. Acknowledgement

The results presented here originate from two different research projects. Both projects AiF No. 8198 and 8532 were supported by the German Turbine Industry, the Technische Vereinigung der Großkraftwerksbetreiber e. V. (VGB) and sponsored by the German minister for economic affairs (Bundesminister für Wirtschaft).. The last one was also supported by the DECHEMA (Deutsche Gesellschaft für Chemisches Apparatewesen, Chemische Technik und Biotechnik. e.V.).

References

1. Stahl-Eisen Prüfblatt 1260: Relaxationsversuch bei erhöhten Temperatur mit Schraubenverbindungsmodellen (Entwurf 12/93).
2. H. Wiegand, K. H. Kloos and W. Thomala, *Schraubenverbindungen*, Springer Verlag, 1988.
3. K. G. Maile, G. Hänsel and H. Purper, Langzeituntersuchungen zum Relaxationsverhalten warmfester Schraubenverbindungen, Abschlußbericht, AiF-Forschungsvorhaben Nr. 6841, MPA Stuttgart, 1994.
4. Relaxationsverhalten warmfester Stähle fur Schrauben, KEG-Forschungsvohaben, Dok. Nr. 6210-KF/1/101 F 6.2/74, VDEhDüsseldorf, 1978.

SESSION 4

Austenitic Bolting Steels

Chairman: H Everson

UES Steels, Stocksbridge, UK

15

Warm Worked Esshete 1250: A High Strength Bolting Steel

J. ORR, H. EVERSON* and G. PARKIN†

British Steel Technical, Swinden Laboratories, Rotherham, UK
*UES Steels, Stocksbridge, UK
†Spencer Clark Metal Industries, Rotherham, UK

ABSTRACT

Esshete 1250, established in British and ASTM Standards as a tube/pipe steel, is also available in the warm worked condition for high strength applications such as bolts and pump shafts. This paper reviews the applications for the steel and describes the principles and practicalities of warm working austenitic stainless steel and the mechanical properties achieved for warm worked Esshete 1250 at ambient and elevated temperatures. Particular emphasis is given to those properties influencing the performance of warm worked Esshete 1250 as a bolting material. Details are given on tensile properties (up to 700°C), stress rupture and creep properties (550–700°C), stress relaxation (0.15% strain at 600–700°C), and physical properties. Toughness data are presented to describe its resistance to embrittlement at 625°C. The test durations for stress rupture and stress relaxation extend to over 70 000 h. Comparisons are drawn between warm worked Esshete 1250 and other strengthened austenitic steel and nickel base alloys.

1. Introduction

The drive towards higher efficiency power generation requires higher temperature and steam pressures than the current 'standard' values of 565°C and ~200 bar, respectively (in the UK). This in turn requires materials to operate at such higher temperatures for which a current industry target is 650°C. Thus, austenitic stainless must be considered, since available ferritic steels do not have sufficient strength for service at ~650°C. The subject of this paper is austenitic stainless steel bolting material.

Esshete 1250, produced by UES Steels, is a fully austenitic stainless steel grade specified in British and ASTM Standards for tube and pipe products [1–3]. It is well characterised by the results from an extensive data collection programme [4]. In the solution treated condition Esshete 1250 is used for superheater tubing and steam pipes in many UK, and some worldwide, power stations on account of its high strength at elevated temperatures, stability of microstructure. and ease of weldability relative to other conventional austenitic stainless steels [4].

However, in common with other austenitic stainless steels, the strength of Esshete 1250 at room temperature (typical 0.2% proof strength $R_{p0.2}$ = 350 MN m^{-2}) is much lower than that for hardened and tempered alloy steel bolting steels (typical $R_{p0.2}$ = 750MN m^{-2}). For temperatures above ~580°C. hardened and tempered alloy steel bolting materials do not have sufficient strength. There are some materials available, e.g. Nimonic alloys, but these are expensive and do not always have suitable properties for the full service range.

Warm worked austenitic stainless steels, with a lower cost than nickel base materials, have significantly higher strengths than solution treated grades and, thus, offer an opportunity for some bolting applications. Other applications where warm worked austenitic stainless steels have been used are pump shafts and turbine blades.

The procedure developed for warm worked Esshete 1250 and the attractive ranges of properties which are developed for ambient and elevated temperatures are described in this paper. A brief description of the steelmaking and rolling routes used for the production of Esshete 1250 is included. Further information on steelmaking and rolling of this and other bolting materials is given in Ref. [5].

2. Steelmaking and Rolling of Esshete 1250

Esshete 1250 is produced in billet form by UES Steels using the route shown schematically in Fig. 1.

At the steelmaking stage selected raw materials are melted in an electric arc furnace (EAF) to provide a base containing the bulk of the alloying elements.

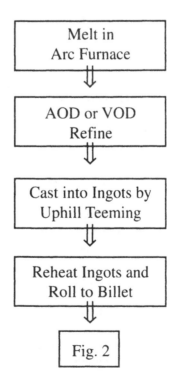

Fig. 1 Esshete 1250: steelmaking and rolling process route.

The liquid metal is transferred into an argon oxygen decarburisation or a vacuum-oxygen decarburisation unit for subsequent refining. During this stage, carbon is removed to the desired level and lime and fluxing agents are added to produce slags of optimum composition for desulphurisation to reduce sulphur to low levels. The aim composition, within the specified range given in Table 1, is achieved by making any necessary alloying additions. Finally, the temperature of the liquid steel is adjusted to the optimum temperature for casting as ingots.

Ingots of appropriate size and configuration are produced by teeming the liquid metal from a ladle using uphill pouring techniques.

After solidification the ingots are removed from the moulds, inspected, and charged to a reheating furnace. The ingots are soaked at a specific temperature within the range 1100–1200°C and rolled to billets of appropriate size and section for further processing, in this case for warm working by a reroller.

3. Principles and Practice of Warm Worked Austenitic Stainless Steels

Warm working can be described as the deformation of material at a suitable elevated temperature, sufficient to increase the strength by increasing the dislocation density which is not significantly decreased by recovery or recrystallisation on cooling to ambient temperature. One other constraint is that the warm working should be carried out at specified temperatures above the service temperature range so that no significant recovery or recrystallisation occurs during service.

Warm working is carried out most commonly for bars using rolling mills of the merchant bar design because of their flexibility in terms of hot reduction and pass sequence relative to the more modern automatic/continuous repeater mills. Other component forms are made using presses, forges, and sheet mills. Further advantages of bar mill processing is that ribbed/profiled bars can be produced and short order runs can be accommodated.

The material to be processed by warm working must have the required combination of strength and ductility at the warm working temperatures, usually in the range 500–1000°C. Both these parameters are necessary to enable the material to be rolled within the loading limit and power capability of the mill, such that the unsupported areas in the roll gap do not crack, burst, or overextend. The rolling process must also generate the required metallurgical structure and properties. The amount of strain which remains after warm working and recovery has to be above a certain minimum, such that critical grain growth does not occur during service at elevated temperatures.

Because of the relatively small amount of deformation (per rolling pass) and limits im-

Table 1. *Esshete 1250: specified composition ranges, wt%*

Value	C	Si	Mn	P	S	Cr	Mo	Ni	B	Nb	V
Minimum	0.06	0.20	5.50	-	-	14.0	0.80	9.00	0.003	0.75	0.15
Maximum	0.15	1.00	7.00	0.040	0.030	16.0	1.20	11.00	0.009	1.25	0.40

posed by mill loads and power, warm working is limited to a maximum bar size of 90 mm dia., as determined by the penetration of deformation to the bar centre.

Warm working is employed most typically for austenitic stainless steel bars. One example of successful commercial exploitation is warm worked Hiproof 304 and 316 steel (containing ~0.25% N), with proof strength values 200–300% higher than for solution treated material, used for civil engineering projects such as concrete reinforcement where high strength and corrosion resistance are required.

The example of interest to this conference is warm worked Esshete 1250 steel used for bolting/fastener applications (see specified composition ranges in Table 1).

A process route for the manufacture of warm worked bar sizes up to 90 mm dia. in Esshete 1250 is shown schematically in Fig. 2.

Preliminary experimental studies involved rolling Esshete 1250 bars at 600–800°C with up to 22% deformation (Fig. 3). This showed that rolling at ~700°C with a deformation of 10–15% gave a significant and fairly consistent increase of strength relative to the solution treated condition, which was subsequently confirmed in material rolled in commercial mer-

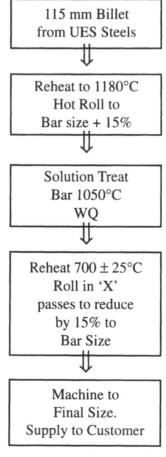

Fig. 2 *Process route for warm worked Esshete 1250 bar (pass sequence and reductions depend on required final bar size).*

chant mills (Fig. 3(a)). It is on this that the commercial route is based (Fig. 2), with optimum conditions of 700 + 25°C and ~15% deformation. Creep data show that such material has a creep rate 10 times lower than that of solution treated material (Fig. 3(b)).

Furthermore, the centre to surface difference in 0.2% proof strength for a 75mm dia. bar rolled with 11.5% deformation at ~650°C was only ~15 MN m^{-2} for a strength level of ~500 MN m^{-2}.

Thus, typical commercial warm working of Esshete 1250 is carried out at about 700°C and with ~15% deformation since this is considered to give consistent properties (Fig. 3(a)).

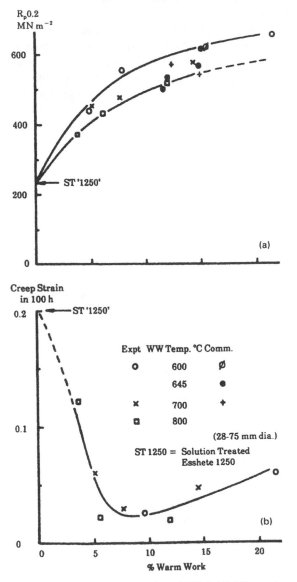

Fig. 3 *Effect of warm work on properties of Esshete 1250. (a) 0.2% proof stress at RT; (b) creep strain (247 MN m^{-2} at 650°C).*

4. Properties of Warm Worked Esshete 1250
4.1. Microstructure and Hardness

Esshete 1250 solidifies to austenite–austenite/NbC eutectic which after hot working and solution treatment becomes dispersed niobium carbide particles in austenite [6]. The basic microstructure is not altered by warm working, except for some grain elongation, which remains stable for at least 77 000 h at 600°C (Fig. 4). The grain size of warm worked Esshete 1250 bar depends on bar size, degree of deformation, and location of metallographic sample within the bar section, as shown in Fig. 4. The grain sizes, in general, are finer that those in the solution treated condition.

The hardness of solution treated Esshete 1250 is typically about 150 HV30, whereas that of warm worked material can exceed 300 HV30. However, the rolling deformation typically 15% at ~700°C, achieved over several passes through the mill, is not uniformly distributed across the bar section, leading to the typical hardness depth profile for bars of 35–75 mm dia., as shown in Fig. 5. Although there is a significant change of hardness from the outside

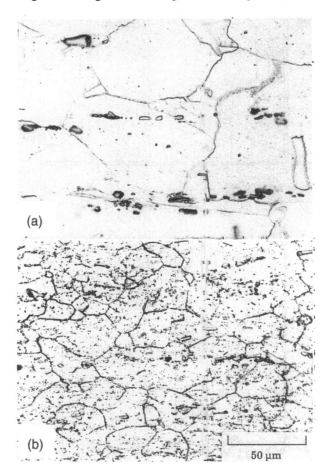

Fig. 4 Microstructures of warm worked Esshete 1250 bars. (a) 22 mm dia. bar tested for 6703 h at 550°C: (b) 45 mm dia. bar tested for 77 087 h at 600°C.

to the centre, the central hardness is significantly higher than that of solution treated material, demonstrating that warm work penetrates throughout bar sections at least up to 75 mm dia.

4.2. Tensile Properties

Room and elevated temperatures test data for warm worked Esshete 1250 show that the strength of this material is significantly above that for solution treated material up to at least 700°C (Fig. 6(a)).

The data available have been subjected to an assessment technique under development in ECISS [7], giving the minimum values shown in Table 2.

In some applications warm worked material may be reheated for a short period as part of the process route. The strength of warm worked Esshete 1250 is retained on reheating up to ~700°C. Above about 750°C, some softening occurs as the material recovery rate increases at temperatures above that used for warm working (Fig. 6(b)).

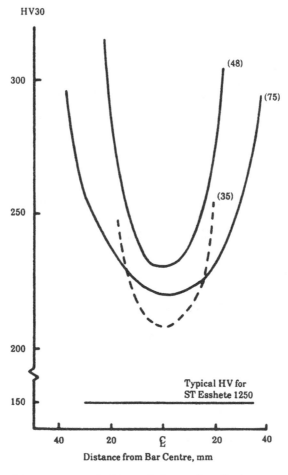

Fig. 5 *Hardness profiles of warm worked bars; ( ) denotes bar dia. in mm.*

Table 2*. Warm worked Esshete 1250: mechanical property values*

Temp., °C	Minimum 0.2% proof strength, MN m^{-2}	Average stress to rupture, MN m^{-2}		Average residual stress, MN m^{-2} (for 0.15% strain)	
		10 000 h	100 000 h	10 000 h	100 000 h
RT	510	...	...	...	...
100	475	...	...	..	...
200	430	...	...	...	...
300	400	...	...	...	...
400	385	...	...	...	...
500	375	...	...	...	...
550	370	402	370	200	160
600	360	340	280	165	130
625	355	300	232	140	115
650	345	260	185	122	...
700	315	...	...	55	...

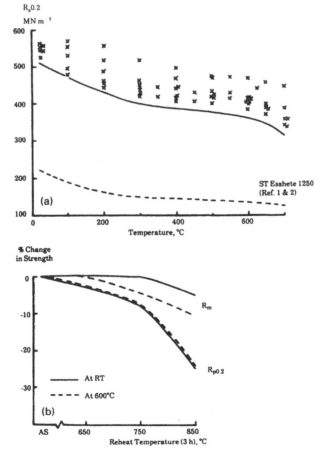

Fig. 6 *(a) 0.2% Proof stress values and (b) stability of warm worked strength.*

4.3. Time Dependent Properties

In common with other elevated temperature service materials, the usefulness for service depends on the level and consistency of strength at and around the service temperatures.

For Esshete 1250 the service temperature is usually within the range 600–650°C [4]. Since warm working is carried out at temperatures as low as 700°C, this same temperature range may be assumed to apply for warm worked Esshete 1250 (see Fig. 6(b)).

The principal application for this grade is anticipated to be bolts/fasteners for which stress relaxation property data are the most appropriate, but other applications require stress rupture and creep stress-strain data.

4.4. Stress Rupture/Creep Strain

Short term data of ~2000 h at 625°C, for warm worked Esshete 1250 46 and 75 mm dia. bars, show a variation of strength across bar diameter but only a small variation of ductility (Fig. 7). This strength profile, being for short term tests, is a reflection of the variation of hardness (Fig. 5).

So that these results for warm worked material can be placed in context, it should be noted that for Esshete 1250 in the solution treated condition the same test condition, 370 MN m^{-2} at 625°C, would cause rupture in a very short duration, typically << 10 h.

Thus, the stress rupture strength of warm worked Esshete 1250 is significantly above that of solution treated material, even in the softest region of the bar, and also good ductility values are retained.

Of more significant interest are longer term stress rupture data characteristics, which are summarised in Fig. 8. These data were derived from the results for 20 batches of warm worked Esshete 1250 tested at 550–700°C for up to 77 000 h, giving the results summarised in Table 2. For example, the stress rupture strength of warm worked Esshete 1250 is 232 MN m^{-2} for 100 000 h at 625°C, which is ~90 MN m^{-2} above that for solution treated material. This is a fairly consistent margin across the relevant temperature–stress range. A similar assessment exercise for 1% creep strain data gave the average stress curve included in Fig. 8, indicating stress for 1% creep to be ~80% of that for rupture. However, the rupture ductility of warm worked Esshete 1250 is also high, typically 15–25%. A further reflection of the high ductility of warm worked Esshete 1250 is that the strength of notched specimens is at least equal and more often superior to that of plain specimens.

4.5. Stress Relaxation

Stress relaxation data, for 0.15% strain tests at 600–700°C for durations up to 70 000 h, are available for warm worked Esshete 1250, as shown in Fig. 9 for residual stress values at 625°C. It will be noted that the scatterband (+20%) is similar to those for heat treated steels, despite the inherent variations in warm worked bars from differences in rolling temperatures, deformations used, and the distribution of deformation, particularly through the larger section bars.

Average residual stress values for 0.15% strain tests for warm worked Esshete 1250 are included in Fig. 8.

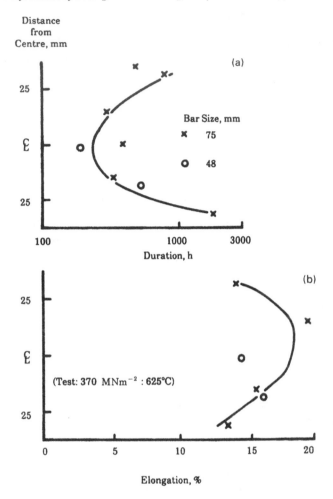

Fig. 7 *Short term stress rupture data for warm worked Esshete 1250 bars. (a) Strength; (b) ductility.*

Cyclic stress relaxation tests based on 015% strain at 650 and 675°C with cycle durations of ~3000 and ~2300 h, respectively, show that warm worked Esshete 1250 maintains or even increases its stress relaxation strength over a significant number of cycles (Fig. 10(a)).

Further data show that strain values up to 0 25% result in residual stress value differences being maintained for up to 30 000 h at 625°C (Fig. 10(b)).

4.5.1. Impact toughness

The impact toughness of austenitic stainless steels is usually very high, typically over 100 J at ambient temperature, but after exposure over particular temperature ranges can, in some cases, become quite low. For Esshete 1250, although some deterioration in toughness does occur, this is not as marked as, for example, type 316 [4].

Warm worked Esshete 1250, by virtue of the higher dislocation density, has an initial

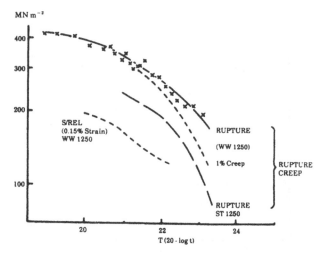

Fig. 8 *Stress rupture, creep, and relaxation strength of warm worked Esshete 1250.*

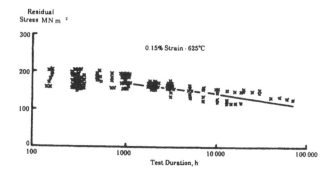

Fig. 9 *Stress relaxation strength of warm worked Esshete 1250.*

toughness lower than that of solution treated material. However, after ageing from 10 000 h at 625°C, the toughness values in both conditions are similar (Fig. 11).

5. Discussion

The results presented in this paper show that the property values claimed for warm worked Esshete 1250 (Table 2) are based on an extensive databank.

The significance of these property values can be assessed by reference to those for other bolting materials. For this purpose, three other materials have been selected namely, Durehete 1055 and Nimonic 80A representing well established high strength ferritic and nickel base bolting materials and a German austenitic steel X8CrNiMoBNb16.16 (1.4986), also supplied and used in the warm worked conditions [8] (Table 3). The data used for these materials are drawn from published information [8–10].

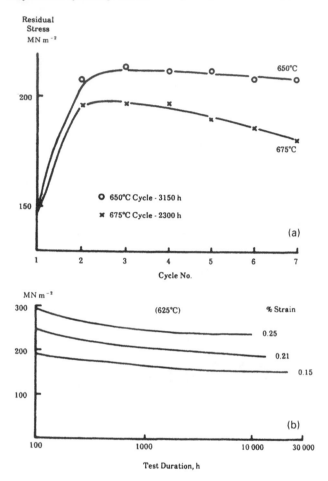

Fig. 10 *(a) Warm worked Esshete 1250 cyclic stress relaxation and (b) effect of strain on stress relaxation.*

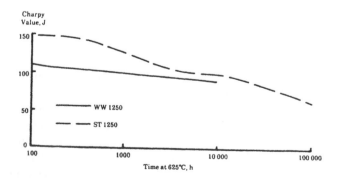

Fig. 11 *Charpy data after ageing at 625°C.*

5.1. Elevated Temperature Strength

The strengths at elevated temperatures of the selected materials are shown in the classical design type diagram in Fig. 12. The data used are the minimum 0.2% proof strength and average 100 000 h stress rupture values, which in the case of warm worked Esshete 1250 are available from Figs. 6(a) and 8.

The strength values for Durehete 1055, which are described elsewhere [11], are the highest for this group of materials up to ~500°C. However, it is at higher temperatures where austenitic steels are likely to be required, especially with the current drive towards higher efficiency power generation, through the adoption of higher steam pressures and temperatures.

Nickel base materials are expensive and can produce operational problems because of their order disorder characteristics which occur over particular ranges of stress and temperature. However, as indicated in Fig. 12, where strength at elevated temperature is the determining factor Nimonic 80A does offer attractive properties up to 600°C. Where other materials and characteristics have to be taken into account, also for temperatures up to 600°C warm worked Esshete 1250 has superior strength to that of the X8CrNiMoBNb16.16 (1.4986) steel.

Above 600°C, although the strength differences are smaller than for the regime where tensile properties rule for design, the proportional differences in strength are still considered to be significant. An enlargement of this region of Fig. 12 demonstrates that, e.g. for 100 000 h at 650°C warm worked Esshete 1250 has a strength advantage of 18% over both Nimonic 80 and the X8CrNiMoBNb16.16 warm worked steel. These two materials are indicated to have similar strengths above 580°C [8].

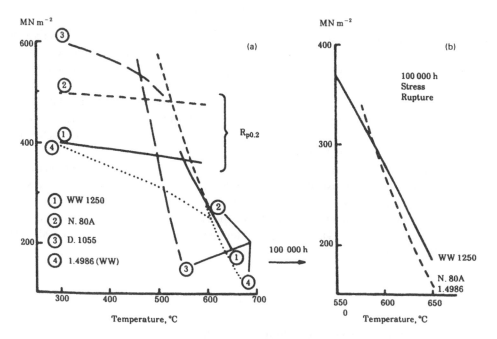

Fig. 12 *Design 'strength' data for bolting materials.*

5.2. Stress Relaxation

The stress relaxation strength, for 015% strain, of warm worked Esshete 1250 is intermediate between that of Durehete 1055 and Nimonic 80A (Fig. 13(a)). These data also indicate that above ~ 600°C, the stress relaxation strengths of warm worked Esshete 1250 and Nimonic 80A become similar.

Furthermore, reloading of Esshete 1250 leads, at least over 7 cycles, to increases of residual stress (Fig. 10(a)). Supplementary data for such tests indicates that, after 7 cycles at 650–675°C, the total strain accumulated is only 0.25–0.3%, which is well below the rupture ductility of the material and, therefore, there should be no risk of cracking of reloaded bolts in warm worked Esshete 1250. Also strain values higher than 0.15% can be used effectively without the stress decay often seen for ferritic steels (Fig. 10(b)).

An empirical relationship appears to exist between the average stress values remaining after a single cycle stress relaxation test and that for rupture (Fig. 13(b)). These data indicate that higher stresses are required for low strain values (e.g. 0.15%) for warm worked Esshete 1250 than for Durehete 1055 or Nimonic 80A, which may be beneficial for bolt applications.

The significant relaxation strength advantage of warm worked Esshete 1250 over the ferritic steels (Fig. 13(a)) and lower cost than nickel base alloys indicate this material to be very attractive for bolting applications at 600–650°C. Service temperatures higher than 650°C are not recommended because of the significant decrease of relaxation strength as the temperature–time combination approaches that of the warm working regime (Table 2).

5.3. Other Service Properties

The physical properties of austenitic stainless steels apply generally to warm worked Esshete 1250. Therefore, the known requirements relating to the relatively high thermal expansion and lower thermal conductivity factors which apply to the use of such steels also apply to warm worked Esshete 1250.

There are some variations, particularly in elastic modulus values, which relate to the variation of stored energy across bar sections. with a gradient from edge to centre. However, the variation is only *ca.* 6% in 75 mm dia. bars, as shown in Fig. 14. Data for Durehete 1055 and Nimonic 80A are included for comparison purposes.

The toughness of warm worked Esshete 1250, although lower than that for solution treated material, is well above that usually required by bolting material specifications. The slow deterioration during service is related to carbide precipitation (Fig. 4(b)) offset by some recovery of the dislocated microstructure. Sigma phase forms only slowly in Esshete 1250 and only above about 700°C [4]. The projected toughness value of ~50–60 J after 100 000 h at 600–650°C also indicates satisfactory performance characteristics.

6. Conclusions

Warm worked austenitic steels have significant strength advantages at ambient and elevated temperatures over solution treated materials, thus making them candidates for power gen-

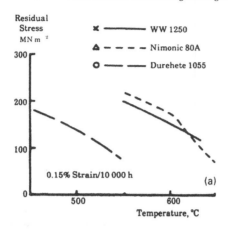

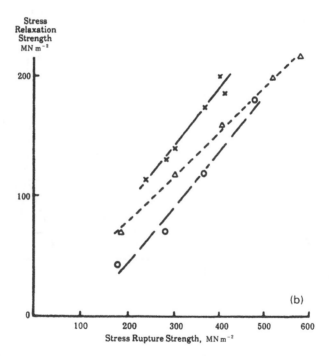

Fig. 13 *(a) Stress relaxation strengths for 0.15% strain for 10 000 h and (b) stress relaxation–stress rupture relationships.*

eration service applications, such as bolts, blading, and shafts. Esshete 1250, a fully austenitic stainless steel with a balanced and stable microstructure, is one such candidate material.

The process route used to produce steel, billets, and warm worked bars in Esshete 1250 up to 90 mm dia. have been described in this paper. Although there are some variations of hardness/strength across such sections because of the limited amount of deformation used, typically 15% at 700°C, the strengths at the centre of such bars is significantly higher than in solution treated material.

Table 3. Bolting material grades

Grade	Composition, wt%															
	C	Si	Mn	P	S	Cr	Mo	Ni	Al	As	B	Cu	Nb	Sn	Ti	V
Durehete 1055	0.17	—	0.35	—	—	0.90	0.90	—	—	—	0.001	—	—	—	0.07	0.60
	0.23	0.40	0.75	0.020	0.020	1.20	1.10	0.2	0.08	0.02	0.010	0.20	—	0.20	0.15	0.80
Nimonic 80	—	—	—	—	—	18.0	—	Bal.	1.00	—	—	—	—	—	1.8	Co≥2.00
	0.10	1.00	1.00	0.030	0.015	21.0	—	≥65.5	1.80	—	—	—	—	—	2.7	Fe≥3.00
X8CrNiMoBNb 16.16	0.04	0.30	—	—	—	15.5	1.60	15.5	—				10XC		—	
(1.4986)	0.10	0.60	1.5	0.045	0.030	17.5	2.00	17.5	—				1.20		—	—

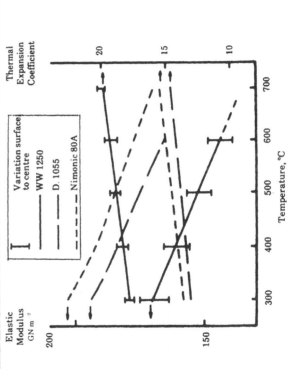

Fig. 14 Physical property values.

The superior strength of warm worked Esshete 1250 is shown to be stable, at least up to 650°C, by the results from stress rupture and stress relaxation tests — some extending to > 70 000 h; this leads to a strength advantage, particularly above 600°C, over other candidate materials such as Nimonic 80A and other warm worked austenitic steels.

Furthermore, the notched strength and rupture ductility remain high, thus meeting the one significant criterion for bolt applications which can contribute to premature failure in other alloy materials.

The toughness of warm worked Esshete 1250 is similar to that for solution treated stainless steel and is predicted to remain above ~50 J after 100 000 h at 600–650°C.

Therefore, warm worked Esshete 1250 is recommended for applications as bolts/studs up to 650°C — above this temperature the material begins to soften as the dislocated structure induced by working at 700°C begins to recover and recrystallise.

7. Acknowledgements

The authors acknowledge the assistance given by other staff in their organisations in preparing data for this paper. They would like to thank Dr M. J. May, Manager, Swinden Laboratories and Product Technology and Dr R. Baker, Director, R&D, both British Steel, Dr I. G. Davies, Technical Director, UES Steels, and the Directors/Technical Management of Spencer Clark Metal Industries Ltd and associated companies for permission to publish the paper.

References

1. BS 3059: Part 2: 1990: 'Steel boiler and superheater tubes: Part 2 — specification for carbon, alloy and austenitic steel tubes with specified elevated temperature properties', 1990, British Standards Institution. London.

2. BS 3604: Part 1: 1990: 'Specification for seamless and electric resistance welded tubes', 1990, British Standards Institution, London.

3. ASTM A213: 'Seamless ferritic and austenitic alloy steel boiler, superheater and heat exchanger tubes', 1992, ASTM Standards, Philadelphia, PA.

4. J. Orr and V. Nileshwar, in *Proc. Conf. Stainless steels '84*, Gothenburg, Sweden, September 1984.

5. H. Everson and G. Oakes, this volume, pp. 389–398 and 399–409.

6. J. D. Murray, J. Hacon and R H. Wannell, in *Proc. High Temperature Properties of Steel*; 1967, London, The Iron and Steel Institute.

7. Report ECISS TC22 WG1, Working Group on 'Assessment of tensile properties' to ECISS Technical Committee TC22, 1994.

8. DIN 17240: 'High temperature and extra high temperature materials for nuts and bolts: quality specifications', FES im DIN, July 1976.

9. 'Nimonic alloy 80 data sheet', Publication 3663, April 1975, Henry Wiggin and Co. Ltd.

10. 'Durehete 1055 data sheet', UES Steels, Stocksbridge Engineering Steels, 6/MM, 5 June 1989.

11. J. Orr, D. Burton, H. Everson and J. Beardwood, this volume, pp. 138–149.

16

Titanium Nitride Strengthened Steels for Bolting Applications

R. G. HAMERTON, D. M. JAEGER and A. R. JONES

AEA Technology, Risley, Warrington, Cheshire, UK

ABSTRACT

Reactive powder metallurgy is being used in the development of enhanced performance austenitic and ferritic stainless steels which are internally nitrided to give a fine dispersion of strengthening TiN precipitates. A key step in the manufacturing route is the combination of the Ti-containing steel with a solid-state nitrogen donor (chromium nitride). This lower stability nitride dissolves during thermomechanical treatment, releasing nitrogen to react with dissolved titanium. Mechanical properties are given for a developmental TiN dispersion-strengthened austenitic steel and the promising potential for the further development of both austenitic and ferritic alloys for use as high temperature bolting materials is discussed.

1. Introduction

One method used to improve high temperature strength in metallic alloys is to introduce a dispersion of fine second-phase particles to impede dislocation movement. Austenitic stainless steel containing a fine dispersion of TiN precipitates emerged from a search during the late 1960s and early 1970s within the UK nuclear industry for high strength candidate alloys for application in advanced gas-cooled reactor (AGR) fuel cladding. The materials were produced by internally nitriding a thin wall 20Cr–25Ni steel tube containing up to 2 wt% Ti in a N_2/H_2 gas mixture at atmospheric pressure and a temperature of *ca.* 1150°C. This process typically yielded a 0.5–2 μm dispersion of TiN precipitates [1].

Short term strength can readily be improved through the introduction of a finely dispersed second phase. However, in order for that strength to be maintained in the long term, the second phase dispersion must be resistant to coarsening. This was the principal attraction of a TiN dispersion-strengthened variant over other alternatives such as Nb-containing austenitic stainless steels, which can be strengthened by precipitation of lower stability niobium carbo-nitrides [2]. The high resistance of TiN precipitates to coarsening in austenitic steels, as suggested by the high thermodynamic stability of TiN, has been demonstrated by Ferguson *et al.* [3] at temperatures up to 1000°C. The creep performance of the nitrided 20Cr–25Ni–Ti steels showed substantial improvement compared to the 20Cr–25Ni–Nb standard cladding alloy. The steady-state creep rate at 750°C was typically reduced by a factor of 10^4 for alloys containing up to 3 wt% Ti [4].

The performance of the gas nitrided austenitic steels led to an investigation of the potential of similarly nitrided ferritic steels for application as fuel cladding within fast reactors, where high strength was demanded at temperatures exceeding 550°C. Initial work was performed on modified Ti-containing variants of the ferritic:martensitic alloys FV448 and FI [5]. The nitriding characteristics of these steels are complex due to the effects on phase stability of changes in matrix composition as nitrogen is introduced and TiN is precipitated. The results suggested that the TiN precipitate sizes were in general coarser than for the equivalent austenitic steel case, and this led to attempts to design alloys which were austenitic during the elevated temperature nitriding treatment but ferritic at the operating temperature [6].

2. Powder Metallurgy Routes for the Production of TiN Dispersion-Strengthened Steels

2.1. Reasons for Using a Powder Route

One drawback of the nitriding process described previously is the effect of nitrogen diffusion distance on the size of the TiN precipitates and, hence, their spacing. As the nitriding front penetrates the steel, its speed of progress decreases and the particles precipitating at the front increase in size. This is a characteristic feature of internal oxidation or nitridation processes [7]. A typical finding from the gas nitriding of AGR cans was that TiN precipitates as fine as 30 nm could be produced at the surface of the can wall, but at the centre of the wall this precipitate diameter was more typically *ca.* 1 μm, corresponding to a diffusion distance of *ca.* 0.2 mm. This feature precludes the fabrication of bulk components with a fine dispersion of TiN precipitates using a gas nitriding route.

One way to overcome this limitation is to use a powder metallurgy process to manufacture a bulk component. In this way, the nitrogen diffusion distance is at most equal to the radius of a powder particle, typically 50 μm or less. Initial trials used a fluidised bed to nitride 20Cr–25Ni–Ti gas atomised powder which was then consolidated into bar product by extrusion [1].

In gas nitriding, high temperatures (typically 1100°C) must be used to avoid the formation of lower stability chromium nitrides in addition to the desired TiN dispersion. There are advantages to be gained from an alternative approach to gas nitriding which is being developed and which involves the deliberate introduction of chromium nitride into the steel as a nitrogen donor. The nitrogen donor is forced to dissolve during thermo-mechanical treatment, releasing nitrogen to react with dissolved titanium. At elevated temperature, the thermodynamic driving force for the dissolution of the chromium nitride donor particles is sufficiently high that the nitrogen activity in the steel is equivalent to that produced by nitrogen gas at a pressure significantly higher than 1 atm (e.g. 15 atm at 1150°C) [1]. This increased nitrogen activity further promotes the formation of a finer dispersion of TiN precipitates.

2.2. Powder Route Steels Using a Nitrogen Donor

A number of methods have been investigated for combining the nitrogen donor (chromium nitride) with the steel. These include:

Mechanical Alloying [1]. Chromium nitride particles are uniformly dispersed within the parent steel powder during high energy ball milling.

Gas Nitriding in Ammonia. The steel powder is nitrided in ammonia at relatively low temperature (e.g. 700°C) to produce a chromium nitride layer at the powder surfaces.

Mechano-Fusion. This is a mechanical powder processing technique which can be used to modify powder shape, to coat one type of powder with another or even to mechanically alloy. In the present case, the process has been used to coat steel powder with chromium nitride particles, simulating the nitride layer produced by ammonia nitriding.

The Osprey Process [8]. In this, a partially liquid spray of gas atomised steel droplets is collected onto a substrate enabling a solid pre-form to be fabricated. A second phase powder, in this case chromium nitride, can simultaneously be injected into the atomisation zone to become uniformly incorporated into the pre-form.

In each case, the powder or pre-form intermediate must be thermo-mechanically processed to consolidate the material and to achieve the transformation from the relatively coarse dispersion of chromium nitride to the desired fine TiN dispersion through a process of donor dissolution and internal nitriding. The Mechano-Fusion and Osprey routes currently offer the most promising options for the formation of bulk components on a commercially attractive scale and cost.

3. Microstructure and Properties

Table 1 gives the mechanical properties of two developmental powder route TiN dispersionstrengthened austenitic stainless steels based on a 20Cr–25Ni composition, demonstrating the marked improvements in hardness and both room and elevated temperature (600°C) strength which have been achieved compared to conventional austenitic stainless steels. The improved mechanical properties probably have a contribution both from finely dispersed TiN particles and a comparatively high nitrogen content in the matrix. Figure 1 demonstrates the fine scale of TiN dispersion which is achievable: particle sizes in this steel, manufactured following the Mechano-Fusion route for combining steel powder with nitrogen donor, were found to be as low as *ca.* 20 nm. Wilson [1] has shown that, for such particle sizes in a steel containing 1.8 wt% Ti, an increment of about 400 MPa on the yield stress is predicted by a theoretical analysis of the effects of an array of spherical particles.

As well as contributing a significant solid solution strengthening to the matrix, the introduction of nitrogen in excess to that required to transform the titanium present to TiN, may have the additional benefit of improving the corrosion resistance of both austenitic and ferritic steels. In particular, the resistance of austenitic stainless steels to pitting attack is known to be improved by dissolved nitrogen [9].

The increased hardness of these steels is accompanied by a marked improvement in wear performance. Figure 2 shows how the wear rate (as measured in a simple pin-on-disc test under a range of conditions) is lower by an order of magnitude in the TiN dispersion-strengthened austenitic steel compared to the conventional 321 type steel.

Table 1. *Room and elevated temperature mechanical properties of the posder route reactively processed austenitic stainless steel compared with conventional stainless steels (PS=Proof Strength, UTS=Ultimate Tensile Strength, El=Elongation).*

	Room Temperature				600°C	
	0.2% PS (MPa)	UTS (MPa)	Hardness/ (Hv)	El. (%)	0.2% PS/ (MPa)	UTS/ (MPa)
TiN-strengthened austenitic (variant 1)	657	978	354	18	409	592
TiN-strengthened austenitic (variant 2)	890	1078	384	5.5	374	582
316 L (annealed)	170	450	190	40	108	359
Commercial high nitrogen austenitic	345	655	205	35	–	–
Duplex	550	760	320	15	–	–
Martensitic (410)	620	825	350	12	306	436

4. Application as Bolting in Advanced High Temperature Plant

The need to improve plant efficiency demands the development of materials which can offer extended life under existing operating conditions or which can enable operating conditions to be extended. The TiN dispersion-strengthened steels originally developed for application under the demanding conditions of nuclear power plant may offer significant property benefits over steels currently used in other high temperature plant. Both austenitic and ferritic variants are currently being developed for a number of applications, including bolting for advanced gas and steam turbines and tubing for advanced boilers [10, 11].

The markedly increased hardness of the developmental powder route austenitic stainless steel compared to conventional austenitic stainless steels promises a bolting material with an increased resistance to galling/seizure. Furthermore, the investigations of the properties of gas nitrided stainless steels clearly show that substantial improvements in the creep performance of austenitic stainless steels can be expected upon introduction of a fine dispersion of TiN particles.

A high nitrogen ferritic steel is also under development for use as bolting in advanced steam turbines. As previously noted, consideration must be given in this case to the changes in matrix phase stability which result from compositional changes during nitriding. This is further complicated when a nitrogen donor is used; during nitriding, the chromium content

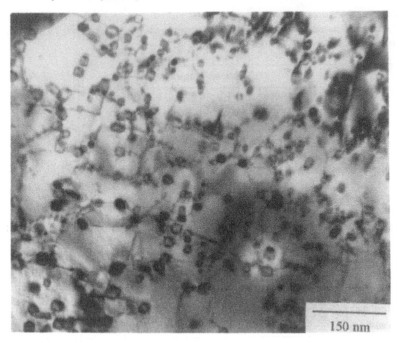

Fig. 1 *Fine TiN precipitates in a TiN dispersion-strengthened austenitic stainless steel produced by a powder route which involves coating steel powder with chromium nitride particles using Mechano-Fusion (TEM micrograph).*

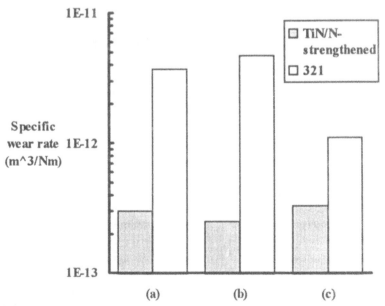

Fig. 2 *Specific wear rate for a powder route TiN dispersion-strengthened austenitic stainless steel compared with a conventional 321 steel under three sets of conditions of load and speed: (a) 5.3 MPa, 11 mm s⁻¹; (b) 5.3 MPa, 1 mm s⁻¹; (c) 0.5 MPa, 11 mm s⁻¹.*

of the matrix changes as well as the titanium and nitrogen contents. The accomplishment of a fine dispersion of TiN particles will (a) impart stress relaxation resistance accompanied by improved long term microstructural stability, and (b) improve resistance to the precipitate coarsening processes, which ultimately limit the life of conventional martensitic bolting steels.

References

1. E. G. Wilson, in *High Nitrogen Steels*, The Institute of Metals, London, 1989, 305–309.
2. G. Knowles, *Met. Sci.*, 1977, **11**, 117–122.
3. P. Ferguson, J. H. Driver and A. Hendry, *J. Mat. Sci.*, 1983, **18**, 2951–2956.
4. A. C. Roberts and H. E. Evans, *Proc. Conf. on Mech. Behaviour and Nuclear Applications of Stainless Steels at Elevated Temperatures*, Metals Society, London, 1982, 51–57.
5. A. M. Wilson, in *High Nitrogen Steels*, The Institute of Metals, London, 1989, 392–397.
6. K. C. Laing and A. Hendry, in *High Nitrogen Steels*, The Institute of Metals, London, 1989, 353–357.
7. N. Birks and G. H. Meier, in *Introduction to High Temperature Oxidation of Metals*, Ch. 5, Edward Arnold, London.
8. A.G. Leatham, A. Ogilvy and L. Elias, in *P/M in Aerospace, Defence and Demanding Applications — 1993*, Metal Powder Industries Federation, Princeton, NJ, 1993, p.165.
9. A. Hendry, in *Proc. Europ. Technical Conf. of the Wire Assoc.*, Berlin, 1993.
10. Titanium Nitride and Nitrogen Strengthened Austenitic Stainless Steels, DTI LINK Project.
11. A Novel TiN-Strengthened, Creep Resistant 9–12% Cr Ferritic Steel for Advanced Steam Power Plant, Brite-EuRam Project 7500.

17

The Role of Molybdenum Disulphide-Containing Lubricants in the Stress Corrosion Cracking of High Strength Bolting Materials

A. D. HOPE and E. B. SHONE

Shell Research Arnhem, P.O. Box 40,6800 AA, Arnhem, The Netherlands

ABSTRACT

Molybdenum disulphide-containing lubricants are commonly used as anti-seize compounds for bolting applications. Such lubricants can decompose to cause the stress corrosion cracking of high strength bolting materials. This paper describes the decomposition of molybdenum disulphide under a range of conditions and illustrates the consequences for high strength bolting materials of the type commonly used in the petroleum industry.

Inductively coupled plasma optical emission spectrometry has been used to demonstrate that H_2S can be liberated from the lubricant if it is in contact with water and the temperature approaches 100°C.

It is shown that the failure of high strength bolting steels (AISI C4140) of the type used in the high pressure reactor piping system of a hydrocracking unit may be attributed to sulphide stress corrosion cracking. Slow strain rate testing has been used to demonstrate that the H_2S liberated from molybdenum disulphide containing lubricants in the presence of water at 100°C can cause such cracking. The influence of the water chemistry (demineralised water, artificial sea water and acidified artificial sea water) on the liberation of H_2S is given.

As a second example, the stress corrosion cracking of a high strength copper–nickel alloy often used as a bolting material for offshore applications is discussed. Cracking occurred in the presence of another commercially available molybdenum disulphide containing lubricant and moisture. It is shown that stress corrosion cracking can occur even at ambient temperatures. The influence of stress is also discussed.

1. Introduction

Molybdenum disulphide containing lubricants are frequently used in the oil and gas industry to reduce friction between highly loaded contacting surfaces, prevent seizure and the possibility of fretting damage. In most instances the use of such lubricants is satisfactory, however, recently the incidence of failures attributed to with the use of such lubricants has increased. As a result of this a brief literature review has been carried out which has shown that similar failures have been reported by several other workers. Berge [1] has reported that failure of bolts in steel isolation valves (AISI 4340, ASTM 193 B7) by stress corrosion cracking was associated with the use of molybdenum disulphide containing lubricants. Czajkowski has reported similar failures of ASTM A471 discs steels [2] and bolts made of ASTM A540 B24 steel [3] used in the nuclear industry. Rungta [4] has reviewed the materials aspects of bolting applications in the nuclear industry and reported that a significant number of stress corrosion cracking failures have occurred which had been attributed to the presence hydrogen sulphide. He suggests that the hydrogen sulphide may be formed by the decomposition of a molybdenum disulphide containing lubricant that was present on the bolts.

During a recent inspection of the piping in a refinery hydrocracking unit, bolts and spring washers made from a high strength C4140 steel were found to be cracked. Molybdenum disulphide containing lubricants had been applied to these components during their installation to avoid seizure. Additionally, high strength precipitation hardening copper–nickel alloys that were being used in production choke valves on offshore installations have failed after short times in service. Again, molybdenum disulphide containing lubricants had been applied to the bolts. In this paper we will describe the work which has been carried out to establish the cause of these failures and the role played by the molybdenum disulphide containing lubricant.

2. The Failure of C4140 Steel Bolts and Spring Washers
2.1. Background

During a shutdown inspection of a high pressure reactor piping system of a hydrocracker unit, bolts and spring washers made of a high strength C4140 steel (HV 350–400) were found to have cracked. The composition of C4140 material is given in Table 1. During their installation a molybdenum disulphide containing lubricant had been applied to the threaded portion of the bolts.

In service the bolts would be exposed to temperatures ranging from ambient to perhaps 400°C for short periods although for most of the service life the temperature would be in the range 90–110°C. The bolts were used in flanges and crevice like conditions could readily exist. It is possible that water could seep into these crevices which may also contain the grease. The quantity of water could vary but could also contain chloride ions especially if the plant is located at a coastal location.

Since, failures associated with the use of molybdenum disulphide containing lubricants had been reported in the literature [1, 3, 4] a small experimental programme was conducted to establish the susceptibility of C4140 high strength steel to cracking when it is in contact with molybdenum disulphide containing lubricants and water vapour.

2.2. Slow Strain Rate Testing

From the literature it is apparent that a molybdenum disulphide containing lubricant may decompose in the presence of water at elevated temperatures to liberate hydrogen sulphide [4, 5]. The decomposition by hydrolysis of a commercially available molybdenum disulphide containing lubricant was confirmed using inductively coupled plasma optical emission

Table 1. Composition of C4140 in wt%

Alloy	C	Cr	Cu	Mn	Mo	Ni	P	S
C4140	0.41	1.008	0.18	0.879	0.178	0.073	0.013	0.01

spectrometry techniques. H_2S was liberated from the lubricant if it was in contact with water at a temperature approaching 100°C. Further tests showed this process could be accelerated in the presence of chloride ions and acidification both of which could be present in a crevice (such as a bolted flange) in a coastal or offshore installation.

As there is no doubt that the bolts could have been subjected to such an environment a series of slow strain rate tests (SSRT) were carried out to assess the susceptibility of AISI C4140 steel to cracking in the presence of a molybdenum disulphide containing lubricant at a temperature of about 100°C in a range of aqueous environments. The SSRT samples were prepared by polishing to a 600 grit finish and then ultrasonically cleaned in an acetone bath. Following this a film of the lubricant was applied to the samples which were then wrapped in an inert insulating material (glass wool) and soaked in the aqueous solution to be used during the test. It was anticipated that the insulating material would be stable in hot water and therefore would not contribute to the corrosivity of the aqueous environment. The prepared sample was then placed into the SSRT rig and 60 mL of the appropriate test solution was added to the 740 mL capacity autoclave to promote a solution saturated vapour during testing at about 100°C. A list of the various test environments to which samples were exposed is given in Table 2. All samples were exposed for periods of 72 or 96 h after which they were strained at rates of 2.1, 4.2 or 5.2 E-6 s^{-1}. At the completion of each test the samples were examined with the aid of optical and scanning electron microscopy and energy dispersive X-ray analysis (EDAX) was carried out when appropriate. The results of the SSRTs are summarised in Table 2 and demonstrate that a very significant reduction in ductility (as measured by elongation to failure and reduction in cross sectional area assessments) occurs when this steel is stressed in hot aqueous lubricant containing environments.

SEM and metallographic examination of the fractures associated with the presence of the lubricant showed them to be brittle, transgranular and with multiple initiation sites as is typified in Fig. 1, where as those associated with the blank test environment (nitrogen) were

Table 2. *Results of slow strain rate tests on C4140 steel in various environments at 100±4°C*

Environment and strain rate	Reduction of area (%)	Elongation to failure (%)
Blank test in N_2 (no lubricant)	51	13.90
Seawater and lubricant SR = 4.2E-6/s^{-1}	7	9.27
Demineralised water and lubricant SR = 4.2E-6/s^{-1}	16	8.53
Demineralised water and lubricant SR = 5.2E-6/s^{-1}	53	13.90
Demineralised water and lubricant SR = 2.1E-6/s^{-1}	52	13.90
Acidified seawater and lubricant SR=4.2E-6/s^{-1}	23	9.44

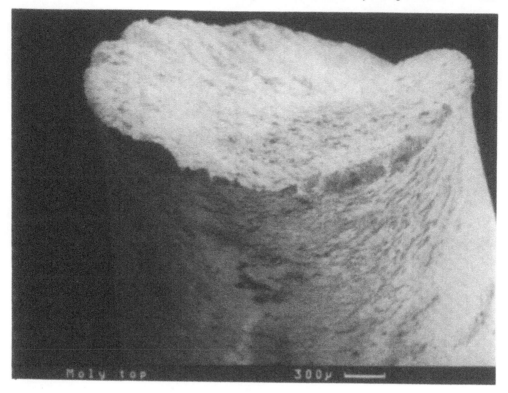

Fig. 1 *Typical fractograph of a brittle SSRT sample.*

ductile (Fig. 2). EDAX analysis of corrosion product on the fracture surfaces and in secondary cracks of bolts from molybdenum disulphide containing test environments showed that they were rich in sulphur.

From this work it could be concluded that AISI C4140 steel (HV 350–400) is susceptible to sulphide stress corrosion cracking in the presence of molybdenum disulphide containing lubricants and water at about 100°C. It apppears that the molybdenum disulphide decomposes in the presence of moisture and generates H_2S which causes cracking.

3. The Failure of High Strength Copper–Nickel Alloy Bolts
3.1. Background

Recently it has been reported that 12 high strength copper–nickel bolts which were part of a choke valve used in gas production had cracked and fractured a short time after entering service. Further inspections revealed that other bolts of a similar material in similar applications also contained cracks and were susceptible to fracture. In addition, an inspection was carried out on bolts in the 'as-received' condition (i.e. the bolts had never been in service) and this showed that they also contained cracks. Extensive metallurgical examinations have been carried out on representative samples of all fractured and cracked bolts and this is briefly described below.

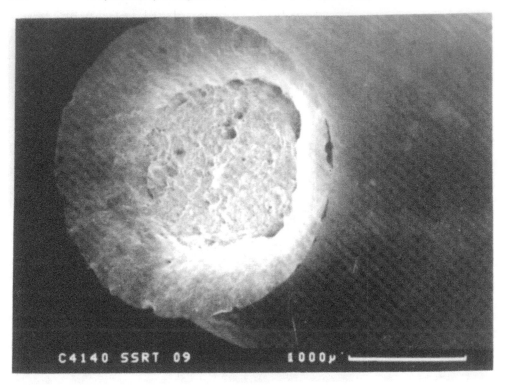

Fig. 2 *Typical fractograph of a SSRT sample from the blank test environment (nitrogen) with no molybdenum disulphide.*

3.2. Material and Service Conditions

The bolting material was a forged high strength precipitation hardening copper–nickel alloy of the composition given in Table 3. These bolts were inserted in duplex stainless steel Choke valves and torqued during manufacture. They were subsequently hydrostatically pressure tested at ambient temperature prior to transportation to the location of use. During the assembly of the valves molybdenum disulphide containing lubricants are applied to all bolting materials and after pressure testing the assemblies are sealed to prevent the ingress of potentially corrosive environments. During service the bolts experience an operating temperature of about 75°C.

Note that the applied molybdenum disulphide containing lubricants were produced by a different manufacturer to the lubricant used on the C4140 bolts.

Table 3. *Inductively Coupled Plasma analysis of the composition (wt%) of the high strength copper–nickel alloy bolts*

Ni	Fe	Al	Mn	Si	Nb	Cr	Mo	Cu
15.5	0.96	1.78	4.31	0.005	0.75	0.41	0.02	Rem

3.3. Investigations

The results of mechanical tests carried out on samples of as-received, cracked and fractured bolts are given in Table 4. These tests along with chemical analysis and metallographic sections confirmed that the material was within specification and that the bolts had been manufactured correctly.

Examination of cracked and fractured bolts from service showed that the damage could be attributed to transgranular stress corrosion cracking that had been initiated at stress raisers such as in the roots of the threads, at the joint of the bolthead and shank or at small corrosion pits on the bolt surface. Figure 3 shows some typical fracture surfaces whilst Fig. 4 (p.233) shows a metallographic section through a bolt illustrating how the branched transgranular cracks typical of stress corrosion cracking propagating corrosion pits. Analysis of deposits on the fracture surface was inconclusive due to significant contamination which occurred after fracture.

Examination of bolts in the as-received condition and which had never been in service showed them to contain similar but less extensive cracking (cracks were typically 50 µm in depth). Figure 5 (p.234) shows a typical cross section illustrating how the cracks initiate from pits. EDAX analysis of the deposits associated with the pits showed them to contain primarily sulphides.

From the above examination it would appear that the cracking could be attributed to sulphide stress corrosion. A detailed analysis of the bolt manufacturing route and of the service conditions led to the conclusion that the lubricant used was the only possible source of sulphur and the damage must be associated with its use. As the lubricant is known to contain molybdenum disulphide and that in the presence water this can decompose by either hydrolysis at temperatures around 100°C to produce hydrogen sulphide [5] or by oxidation at room temperature to form sulphuric acid [3, 4, 6]. If the later mechanism was operative it could cause localised acidity and localised corrosion such as pitting. The operating temperature of the bolts in service is around 75°C. It is not known if this temperature is high enough to cause hydrolysis of molybdenum disulphide and without further testing it is not possible to say in this case if the molybdenum disulphide decomposed by hydrolysis or oxidation. It is unlikely that the as-received bolts have been exposed to elevated temperatures. The source of water was likely to have originated during valve assembly. It is likely however that the service temperature during operation accelerated crack growth in service.

In summation it appears that the cracking and fracture of the bolt material can be attributed to stress corrosion cracking associated with presence of sulphides. The precise details of the operative mechanism are still uncertain and further research would be required to clarify them. However, there is little doubt that the damage could be caused by the presence of decomposition products from molybdenum disulphide.

Table 4. Mechanical properties of the high strength copper nickel alloy bolts

Material	Hardness (Brinell)	0.2% Proof Stress (MPa)	Ultimate tensile strength (MPa)
Specification	280	750	930
Intact bolt from service	269	745	958

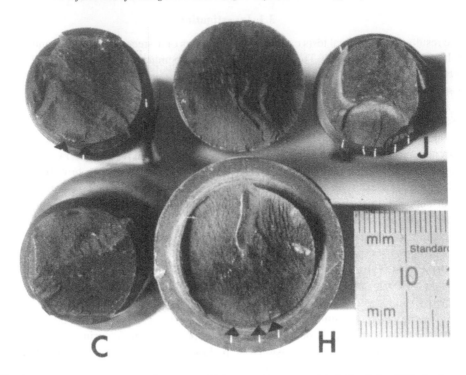

Fig. 3 *Typical fracture surfaces of ex-service high strength copper–nickel alloy bolts. ('i' indicates initiation site).*

4. General Discussion

A combination of three factors are generally required to cause sulphide stress corrosion cracking. If any one of these factors can be prevented then stress corrosion cracking failures can usually be avoided.

These factors are:

(i) The stress in the bolts is beyond the yield strength of the material.

(ii) The material is susceptible to stress corrosion (e.g. has a high hardness).

(iii) A corrosive environment exists.

The application of lubricants onto the bolts during assembly can result in the introduction of tensile stresses that exceed the yield point into the material as the level of friction is reduced and torque values are easily exceeded. If this occurs then the material may be prone to stress corrosion cracking.

In view of the failures there seems little doubt that the materials considered were susceptible to stress corrosion cracking and that the level of stress in the materials exceeded their yield strengths. Whilst the application of lubricants onto the bolts during assembly may have

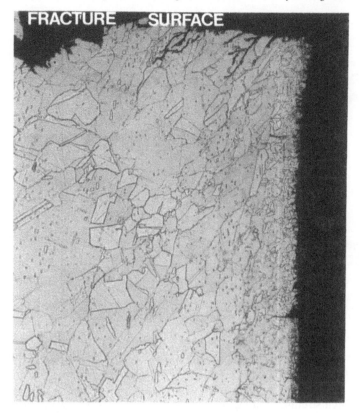

Fig. 4 *A metallographic section through an ex-service bolt showing a portion of the fracture surface and secondary cracks. Note also secondary cracks initiating from pits on the bolt surface (× 500 magnification, etch: 5% ferric chloride solution).*

resulted in torque values being exceeded and yielding occurring this is likely to happen in many practical situations. For this reason care must be taken in the use of high strength materials in applications where the environment is conducive to corrosion and cracking.

In these failures the environment has been characterised by the presence of water and molybdenum disulphide containing lubricants. It would appear that if the corrosive nature of the environment could be eliminated then damage could be prevented. In practice, the complete elimination of water from the immediate bolting environment may not be possible however it would seem fairly simple to avoid the use of molybdenum disulphide containing lubricants. It is recommended that molybdenum disulphide containing lubricants should not be used in conjunction with high strength materials of the type considered in this brief paper.

5. Conclusions

1. It has been shown that commercially available molybdenum disulphide containing lubricants can be hydrolysed in the presence of water at temperatures of around 100°C produce hydrogen sulphide.

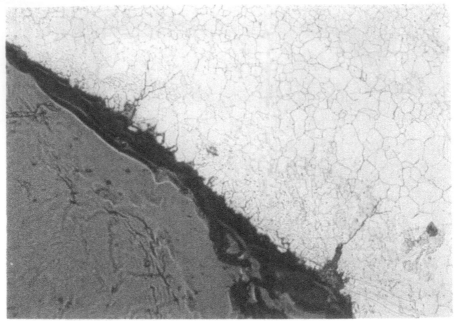

Fig. 5 *Finely branched transgranular cracks typical of stress corrosion cracking initiating from pits in as-received material which had never been in-service (× 500 magnification, etch: 5% ferric chloride solution).*

2. C4140 high strength bolting steel has been shown to be susceptible to sulphide stress corrosion cracking associated with the hydrolysis of molybdenum disulphide containing lubricants. This can be used to explain cracking of bolts observed in the high pressure piping systems of a hydrocracking unit.

3. The stress corrosion cracking of high strength copper-nickel alloy bolts at ambient temperatures can also be attributed to the decomposition of a molybdenum disulphide containing lubricant.

6. Recommendations

It is recommended that molybdenum disulphide containing lubricants should not be used in conjunction with high strength materials of the type considered in this paper.

7. Acknowledgements

The experimental work of J. C. Nava Paz and B. Deurhof is gratefully acknowledged. The authors would also like to thank Shell Research for permission to publish this paper.

References

1. P. J. Berge, Corrosion of power plant components in the presence of sulphur and chloride compounds, *Mat. Perform.*, 1986, **25**, 33–36.

2. C. J. Czajkowski and J. R. Weeks, Examination of cracked turbine discs from nuclear power plants, *Mat. Perform.*, 1983, **22**, 21–25.

3. C. J. Czajkowski, Corrosion and stress corrosion cracking of bolting materials in light water reactors, in *Proc. Int. Symp. on Environmental Degradation of Materials in Nuclear Power Systems Water Reactors*, NACE, Myrtle Beach, South Carolina, 1983, p.192–208.

4. R. Rungta and B. S. Majumdar, Materials behaviour related issues for bolting applications in the nuclear industry, in *Int. Conf. on Improved Technology for Critical Bolting Applications*, Chicago, IL, USA, 20–24 July 1986, ASME Publishers, p. 39–48.

5. A. R. Lansdowne, Molybdenum disulphide lubrication, Report No. ESRO CR-402, European Space Research Organisation, Neuilly, France, 1974.

6. F. G. Claus, *Solid Lubricants and Self-Lubricating Solids*, Academic Press, London, 1972.

18

Experience with Stainless Steel Bolts in the Helium Circulators of the Fort St Vrain Reactor

J. R. LINDGREN, H. D. SHATOFF, R. H. RYDER and R. L. HELLNER*

General Atomics, San Diego, Ca, USA
*Public Service Company of Colorado, Denver, Co, USA

ABSTRACT

Failures were experienced in some of the bolts used in the steam-side ducting, rotors, and stators of the helium circulators of the Fort St Vrain (FSV) High Temperature Gas-Cooled Nuclear Reactor. The failures were from stress corrosion cracking caused by impurities in the water/steam used to drive the circulators. The bolts were replaced and more stringent water chemistry controls applied to the circulator water system. This paper describes the investigation carried out to provide a basis for repair, replacement, and continuing operation.

The FSV electricity generating plant was operated between 1978 and 1989. Helium was used as the heat transport fluid between the core and the steam generators. Helium temperatures ranged from 750°C at the outlet from the core to 400°C at the inlet to the circulators. The bolts operated in a flowing water/steam environment at up to 400°C.

The finite element analysis of the configuration made as part of the original design process showed that the stresses in the bolts met requirements similar to those in the ASME Boiler and Pressure Vessel Code requirements for maximum average stress, maximum membrane plus bending stress combined with shear stress, and the fatigue-controlled limits. After the failures, the analysis was updated using as-built dimensions and as-operated loading conditions. The results of this second analysis also showed compliance with the stress requirements. In-service and residual stresses were nevertheless high enough to promote stress corrosion cracking in an aggressive environment. The source of the aggressive environment was identified as the ingress of caustic compounds into the auxiliary water/steam used to drive the circulators under certain conditions.

Of the two materials used for the stainless bolts one, A-286, is a precipitation-hardened austenitic steel originally developed for aircraft engine applications where high tensile strength and rupture properties are required. However, its good corrosion resistance has also made it an ideal choice for lower temperatures applications requiring moderate strength coupled with corrosion resistance. The other material was AISI-410 martensitic stainless steel.

Since the original A-286 bolts performed satisfactorily for 60 000 h in spite of the detrimental environment, it was decided, because high strength was required, to utilise the same material for replacement. The AISI-410 bolts were replaced with bolts made from Inconel X-750 for better corrosion resistance. More stringent water chemistry requirements were also introduced to hold down the level of caustic in the auxiliary water/steam supply system.

1. Introduction

The Fort St Vrain high-temperature gas-cooled power reactor (HTGR) was operated by Public Service Company of Colorado until 1989. It is currently being decommissioned. The 330

MWe plant was designed by General Atomics and was a logical development from the successful 45 MWe demonstration plant built and operated at Peach Bottom, Pennsylvania, from 1967 to 1974.

In the FSV plant, helium at approx. 5 MPa was used as the primary coolant. The helium flowed downward through the graphite moderated reactor core and thence down into two identical parallel loops each containing six steam generators and two helium circulators. The helium was returned upwards to the upper plenum of the core around the outer periphery of the core reflector. Figure 1 shows the arrangement of the reactor.

The unique design of the circulator machine consisted of a single stage axial flow helium compressor driven by a single stage steam turbine. A single stage pelton wheel water turbine provide a back-up drive and could be operated on feedwater, condensate, or emergency fire water. The layout of the machine is shown in Fig. 2. Although a considerable amount of developmental testing was accomplished before the machines went into service, during plant operation they were found to be sensitive to the performance of the complex support auxiliary systems necessary for the proper operation of the units. These auxiliary systems provided water for the bearings and buffer helium to prevent leakage in either direction between the steam and the helium. A number of problems, including leaks and component failures, were caused by improper performance of the auxiliaries. In particular, this paper deals with one incident resulting from stress corrosion of stainless steel bolts. The impurities causing the corrosion may have been introduced into the secondary coolant (steam/water) demineralisers as a result of regeneration procedures over a fifteen year period or from

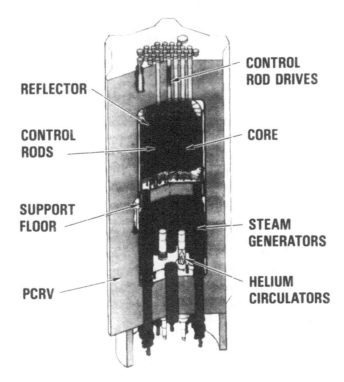

Fig. 1 Fort St Vrain reactor arrangement.

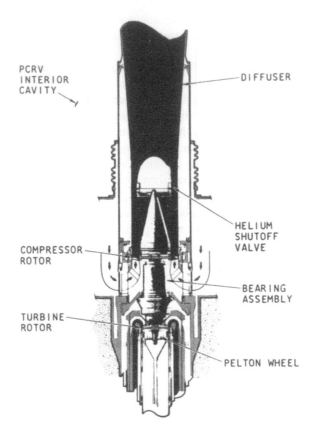

Fig. 2 Fort St Vrain helium circulator.

limited use that was made of sodium hydroxide for auxiliary water treatment some five years before the incident discussed in this paper.

In July 1987, during operation at approximately 70% power, a circulator trip and a shut-down of loop No. 2 occurred. Upon re-establishment of loop operation at low power, high shaft wobble and helium leaks were found on one of the circulators. The plant was shut down and the circulator replaced with a spare.

In the removed circulator, pieces of failed hardware were found residing in the steam outlet plenum. Components associated with the labyrinth seal and insulation had dropped onto the turbine wheel and significant damage had been caused to the steam ducting and steam turbine rotor. Failure and damage had also occurred in a number of other components, including stainless steel bolts. The materials in the failed or damaged components were A-286 and Type 410 stainless steels, Monel 400, Inconel 600, 5% Cr steel, 4140 low alloy steel and carbon steel.

Judging by the number of items and differing materials affected, it appeared that some common mode failure or common initiating event had occurred.

The objective of the investigation subsequently carried out was to ascertain the cause (or causes) of the failures and to determine what, if any, remedial action should be taken other than repairing and returning the components to service. The components in the vicinity of the failures were examined. Some had obviously been damaged as a result of the failure of other components. At the end of the investigation it was clear that the initiating failure mechanism was stress corrosion cracking.

The bolts in particular had suffered severely from this phenomenon and in many cases failed. However, it was not clear from the results of the study which items had failed first and precipitated the overall failure. This paper will be concerned only with the evaluations done on the A-286 and AISI-410 stainless steel bolting listed in Table 1.

2. Stainless Steel Bolting Materials Used in the FSV Helium Circulators

2.1. A-286 (Austenitic)

A-286 is a precipitation hardening austenitic stainless steel of typical composition shown in Table 2 [1]. The material was originally developed for aircraft engine applications requiring high tensile strength and rupture properties up to 660°C. Over the last two decades it has been one of the most widely used materials in that type of application. Its composition gives it good corrosion resistance and it has also been used extensively for ambient temperature applications requiring moderate strength coupled with good corrosion resistance. In the solution treated and precipitation heat treated condition (aged) A-286 fasteners possess a minimum tensile strength of 900 MPa. Strength levels significantly above this can be achieved using a combination of cold work and accelerated aging. By this technique fasteners with a minimum tensile strength as high as 1380 MPa can be produced and are in use in the aerospace industry. The material has also been used for bolts in light water nuclear power plants [2].

Table 1. Stainless steel bolting used in the FSV helium circulator

Part Name	Material	Operating Conditions		
		Environment	Temperature °C	Preload MPa
Labyrinth seal mounting bolts	AISI-410	steam	400	390
Steam ducting pressure tap bolt	A-286	steam	260	910
Steam ducting bolts	A-286	steam	260	790
Stator alignment sleeve cap screws	AISI-410	steam	400	345
Stator water deflector screws	AISI-410	steam	400	345
Stator installation cap screws	AISI-410	steam	400	345

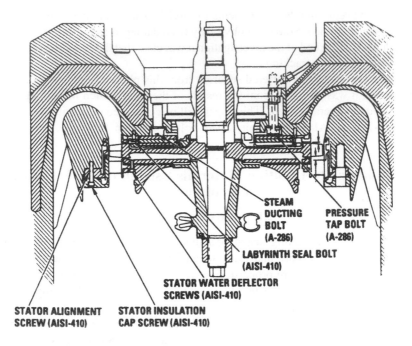

Fig. 3 *Fort St Vrain circulator cross section.*

Table 2. *Chemical composition by weight (%) of bolting materials*

A-286			**AISI-410**		
	Cr	15		C	0.15 max.
	Ni	25.5		Cr	12.5
	Mo	1.3		Mn	1.0 max.
	V	0.3		Si	1.0 max.
	Ti	2.1		P	0.04 max.
	B	0.006		S	0.03 max.
	Fe	bal.		Fe	bal.

A-286 in a high strength condition was used for bolting in the steam ducting of the FSV helium circulators [3] in the locations shown in Fig. 3.

2.2. AISI-410 (Martensitic)

AISI-410 is a martensitic stainless steel which, when used for fasteners, is normally ordered to specifications ASME SA193-B6 or ASTM A193-B6. Typical composition is shown in Table 2.

The material is widely used for bolting and fasteners. However, when heat treated to strength levels above a certain value, it is sensitive to stress corrosion cracking in moist caustic, chloride, or sulphur containing environments. This critical strength value depends on service temperature [4]. Although a tempering heat treatment of one hour at 620°C was applied to the AISI-410 fasteners used in the FSV helium circulators, many showed damage from stress corrosion cracking when examined as a result of the incident described.

The AISI-410 material was used for the labyrinth seal and stator bolting as shown in Fig. 3.

3. Metallurgical Evaluation
3.1. Examination of A-286 Bolts

The steam ducting bolts were 0.75 in (20 mm) dia. by 3.5 in (89 mm) long and their locations are shown in Fig. 3. These bolts operated at approximately 260°C and were initially torqued to a preload of 790 MPa. The hollow pressure tap bolt (No. 9) was torqued to 910 MPa preload. Upon disassembly of the steam duct the heads of these bolts were observed to have been worn down by as much as 6 mm on a diameter and the pressure tap bolt had totally separated (Fig. 4). Others, notably numbers 8, 10 and 12 were also severely cracked.

Bolt No. 8 was cracked in the shank approximately 25 mm below the bottom of the head (Fig. 5). The fracture surface is shown in Fig. 6. The crescent shaped area indicated by the arrow is the servicecracked area, the remainder is due to overload. The service-cracked area was covered in iron oxide corrosion product but examination by EDS showed no contaminants such as sodium, chlorine, or sulphur. Careful study of the fracture surface showed that the crack was transgranular with some evidence of post-fracture corrosion and extensive secondary cracking in some regions. Faint striation marks were discernible in the transition region between the crescent shaped region and the flat overload region. This indicated rapid acceleration of fatigue crack growth rate prior to overload failure. Cracks were also found initiating from the roots of the threads below the shank. These cracks were transgranular and

Fig. 4 *Photograph of the steam ducting bolts from below. Note debris from failed bolt head jammed in bolt recess in top left corner.*

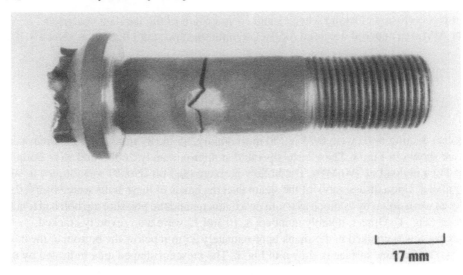

Fig. 5 *Steam ducting bolt No. 8.*

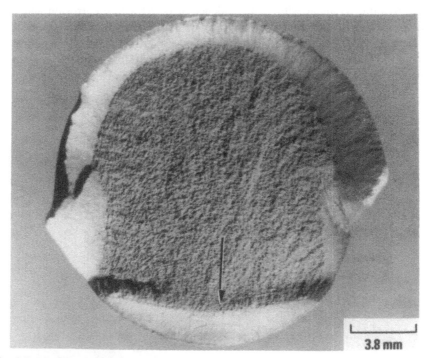

Fig. 6 *Fracture surface of steam ducting bolt No. 8.*

propagating perpendicular to the applied load. It was postulated that the crescent-shaped initial crack was caused by SCC and, in the final stages, propagated rapidly as a result of the loads imposed by the rubbing of the debris just prior to overload failure.

The pressure tap bolt, No. 9, had fractured in the reduced shank above the first thread. The fractured surface had a brittle failure appearance and was covered with corrosion products. SEM/EDS analysis indicated iron oxide again with no evidence of contaminants. However, numerous secondary cracks, typical of stress corrosion cracking, were found initiating from the fracture surface and extending longitudinally in the direction of the axis of the bolt (Fig. 7).

A new bolt (available from the spares stock) was sectioned in the threaded region for comparison with the ex-service bolts. No defects were found in this bolt.

3.3. Examination of AISI-410 Bolts

The labyrinth seal (Fig. 3) of the circulator was held by twelve 6 mm dia., 16 mm long bolts made from AISI-410 martensitic steel. Disassembly of the unit revealed that three bolt holes were empty with no apparent damage to the threads. The remaining holes contained remains of fractured bolts. The majority of the fractures had occurred approximately one thread

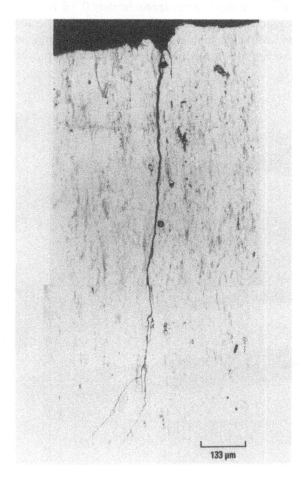

Fig. 7 *Secondary cracks from fracture surface of steam ducting bolt No. 9.*

below the mating surfaces. Six of these were removed for examination. Chemical analysis showed that the bolts met the ASTM specification for 410. The fractured surfaces had a brittle appearance suggesting a progressive type of failure typical of stress corrosion cracking. Corrosion products analysis showed the presence of iron oxides but no contaminants such as Na and Cl. The six bolts and two bolt sections all experienced cracking in the root of the threads, typically the first three below the mating surfaces (Fig. 8). Crack propagation was transgranular and initially perpendicular to the direction of the applied load (Fig. 9) The microstructure was spheroidised which is not typical of a quenched and tempered 410 material.

During removal, one of the AISI-410 stator alignment sleeve cap screws fractured approximately 9 threads below the screw head. Cracking at the root of adjacent threads was also evident. The heavy corrosion products found on this bolt were analysed using SEM/EDS and found to contain iron oxides but no contaminants. Metallographic examination revealed multiple stress corrosion cracks initiating at the thread roots (Fig. 10).

Heavy corrosion products and cracking at the thread roots were also observed in two other cap screws. However, in these cases, approximately 0.3% by weight of sodium was found in the corrosion products. Molybdenum disulfide (up to 10%) and silicon dioxide (up to 10%) were also found. Metallographic examination revealed intergranular cracking along prior-austenite grain boundaries in the material of these bolts typical of stress corrosion cracking (Fig. 11).

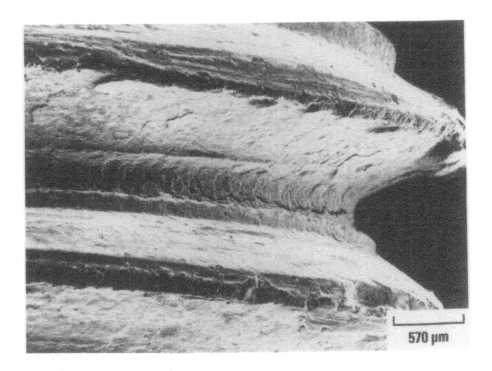

Fig. 8 SEM photograph of typical crack in thread root of labyrinth seal mounting bolt.

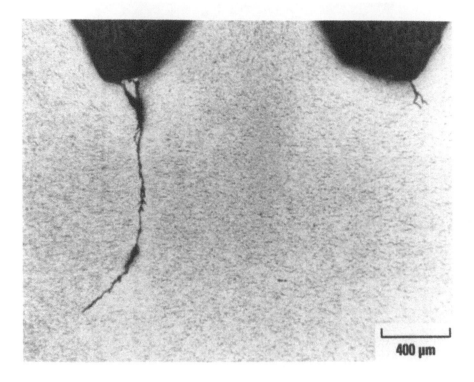

***Fig. 9** Stress corrosion cracks from thread roots in labyrinth seal mounting bolts.*

Cracks were also observed initiating from the face of the threads. These cracks may have been from the original manufacture but some growth had taken place during service as indicated by the branching found.

3.4. Other Evidence

In addition to the bolt damage indicated above, there was significant evidence of contamination, particularly of sodium, and stress corrosion cracking in other components adjacent to the bolts. In particular, the bellows at the lower end of the assembly were cracked and a significant amount of sodium was detected on the crack surfaces.

Some four years earlier, the presence of boiler tube corrosion had led plant operators to introduce a chemical treatment for the auxiliary boilers consisting of sodium hydroxide and trisodium phosphate. This was discontinued in 1985 when large build-ups of sodium salts and iron phosphate caused valve closure and pipe blockage problems. The auxiliary steam was found at that time to contain sodium concentrations ranging from 9 to 820 ppb and phosphate up to 100 ppb. It was probable, therefore, that the stress corrosion problems in the circulator steam turbine region had initiated in 1983 and taken approximately four years to progress far enough to cause failures.

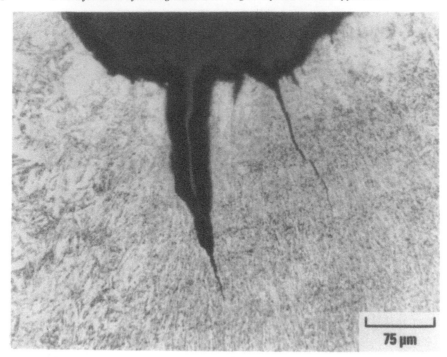

Fig. 10 *Transgranular cracks from root of a stator alignment sleeve screw.*

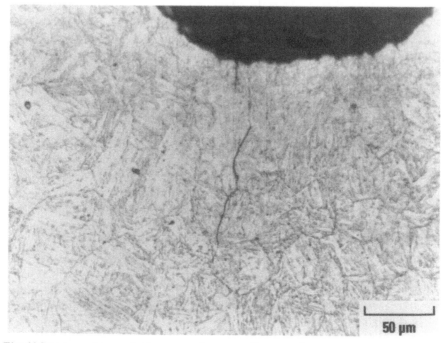

Fig. 11 *Intergranular cracks from root of a stator alignment sleeve screw.*

4. Stress Analysis
4.1. Design Analysis of A-286 Steam Ducting Bolts

The original design analysis was carried out in 1969 utilising the SAFESHELL finite element program. This program is a tool for calculating elastic deformations and stresses in axisymmetric shells of revolution and was developed by General Atomics in the mid-1960s. The analytical model, part of which is shown in Fig. 12, consisted of 176 conical shell segments. The A-286 steam ducting bolts are identified in the figure. These were simulated by cylinders with the same cross-sectional area as that of the bolt. A series of unit load cases were run with this model to separately calculate stresses due to thermal and mechanical loads. Resulting stresses were factored and combined appropriately following the approach of the ASME Boiler and Pressure Vessel Code, Section III. Steady state and transient operating conditions were evaluated.

In 1987, using actual conditions from the plant operation, the thermal part of the loadings was updated using an in-house thermal analysis code (TAC2D) and the resulting temperatures used again in the original SAFESHELL model to recalculate thermal stresses. The unit load cases were then combined to cover the following operating conditions:

- Initial bolt up to 634 MPa (lower casing bolts);
- 5 MPa secondary helium, 0.1 Pa steam, and no thermal load;
- 5 MPa secondary helium, 5 MPa steam, 100% steady state thermal load;
- 5 MPa secondary helium, 0.1 Pa steam, and thermal load 430 s after scram;
- 5 MPa secondary helium, 0.1 Pa steam, and thermal load 1260 s after scram.

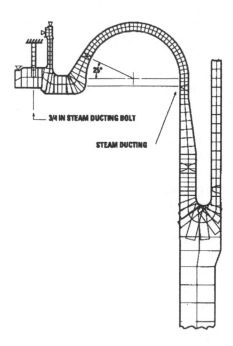

Fig. 12 SAFESHELL finite element stress analysis model of steam ducting.

The yield strength of the bolt material at room temperature was given by the manufacturer as 1260 MPa. A stress intensity of 383 MPa at 177°C was used as an equivalent allowable S_m value. The three limits on bolt stress, following the approach of the 1969 edition of Section III of the ASME Boiler and Pressure Code, were:

1. Maximum average stress in bolt $< 2 \times S_m$,
2. Maximum membrane plus bending plus shear stress at the bolt periphery $< 3 \times S_m$, and
3. Cumulative fatigue damage factor < 1 (The chart for high strength bolting in Section VIII of the ASME B&PV code was used).

All of the bolts, including the pressure tap bolt, met these limits.

5. Discussion and Conclusions

The overwhelming evidence available indicated that stress corrosion cracking due to caustic contamination was the cause of the problem in the circulators. The fact that the stress analysis showed that the bolts were installed and loaded appropriately confirmed that the cause lay in the operating environment. The possibility of material degradation due to molybdenum disulphide lubricant was also considered but, since SCC was found also in unlubricated components, this cause was discounted.

However, the stresses in the bolts were high due to several factors, namely: residual stresses from cold work induced during the heat treatment process to raise the strength of the material; thermal stresses from temperature cycles, and; high installation torques.

Although stress corrosion-like cracks were found emanating from most of the fracture surfaces on the bolts, the EDS results showed little direct evidence of the presence of caustic and chloride contaminants on the fracture surfaces or on the outside of the bolts. One conclusion, therefore, might be that the majority of the cracks investigated occurred as a result of fatigue after an initial bolt failure. The observation of rubbing of bolt heads and damage to the insulation between the steam turbine rotor and the steam ducting would support this contention. However, bolts from other parts of the ducting (e.g. bolts on the stator) showed definite evidence of stress corrosion including significant amounts of sodium. High concentrations of sodium were also found on the crack surfaces in the Inconel 600 bellows expansion joint which is part of the steam circuit.

Because of the wide range of temperature and pressure conditions of the steam, the concentration of caustic in the steam could have ranged from a few ppb to a high precipitate concentration of 80%. Samplings of the auxiliary steam on various occasions showed sodium concentrations in the range 9–820 ppb indicating a wide range of possible caustic concentrations.

Apart from the fact that caustic was known to have been injected into the system between 1983 and 1985, a review of the steam drum level charts indicated that, during startups, water levels in the drum were erratic and could have resulted in sodium hydroxide and trisodium phosphate being carried over into the circulator steam ducting and turbine.

The actions taken as a result of the investigation were as follows:

1. The high strength A-286 material was retained for the stator and steam ducting bolts because it had performed satisfactorily for 60 000 h under adverse conditions and could easily be replaced every 60 000 h during maintenance periods. These were also captive bolts and their failure could not result in damage to rotating parts, only a loss of secondary buffer helium pressure would result from their failure. Preloads were reduced by using a more precise method to install the bolts.

2. More stringent water chemistry controls were implemented to reduce the allowable level of caustics in the auxiliary stem system.

3. The AISI-410 fasteners were replaced with those made from Incoloy Alloy X-750 for improved corrosion resistance. For these bolts high strength was not of paramount importance. The specified heat treatment was 1 to 2 h at 1107°C followed by overaging at 705°C for 20–22 h. This treatment was recommended in EPRI draft specification 2322-16322-HC2 "Alloy X-750 for use in Light Water Reactor Internals". Preloads used were lower than those specified for the original AISI-410 bolts.

4. The molybdenum disulphide lubricant was changed to a nuclear grade compound containing no sulphur compounds.

The failure incident described in this paper was one of a series of problems experienced by the Fort St Vrain helium circulators associated with the auxiliary support systems. In particular, contamination of the steam system and water leakage from the bearings were experienced. These factors have been a dominant influence on the decision to use electric motor driven circulators for current designs of the steam generating high temperature gas-cooled reactor.

References

1. T. R. Roach, Aerospace High Performance Fasteners Resist Stress Corrosion Cracking, *Corrosion '84*, Paper 112. NACE, Houston, TX (1984).
2. C. J. Czajkowski, "Bolting Applications", NUREG/CR-3604, BNL-NUREG-51735, US Nuclear Regulatory Commission, May 1984.
3. M. K. Nichols, "Final Inspection and Repair Summary for Helium Circulator C-2101", GAC19493, November 1988.
4. M. F. Maguire, A. R. Troiano and R. F. Hehemann, SCC of Ferritic and Martensitic Stainless Steels in Saline Solutions, *Corrosion '73*, Vol. 29, No. 7, NACE, July (1973).

19

Experiences with Modern Thread Lubricants

R. HOLINSKI and H. TRAUTMANN

Molykote GmbH, Pelkovenstraße 152, D-80992 Munich, Germany

ABSTRACT

High temperature bolting applications require special alloys. Since such alloys are rather noble, their surfaces are covered only by a very thin oxide layer. Therefore, such metals have a very high tendency to seize or gall during tribological stress. Such bolting application definitely requires a lubricant based on solids, in order to enable a specified assembly of the connection and allow trouble-free disassembly after high temperature exposure. A number of commercially available high temperature thread lubricants have been tested using various bolting alloys. Analytical investigations after thread tests at high temperatures reveal that certain ingredients of the thread lubricants diffuse into the steel surface and cause damage, such as micro cracks and fracture of bolts. Based on this experience a specification for high temperature thread lubricants has been established, excluding a number of elements for use in thread lubricants. Based on current research and development, a high temperature thread lubricant has been developed and tested in bolting applications with various alloys at different temperatures. Lubricating properties and the effects of lubricant on special alloys at high temperatures were tested. Field experiences with this new thread lubricant are also reported.

1. Introduction

A great deal of care is required in the selection of suitable alloys to be used for threaded connections, such as bolts, nuts and washers. The alloy has to provide sufficient yield strength during high temperature service, in order to make the connection a safe one. Bolting materials for high temperature applications require a suitable lubricant, and therefore the selection of suitable lubricants for threaded connections is very important.

In general, high alloyed steels are very corrosion resistant, as the steel surfaces are covered by very thin oxide layers. This implies that such alloys tend to seize easily, when tribologically stressed, i.e. during the assembly of bolt and nut.

Consequently, these machine elements need to be lubricated by solid lubricants, which form a homogeneous solid film on the threaded surfaces. Lubricating films with good adhesive properties protect steel surfaces from seizure and galling during assembly, and separate threads during high temperature exposure. However, this means that lubricating films are in close contact with the metal surfaces and need to be compatible with the various alloys.

2. Composition of Thread Lubricants

For the lubrication of threaded connections pastes are in common use. These products consist of a mineral oil and various solids. Originally, solid lubricants such as graphite and molybdenum disulphide have been used. Later, metal powders such as copper, lead, zinc, tin, aluminum or nickel were part of the paste composition. Also a number of compounds such as metal sulphides, fluorides, and phosphates were used.

The purpose of a thread paste is to enable an assembly of bolt and nut, which leads to a specific clamping force when machine parts are connected. During assembly from the solid particles of the paste, a tightly adhering solid film is formed, which has to have good load carrying capacity in order to prevent seizure and galling. During high temperature exposure the mineral oil evaporates, leaving a dry solid film on the thread surfaces. Again, this film has to have good adhesion on thread surfaces and needs to have good load carrying capacity during high temperatures, which might be in the region of 900°C in the case of gas turbines. After high temperature service, the connection has to be disassembled, damage-free.

In special cases, dry lubricating coatings are also used for thread lubricants, consisting primarily of molybdenum disulphide and a binder. Dry coatings are used for applications in which no oily products are permitted. In order to prevent fretting corrosion, thread surfaces or bolts are covered with a silver layer by galvanic treatment.

3. Failure of Threaded Connections

As previously mentioned, alloys are used with a high amount of chromium, nickel, molybdenum and vanadium for bolting materials exposed to high temperatures. These alloys are very corrosion resistant, which means that surfaces do not react with oxygen to form oxide layers. However, oxide layers are a very good protection against cold seizure once these oxides are tribologically stressed. The best replacement for the lacking oxide layer is a solid lubricant film. If such a film has poor adhesion on thread surfaces or is not homogeneously formed, seizure between thread or bolt and nut will occur during assembly. In these cases, connections cannot be easily dismantled and bolts have to be cut.

Amongst the failures of threaded connections, the greatest threat is presented by a fracture of the bolt (Fig. 1). Such a fracture can lead to a leak of a pressurised vessel or pipe. In 1984, the United States Nuclear Regulatory Commission investigated fractures of bolts in American Nuclear Power Stations. This commission reported that, out of 44 bolt fractures, 19 were caused by stress corrosion. An investigation of the composition of thread lubricants was initiated. It was reported that thread lubricants contained metal powders and also MoS_2. In the presence of humidity, hydrogen sulphide is generated from the sulphur of MoS, which at the same time produces hydrogen. This may cause hydrogen embrittlement, giving rise to cracks or fractures. After these investigations the commission banned all MoS_2-containing lubricants from use in American Nuclear Power Stations.

Metallurgical changes of bolting alloys in contact with thread lubricants have been investigated in a joint project between the Central Electricity Generating Board, Bristol, UK and Molykote, Munich, Germany [1]. A number of commercially available thread lubricants have been used for assembly of threaded connections consisting of a nimonic. These connections were exposed to a temperature of 700°C for several months. After disassembly of bolts and nuts, the metallurgical effects of lubricants on thread surfaces have been investigated on cross-sections of bolts. On examination, thread lubricants containing solder metals,

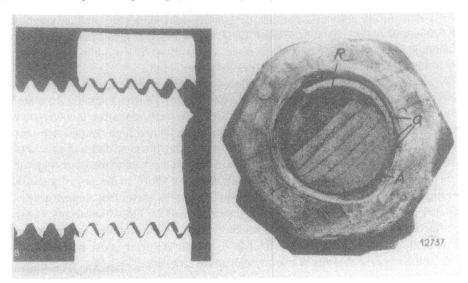

Fig. 1 *Bolt fracture, stress corrosion cracking.*

such as lead, tin or zinc caused destruction of grain boundaries — which may then lead to solder embrittlement (Fig. 2). Ideally, modern thread lubricants should not contain any low melting metals. Sulphur containing solids, such as molybdenum disulphide, caused micro cracks of the alloy in the area of highest stress (Fig. 3).

These findings reveal that certain thread lubricants may have detrimental effects on bolts, leading to cracks or fractures. A safe thread lubricant also must not contain compounds based on phosphorus, arsenic, selenium and antimony. Such species have shown an increase in hydrogen permeation kinetics. Stress corrosion cracking can generally be rationalised in terms of a hydrogen embrittlement mechanism, because certain additives, known to promote hydrogen entry into steel, accelerate stress corrosion cracking. In the presence of water or humidity, certain amounts of halogens, such as chloride, cause corrosive attacks on steel surfaces. Such corrosion will again cause stress corrosion cracking. Based on this knowledge, a number of elements are banned from thread lubricants in order to avoid stress corrosion cracking and fracture of bolts.

4. Development and Testing of a New High Temperature Thread Lubricant

The technical requirements for a thread lubricant are as follows:

(i)   The lubricant must not contain any elements which may cause stress corrosion cracking or bolt fracture during high temperature service.

(ii) A good load-carrying capacity to avoid seizure during assembly.

(iii) A thread friction coefficient $\mu < 0.15$ for bolt alloy 21CrMoNiV57.

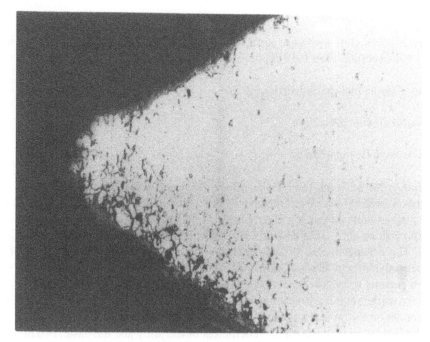

Fig. 2 Cross-section of bolt thread (CrNi alloy). Lubricant: lead paste, 1000 h at 700°C.

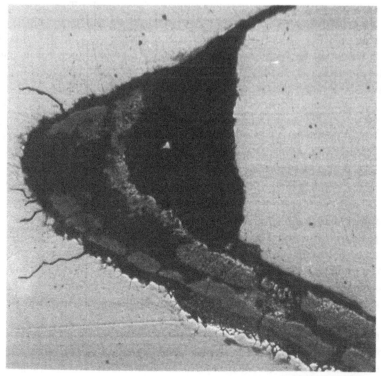

Fig. 3 Cross-section of thread (CrNi alloy). Lubricant: MoS paste, 1000 h at 700°C.

(iv) Good high temperature performance of solids to allow damage-free disassembly of connections after high temperature service.

(v) Good anti-wear properties.

(vi) Corrosion protection.

(vii) No toxic ingredients.

In order to fulfill all these technical requirements, a newly developed thread paste has been evaluated by several test methods. Since a thread lubricant needs to enable trouble-free assembly of bolt and nut at a given torque the friction coefficient has to be determined for various bolting alloys. For friction-value investigations, a testing machine from Erichsen is used. The dependence of the pre-stressing force Fv on the tightening torque Ma which is influenced in turn by the friction value, is determined. The measuring electronics of the test bench permits testing in accordance with practice. The friction coefficient was measured with five different threaded connections to find the mean value. The coefficient of friction is dependent on the type of alloy and on the type of thread paste.

Having determined the friction coefficient of the alloy and the paste, two high temperature stable plates were clamped together by threaded connections. The tightening force was calculated from the previously determined friction coefficient. These clamped plates had been exposed in an oven to various temperatures for 4 weeks. Two different alloys were used at temperatures of 300, 580 and 850°C. After cooling the clamped plates at the end of the test, the break-away torques of the threaded connections were measured by a special electronic torquemeter. Performance of the newly developed thread paste has been compared to a commercially available paste containing lead and molybdenum disulphide. It was found that the ultra-pure paste had about the same performance in three different threaded connections at various temperatures as the commercial paste. All threaded connections could be opened after the 4-week test at high temperatures without seizure. The investigation of the cross-section of a thread after long time high temperature test demonstrates that there is no diffusion of materials of the thread paste into the surface of the steel resulting in deterioration of the thread surfaces (Fig. 4).

The newly developed ultra-pure high temperature paste has extremely low levels of sulphur and halogen. The most important ingredients of a solid are zirconium dioxide and graphite, besides some other solids.

The paste has been tested in exhaust bends and pipes of diesel engines. Requirements of the thread paste include low levels of sulphur and halogens, a low scatter of friction coefficient and a friction level at a value which prevents self-loosening of nuts. The temperature of the bolts has been recorded as high as 600°C. The new paste fulfilled all these requirements and was specified to be used on connecting rods and threads of exhaust pipes and bends.

• In the chemical industry devices are being used to grow single crystals of silicon. The lubricants used on threads have to be ultra-pure and free of any metal. The temperature of the thread was 400°C. The new paste was tested here and was found to be suitable.

Fig. 4 *Cross-section of thread (CrNi alloy). Lubricant: inert paste, 1000 h at 700°C.*

- In pyrolitic devices in the chemical industry, bolt temperatures as high as 950°C are found. In this case, performance at high temperature and high purity level was required. The ultra-pure paste passed these requirements and was approved.

- In aluminium motor blocks a threaded pipe is being screwed into pipe holes. The temperature in this case is 450°C. The lubrication of the aluminium threads posed a problem. The ultra-pure paste was used to successfully lubricate the aluminium threads at this temperature.

- Turbines operating at a temperature of 540°C also have to be lubricated by a high performance ultra-pure paste. Subsequently, threads could be disassembled, without any seizure, after a period of 6 months.

5. Conclusions

The use of a thread lubricant containing solids for the assembly of high temperature stable bolting materials is essential. However, certain solids react with special alloys of the threaded connections which can result in the fracture of bolts during high temperature application.

Based on this knowledge, a new thread lubricant has been developed using inert solids, such as zirconium dioxide. It is essential that the new paste is ultra-pure in respect of sulphur and halogen content, and does not contain any low melting metal powders. This product shows very good tribological properties, and after high temperature service shows no detrimental effects on bolting alloys. The new paste has been tested in the field in various applications and was found to fulfill the requirements of threaded connections in use in various machine designs operating at high termperatures.

Reference

1. R. Holinski, Lubricant diffusion embrittles steel threads, *Tribology*, 1983, (March), 27–30.

SESSION 5

Nickel Based Bolting Alloys

Chairman: RD Townsend
ERA Technology, Leatherhead, UK

SESSION 5

20

25 Years Experience of Nickel Based Bolting Materials

S. M. BEECH

International Research and Development Ltd, Newcastle-upon-Tyne, UK

ABSTRACT

Nimonic 80A has been successfully used as a bolting material for Parsons Power Generation Systems since the mid 1960s, after good experience with this material had been obtained for gas turbine blading. Parsons have currently over 10 000 Nimonic 80A bolts in service, and there have been only 75 bolt failures in total worldwide. A review of the properties of this material has been undertaken, and some of the early problems in using Nimonic 80A for bolting are discussed, including lattice contraction, embrittlement and stress corrosion cracking. The practical remedies that Power Plant manufacturers and operators have provided are highlighted. Finally, the future use of nickel based bolting in the next generation of advanced steam condition Power Plant is addressed.

1. Introduction

Nimonic 80A was first introduced as a material for bolting in the mid 1960s, following on from good experience at Parsons as a high temperature gas turbine blading material. It is very much stronger than the ferritic steel bolting then used for all high temperature applications, such as the high carbon 1% chromium based Durehete series of steels. The need for a much stronger bolting material arose from the large increases in steam turbine unit output that was demanded both in the UK and worldwide at that time. Indeed Parsons went from designing and building 120 and 200 MW output turbines to 500 MW sets in one stage during the early part of the 1960s. The increased jointing stresses that were required for these large units meant that there were considerable difficulties in designing with ferritic based materials, and a stronger substitute had to be found.

At the temperatures of operation, i.e. up to 565°C, Nimonic 80A is microstructurally very stable. In addition it has a high resistance to stress corrosion cracking, and a good fracture toughness. An additional benefit in choosing this material rather than other high strength materials such as high alloy ferritic steels was that the thermal expansion coefficient of Nimonic 80A was similar to that of the low alloy casing materials then employed. This results in easier design methodologies and simpler bolt tightening requirements.

Its use covers HP cylinders, valve covers, steam strainers and loop pipe flanges. In the UK there are in total approximately 15 000 Nimonic 80A bolts, of which Parsons have manufactured 10 000, with approximately 10 000 above 50mm in diameter. Service lives now exceed 150 000 h at up to 565°C and 165 bar. In the 30 years of use there have been only 75 failures, and there has been no failures before 30 000 h use; most of the failures have occurred after 50 000 h service.

2. Nimonic 80A Material and Properties
2.1. Composition and Mechanical Properties

Nimonic 80A is a nickel based alloy containing 20% chromium and 0.5% iron, with small amounts of titanium, aluminium and carbon. A typical chemical composition for this material is given in Table 1. It derives its good high temperature mechanical and long term strength from the formation of carbides and up to 20% volume fraction of gamma prime precipitates. It has a high proof and tensile strength compared to ferritic bolt steels (Table 2), and its thermal expansion coefficient is similar to low alloy bolt steels (Fig. 1).

2.2. Heat Treatment

Nimonic 80A is invariably given a solution heat treatment of 1080°C for 24 h, then ageing treatments at 850°C and 700°C, in each case followed by cooling in air. This heat treatment sequence gives rise to a good elevated temperature strength which arise from two factors: a high density of gamma prime precipitates, and from a dispersion of stable M23C6 and M7C3 carbides. The 850°C solution treatment produces a relatively coarse distribution of gamma prime precipitates (Ni_3(Al,Ti), 200 nm). Experience in the gas turbine industry has shown that although this produces a good creep strength, notch brittleness occurs under creep conditions [1]. In order to produce a more ductile material, a final solution treatment at 750°C is now carried out. This precipitates gamma prime with a size of 80 nm. This two size gamma prime distribution was very successful in producing an alloy much more tolerant to notch brittleness in creep.

It is usually extruded and given a 15% cold stretch to induce cold work prior to heat treatment in order to obtain a smaller and more uniform grain size.

Table 1. Composition of Nimonic 80A

	C	Si	Cu	Fe	Mn	Cr	Ti	Al	Co	Mo	B	Zr
Min.	0.04	–	–	–	–	18.0	1.8	1.0	–	–	0.0015	0.04
Max.	0.10	1.0	0.2	1.0	1.0	21.0	2.7	1.8	2.0	0.3	0.005	0.10

Table 2. Mechanical properties of Nimonic 80A

Tensile strength	1200 MPa mean
Proof strength	750 MPa mean
Hardness	260–290 HV
Room temperature impact strength	40–50 J

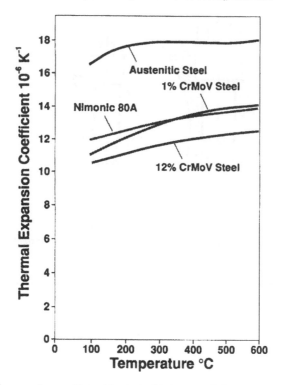

Fig. 1 *Thermal expansion coefficient for bolted joint materials.*

3. Nimonic 80A for Bolting

Jointing stresses between inner casings of steam turbines increased during the 1960s with the change to 500 MW units. Parsons had used Nimonic 80A before as a blading alloy for gas turbines, and it was felt that it would also make a bolting material, as its good high temperature strength could be utilised. The relative sizes between the two types of bolting is shown in Fig. 2. Thorpe Marsh Power Station, which was commissioned in 1963 was the first station to incorporate the new bolting material.

Nimonic 80A has a good combination of characteristics that make it particularly suitable for turbine bolting applications, including the following:

- a creep rupture and creep relaxation strength much greater than low alloy ferritic bolting materials (Fig. 3);

- a high proof strength; and

- a thermal expansion coefficient close to that of low alloy flange steels usually notch insensitive under creep conditions.

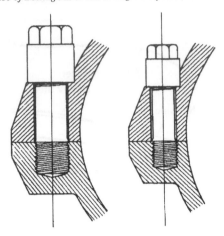

Fig. 2 *Comparison of bolt sizes for Nimonic 80A and low alloy ferritic steel [1].*

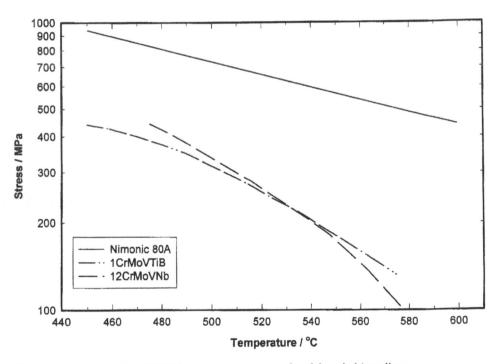

Fig. 3 *Comparison of the 30 000 h creep rupture strengths of three bolting alloys.*

4. Operational Experience

Currently, Parsons have 63 units ranging from 200 to 660 MW capacity in which Nimonic 80A bolts are fitted. The majority of these bolts (8320) are used to secure steam chest and strainer covers, whilst the remainder (576) are used exclusively in the hotter regions of HP inner cylinder flanges. In all Parsons have approximately 10 000 bolts in service at temperatures estimated to be between 490°C and 565°C, some with lives currently out to 125 000 h. Parsons records show that a total of 45 Nimonic 80A bolts have fractured or cracked in service [2], although this may not reflect the complete experience, since failure investigations are sometimes undertaken by the utilities, and the manufacturers are not always informed. There are also four other instances where bolts have failed as a result of arcing damage occurring ill the bore while bolts were being heat-tightened. A summary of the UK service experience of Nimonic 80A [3] is given in Table 3.

4.1. Long Range Ordering

Most (i.e. 34) of the Nimonic 80A bolt failures occurred as a result of overload when long term ordering increased the stress levels in the bolts. During constant strain stress relaxation tests, the stress necessary to maintain the strain in most materials decreases with time because the elastic component of the applied strain is converted to plastic strain, usually by creep deformation processes. In Nimonic 80A, the stress increases as a function of time at 550°C and below (Fig. 4). At higher temperatures, normal stress relaxation behaviour occurs. This is because at lower temperatures there is a contraction of the material due to atomic ordering of the matrix phase (Fig. 5). At temperatures higher than 550°C, normal creep processes dominates. There is some short range ordering following normal 3-stage heat treatment, and at temperatures of less than 550°C, the short range order increases, and then transforms to long range ordering (Ni_2Cr). This reaction is very sluggish, and it takes more than 10 000 h for the beginning of long range ordering to occur at 450°C (Fig. 6). Under stress however, there is a significantly enhanced kinetics [4]. For example it is possible to get up to 0.16% contraction after 50 000 h service and 0.11% in 30 000 h. At temperatures greater than 525°C, the kinetics are increased so that short range ordering is responsible for a lattice contraction of 0.03 % at 550°C.

During prolonged service, the stress on a bolt, tightened to an initial strain of 0.1 or 0.15 %, can increase significantly due to this ordering transformation. Angle of turn on bolt capnuts was used for many years as a measure of initial tightening strain. There is always the possibility for bolts to be incorrectly tightened using this procedure, and for many of the failed bolts, there was the strong suggestion that initial overtightening followed by lattice contraction, was responsible.

Table 3. *Service experience of Nimonic 80A*

Total number of bolts in UK service:	15 000 (10 000 > 50 mm)
Estimated number of failed bolts:	75
Primary failure mechanisms:	
Creep	
Intergranular fast fracture	50%
Stress corrosion cracking	33%
High strain fatigue	8%

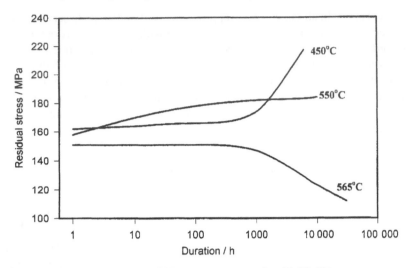

Fig. 4 *Stress relaxation of Nimonic 80A at a constant strain of 0.1% [3].*

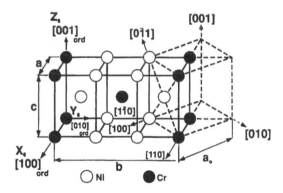

Fig. 5 *Ordered structure of Ni₂Cr. The broken lines refer to the original fcc cell [4].*

Nimonic 80A bolts had traditionally been tightened to a nominal strain of 0.15% but following the failures in the late 1970s that were attributed to lattice contraction, 0.10% tightening strain became the norm. This recommendation was made around June 1981 and from that date utilities have progressively introduced the lower strain during overhauls.

4.2. Embrittlement

Failed bolts made from Nimonic 80A were often found to have very poor charpy impact values, in some cases down to 10 J. This is much lower than virgin stock, and material would recover if using a full three stage re-heat treatment. Further ageing at 600°C again reduces the charpy impact value [5]. Nimonic 80A becomes embrittled upon ageing, with characteristic 'C' curve kinetics (Fig. 7), although the rate of embrittlement varies from heat to heat, and with differing initial values. Auger studies that have been undertaken have rationalised this by finding a broad agreement between the P/Ni ratio with impact energy.

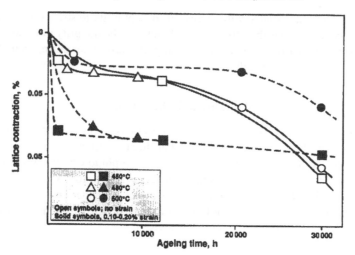

Fig. 6 *Effect of temperature, tlme and strain on lattice parameter contraction in Nimonic 80A* [4].

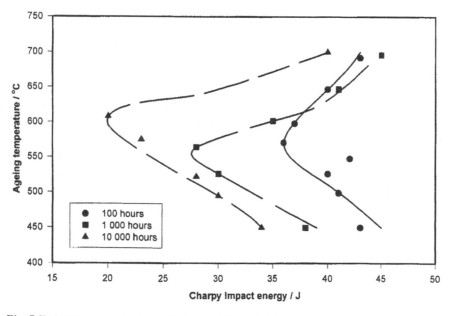

Fig. 7 *Embrittlement behaviour of Nimonic 80A with 30 ppm phosphorus [5].*

4.3. Stress Corrosion Cracking

Stress corrosion cracking, this mechanism has been responsible for a large fraction of the failures within Europe [3]. Sulphur contamination being a common feature on the majority of fracture surfaces. Work carried out by various workers [6–8] has shown that Nimonic 80A is susceptible to stress corrosion cracking in acidic solutions, and it appears that hydrogen ions are responsible for this behaviour, strongly suggesting that hydrogen cracking is occurring, as sulphur is not necessary for cracking to take place [7].

A large grain size is associated with increased susceptibility to stress corrosion cracking, so that the fabrication route is therefore a critical factor in the integrity of nimonic bolting. It is for this reason that extrusion and cold stretching are carried out during manufacture as a more uniform, fine grain size results.

4.4. Thread Lubrication

In view of the propensity of Nimonic 80A to stress corrosion cracking in acidic environments, attention was drawn to the possibility that thread lubricants could give rise to bolt failures. In particular, it is known that molybdenum disulphide which is used in many proprietary lubricants can break down in high temperature steam, for example during a steam leak to molibdic acids. Hence in the early 1980s a recommendation was made which described the problems associated with molybdenum disulphide bearing thread lubricants, and the avoidance of such lubricants was strongly advised.

4.5. Relaxation and Creep Rupture Behaviour

The creep relaxation and rupture strength of Nimonic 80A is much greater than that of the ferritic materials they replaced (Fig. 3), but there has been a tendency for some heats to be notch weakening. This has obvious implications for a threaded bolt. The reasons for this behaviour have been studied, but with inconclusive results. For example in a detailed review of the high temperature properties of Nimonic 80A [9], several aspects of material behaviour requiring further evaluation were highlighted. In particular, the interaction of several variables, notably manufacturing route, notch and specimen geometry (kt), method of notch preparation and machining practice were thought to be responsible for the large amount of scatter seen in notched rupture results from different heats. For example, depending on the manufacturing route adopted for Nimonic 80A bolting applications (Fig. 8) three different surface conditions can occur; if solution treatment is carried out after machining, large oxide fingers can result, even if the surface is protected. If age hardening treatments are carried out after machining, grain growth occurs in the worked region, and if machining is performed as the final step of manufacture, then a worked layer is often seen. These microstructures are shown in Fig. 9. Work is continuing in order to determine, and therefore eliminate the causes for notch brittleness in Nimonic 80A.

4.6. Fractography

It is extremely unusual to find a partially cracked bolt, however this may be the result of untightening procedures. The failures usually occur at the first engaged thread and are characterised by brittle intergranular fast fracture with a single initiation site and little crack branching. Invariably no creep cavitation is seen. Overtightened, embrittled and ordered bolts will all display this fractography.

Stress corrosion cracking is associated with intergranular failure, although some transgranular failures are also observed at low temperatures. These failures are usually have multiple initiation sites, and display some crack branching. Some high strain fatigue failures have been identified, with transgranular or intergranular cracking and beach markings on

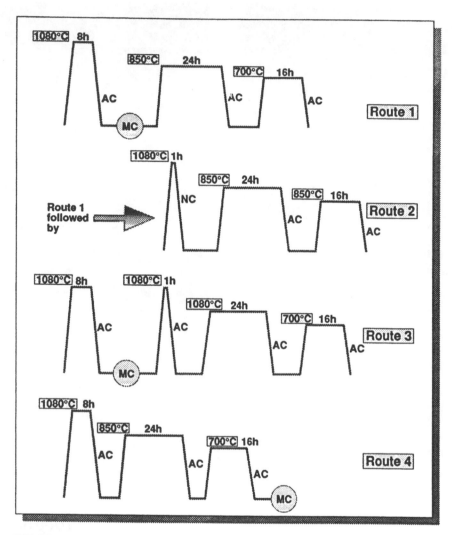

IRD 04519

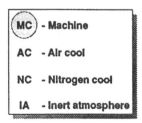

Fig. 8 *Different manufacturing routes used for Nimonic 80A bolting.*

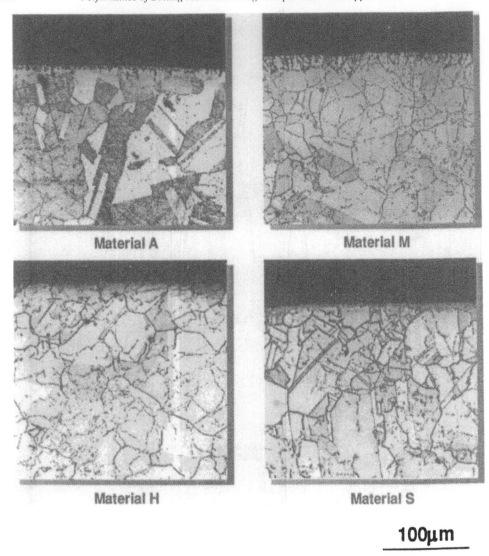

Fig. 9 Surface appearance of untested material after different heat treatment/machining sequences.

the fracture surface. There is usually little crack branching and are characterised by one or two initiation sites. If there are two, they tend to be diametrically opposed.

5. Discussion

The most typical reason for failure in Nimonic 80A bolting is now agreed to be in the following way:

1. The bolts contract in-service due to the ordering reaction, increasing the stress but not to a level that would explain failure by creep rupture if compared with the normal material design data. Exacerbating this is a poor notch ductility when compared with virgin material through embrittlement of phosphorus to grain boundaries. No evidence of creep cavitation is usually found, even in the vicinity of the fractured thread roots. The accumulation of plastic strain at the thread root of the studs is thought to be in important factor in the failure mechanism.

2. In view of the risk of bolts coming into contact with unsuitable thread lubricants, work is being carried out to reduce the susceptibility to stress corrosion cracking [10]. The aim is to utilise the relationship between stress corrosion susceptibility and proof strength. By reducing the proof strength of the material by carrying out a furnace cool from solution temperature to the first ageing treatment, it is hoped that susceptibility is reduced without unduly impairing the relaxation strength.

3. In an assessment of Nimonic 80A as a high temperature bolting for advanced steam conditions [3], the authors concluded that it may be regarded as a suitable candidate. This was due to its superior creep rupture and relaxation properties compared to the ferritic alternatives. Moreover, above about 550°C there is no stress pick-up due to long range ordering (Fig. 6). In order to reduce the embrittling effect of phosphorus, it may be necessary to specify lower bulk levels, as at higher operating temperatures, the material may be more susceptible.

4. Over the last twenty-five years, the various bolt failures that have occurred have allowed a much fuller understanding of Nimonic 80A material. A summary of the factors affecting bolt life is given in Table 4. The result of this has been that effective measures have been taken to minimise the risk of bolt failure. These include avoiding the temperatures over which Nimonic 80A undergoes the ordering reaction, minimising the risk of stress corrosion cracking from control of thread lubricants, and moving away from inaccurate angle of turn

Table 4. Summary of factors influencing Nimonic 80A bolt life

Stress
Initial tightening strain
Notch stress concentration
Negative creep
Thermal transients

Materials and properties
Microstructure (grain size)
Properties after service exposure

Environment
Condensation
Steam impurities and lubricants
Oxygen
Operating temperature

methods for bolt tightening. The success of this is demonstrated in that the last reported failure of a Nimonic 80A bolt in a Parsons' machine was in 1983.

6. Conclusions

There have been very few failures in Nimonic 80A bolting since it was first introduced in the early 1960s, and none since 1983.

There were initial problems with overtightening coupled which ws exacerbated by a lattice contraction in the material.

There were later problems with stress corrosion cracking which were resolved by the removal of molybdenum disulphide as a thread lubricant.

In order to provide even higher integrity bolting, it may be necessary to reduce the phosphorus content, and to carry out a slow cool from the solution heat treatment temperature in order to reduce the overall strength and hence the 2+propensity to stress corrosion cracking.

7. Acknowledgements

The Author would like to thank Mr E. F. Tate of Parsons Power Generation Systems for helpful discussions when writing this paper. Thanks must also be given to International Research and Development Ltd for permission to publish.

References

1. W. Betteridge and P. A. Morgan, *The Nimonic alloys*, Second edition (Eds W. Betteridge and J. Heslop), 1974, pp.7–22.
2. E. F. Tate. Private communication. Parsons Power Generation systems, 1990.
3. S. M. Beech, S. R. Holdsworth, H. G. Mellor, D. A. Miller and B. Nath, 'An assessment of Alloy 80A as a high temperature bolting material for advanced steam conditions', *Proc. Int. Conf. Advances in Material Technology for Fossil Power Plants*, Chicago, 1987, pp. 503–410
4. E. Metcalfe and B. Nath, *Int. Conf. on Phase Transformations*, York UK, 1979, The Institution of Metallurgists, London.
5. B. C. Edwards, D. A. Miller and B. Nath, 'The long term embrittlement characteristics of Alloy 80A', unpublished report.
6. K-H. Mayer and K-H. Keienburg, *Inst. Mech. Engng Creep Conf.*, Sheffield, UK, 1980.
7. J. Parker and T. E. Parsons, CEGB Private communication.
8. D. A. Miller, L. Miles and J. A. Roscow, CEGB Private communication.
9. S. M. Beech, Parsons Internal report, 1981.
10. B. Nath, K-H. Mayer, S. M. Beech and R. W. Vanstone, 'Recent developments in alloy 80A for high temperature bolting applications', this volume, pp. 306–317.

21

Brittle Fracture of Alloy 80A Stud Bolts in a Steam Turbine

P. VINDERS*, R. GOMMANS, G. van OPPEN[†] and K. VERHEESEN

DSM Services, P.O. Box 18 - 6160 MD Geleen, The Netherlands
*Now retired
[†]DSM Hydrocarbons, P.O. Box 24 - 6190 AA Beek, The Netherlands

ABSTRACT

During start-up of a steam turbine 14 out of 34 bolts (material Alloy 80A) ruptured in a brittle manner after only 9 months of service at 475°C. Metallurgical investigations and literature review made clear that longrange ordering (LRO) had occurred. LRO embrittled the material and raised the internal stress level. During start-up the bolts ruptured due to additional thermal loading. Because LRO and subsequent rupture occurred only in the 12 in. bolts, it was concluded that the material composition, the material condition, and the thread application methods were not optimal. In the future Alloy 80A bolting will be purchased using an revised specification.

1. Introduction

In October 1991, during the start-up of a steam turbine 14 out of 34 stud bolts broke in a brittle manner. The bolts were fabricated of Alloy 80A according to the Germans standard DIN-17240. The operating temperature was between 450 and 500°C (850–950°F), while the operating period had been only 9 months (6500 h). Several diameters were installed, ranging from 12 to 3 in., but only the 12 in. dia. bolts broke. Failure investigations and literature review were performed in order to elucidate this uncommon failure mode.

2. Case History

Ethylene plant No. 4 is one of the two ethylene plants that are operated by DSM in the Netherlands. This plant with a name-plate capacity of 560 000 tons of ethylene per year was designed by Lummus, and was commissioned in 1979.

The steam turbine under discussion (No. X551) is driving the propylene compressor. The turbine has a rated output of 19 MW, rotates at 4000 rpm, and was designed and delivered by Turbodyne (now Dresser-Rand).

The high pressure steam (110 bar, 510°C) enters the turbine through the TT-valve and a 4-valve steam chest (see Fig. 1). The valve chest is bolted to the turbine casing. All bolts were originally made of ASTM A193 Gr.B16 material ($0.4C$–$1Cr$–$^1/_2Mo$–$^1/_4V$ type). The following connections are bolted:

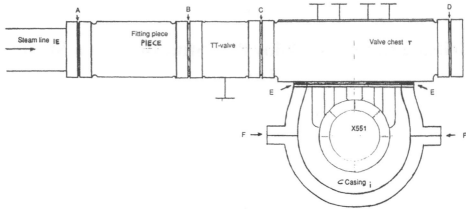

Fig. 1 *Schematic drawing of the X551 steam turbine.*

A. Steam line to fitting piece   D. Blind flange to valve chest

B. Fitting piece to TT-valve   E. Valve chest to turbine casing

C. TT-valve to valve chest   F. Split line in turbine casing.

During the years problems occurred due to leakages. During a scheduled shut-down in 1988 the B16-bolting of the connections A to D were replaced using Alloy 80A (75Ni–20Cr–Al/Ti type). Positive experience had been gained during the following operating period, so it was decided to replace the other bolted connections as soon as possible:

• December 1990 (intermediate shut-down): stud bolts of the valve chest to the turbine casing (E);
• October 1991 (planned shut-down) : stud bolts of the split line in the turbine casing (F).

In October 1991 during start-up of the turbine 14 out of 34 stud bolts of the valve chest to the casing (connection E) failed in a brittle manner. The location of the ruptured bolts is shown in Fig. 2. Most ruptured bolts were located at the turbine side of the valve chest. Just before the failure the critical rotating speed was passed. At that moment the steam temperature was 425°C; while the estimated bolting temperature was *ca.* 300°C. No sufficient spare bolting was available, nor was time available to order new Alloy 80A bolting. Therefore, it was decided to use the original B16 material and the turbine and the plant were re-started within 3 days.

Preliminary failure investigations indicated that the other bolt dimensions did not behave brittle. Extensive failure investigations started, and as a result of that the material specification was revised. The new Alloy 80A bolting material will be installed during the scheduled shut-down of 1995.

3. Failure Investigations

The failure investigations included microscopic examination, fractography, chemical analy-

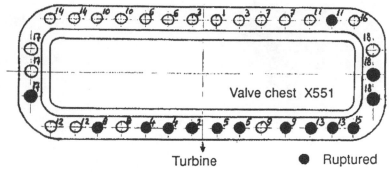

Fig. 2 *Bolting plan of the valve chest to the turbine casing, also indicating the ruptured bolts. Notice that most bolts failed at one side (the turbine side) of the valve chest.*

sis, and mechanical testing. Some new material was exposed to the service temperature (475°C); heat treatments at 700°C were also performed.

The following bolt 'types' were investigated:

- $1^{1}/_{2}$ in.   Broken   6500 h   • 2 in.   Not broken   New
- $1^{1}/_{2}$ in.   Not broken   6500 h   • $2^{1}/_{2}$ in.   Not broken   New
- $1^{1}/_{2}$ in.   Not broken   New   • 3 in.   Not broken   New.

Remark: Not all testing mentioned above were performed on all bolt types mentioned.

The initial axial strain given to the bolts is *ca.* 0.15% (80% of the yield stress). No re-tightenings were performed during the 9 months of service.

3.1. Microscopic investigations

The $1^{1}/_{2}$ in. bolt broke intergranularly. The grain size is very coarse : ASTM $1-1^{1}/_{2}$ (200–250 μm). Precipitates are present on grain boundaries and inside the grains (see photo 2). The hardness of the material is about 340 Hv for the new and broken $1^{1}/_{2}$ in. bolts. It has been observed that the threading must have been applied by cold rolling, and the hardness was 450–540 Hv on the top, and 480–700 Hv at the bottom of the thread. The new $1^{1}/_{2}$ in. bolt has a similar grain size; but the new $2^{1}/_{2}$ in. and 3 in. bolt had a grain size of ASTM 4-5 (60–90 μm).

3.2. Fractography

The fracture surface of the broken $1^{1}/_{2}$ in. bolt shows a completely deformationless intergranular crack (see photo 1). The fracture surfaces of the notch impact test samples of the same bolt look similar. EDX-analysis identified the presence of many carbides on the surface (see Fig. 3(a)) and locally, the presence of antimony (Sb, see Fig. 3(b)); no other elements (as the elements present in the matrix) were detected. Phosphorus (P) was not detected.

Photo 1. *Broken $1^{1}/_{2}$ in. bolt of Alloy 80A in turbine X551.*

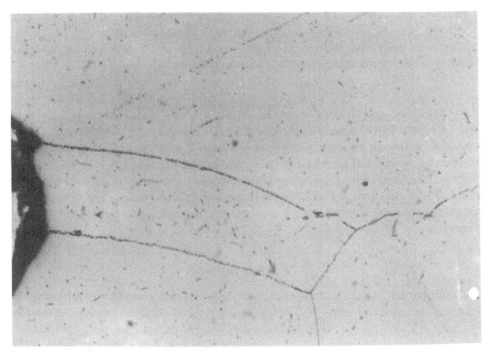

Photo 2. *Intragranular and grain boundary carbides in the broken $1^{1}/_{2}$ in. bolts, which is normally observed in Alloy 80A.*

Photo 3. *Fracture surface of the 1¹/₂ in. bolt, showing a completely deformationless intergranular crack.*

Photo 4. *Detail of photo 3 showing particles identified as carbides.*

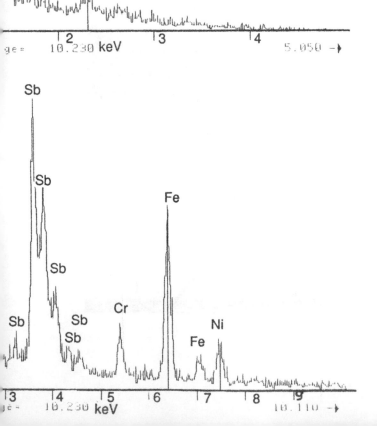

Charpy-V impact test samples of the 2 in. bolt with 30 000 operating hours, have a ductile fracture surface, with some occasional intergranular facets.

3.3. Mechanical testing

The broken $1^1/_2$ in. bolt and the not-broken 2 in. bolt have 'normal' strength at room temperature, as can be concluded from the table given below. The ductility of the broken $1^1/_2$ in. bolt is the lowest. The tensile testing does not show deviations from the DIN-standard. Therefore, notch impact testing was performed at room temperature using Charpy-V specimens (see the table below, impact energy in J).

	$1^1/_2$" bolt broken	2" bolt broken	required DIN 17240
$\sigma_{0.2}$ (MPa)	660	705	> 600
σ_R (MPa)	1045	1195	> 1000
δ_5 (%)	19	24	> 12
RA (%)	19	29	> 12

$1^1/_2$" broken	$1^1/_2$" used	$1^1/_2$" new	2" used	2" new
11	13	20	28	29

For Alloy 80A the room temperature impact value should be higher than 20 J. The broken and used $1^1/_2$ in. bolts have lower impact values, while the new $1^1/_2$ in. bolt just meets the requirement. Both 2 in. bolts exceed the requirements easily.

3.4. Chemical analysis

The melting practice for the broken $1^1/_2$ in. and the 3 in. bolts included electro-slag remelting (ESR). The melting practice for the 2 in. and $2^1/_2$ in. bolting is not known. The chemical composition according to the certificates of the bolts is given below (all elements in wt%; except for P, S, B in ppm).

From the analysis given overleaf the phosphorus content was checked, and was found to be correct.

The heat treatment for the $1^1/_2$ in.-bolting is different from the other bolting. After solution annealing the $1^1/_2$ in. dia. bolts are quenched in oil; the other bolts are air cooled after solution annealing. The temperatures and times for solution annealing and both aging heat treatments are similar.

The thread was applied by rolling **after** all heat treatments had been performed, which is very unusual. Normally the threading is applied (by rolling or cutting) after solution annealing. Sometimes an intermediate solution treatment is given to the bolts. Then the final aging treatments at 850 and 700°C are performed.

Dimension Heat No.	1¹/₂" 28463	2" + 2¹/₂" 718680	2" + 2¹/₂" 718650	3" 29273	Required DIN 17 240
Ni	73.9	73.7	73.6	75.5	bal.
Cr	19.6	19.1	19.2	19.2	18–21
Al	2.68	2.53	2.54	2.59	1.8–2.7
Ti	1.52	1.52	1.47	1.41	1.0–1.8
(Al+Ti) wt%	4.20	4.05	4.01	4.00	—
at.%	7.2	6.8	6.8	6.8	—
C	0.04	0.05	0.05	0.05	0.04–0.1
Mn	0.20	0.39	0.38	0.15	< 1.0
Si	0.26	0.25	0.25	0.22	< 1.0
P	20	50	50	40	< 200
S	10	30	30	0	< 150
Cu	0.04	0.04	0.04	—	< 0.2
Fe	1.50	1.98	1.97	0.56	< 3
B	70	45	44	—	< 80

3.5. Exposure at 475°C

New material of the bolts (with the same heat number as the used bolts) has been aged at 475°C for 3000 and 6000 h. After that impact testing was performed, see below. It is obvious that the impact energy decreases as aging progresses. The impact value of the 1¹/₂ in. bolts after 6000 hours is similar to that of the bolts broken in service. Due to a heat treatment (700°C/3 h) the impact energy can be restored.

	1¹/₂"	2"	2¹/₂"	3"
New	21	29	22	42
475°C/3000 h	14	–	24	38
475°C/6000 h	12	–	18	32
475°C/3000 h + 700°C/3 h	22	–	–	–

4. Discussion

In a recent publication of results from COST-505 is shown that about 50% of the failures occur from SCC, and about 50% from intergranular fast fracture. From the failure investigations made, SCC can be ruled out.

Intergranular fast fracture can be caused by:

– phosphorus segregations on grain boundaries, and
– negative creep caused by Long Range Ordering (LRO).

The bulk content of phosphorus in the 3 in. bolts is higher than in the $1\frac{1}{2}$ in. bolts (40 and 20 ppm respectively). Furthermore:

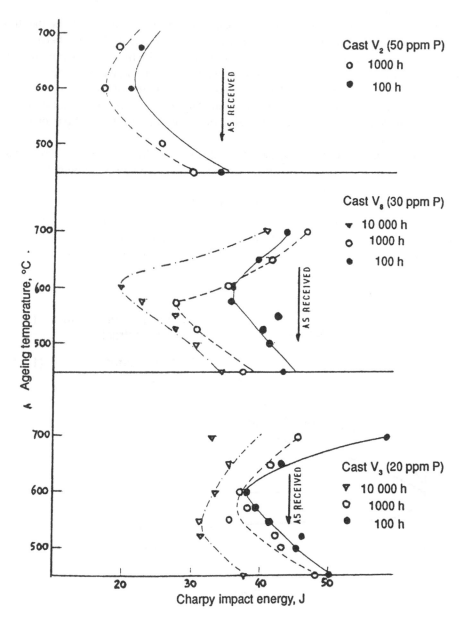

Fig. 4 *Influence of ageing on impact energy for several phosphorus contents, showing that P-segregation is most severe above 500°C [3].*

– phosphorus has not been detected by SEM/EDX on the fracture surface,
– the bulk content is below the 'safe limit' for P-segregation (20 ppm),
– phophorus segregation is less severe at temperatures below 500°C (see Fig. 4).

Therefore, it is very unlikely that segregation of phosphorus to the grain boundaries caused the low impact values.

Below 525°C Alloy 80A shows an unusual behaviour in stress relaxation tests: the stress necessary to maintain a constant strain *increases* with time (see Fig. 5). This behaviour is opposite to the normal behaviour where the stress decreases with time. This is caused by the formation of Ni_2Cr as an ordered phase (also described as long-range ordering (LRO) or negative creep). LRO is time-dependent and causes a contraction of the lattice, and therefore, decreases the volume of the material. Ultimately, because the stress caused by the decrease in volume is larger than the stress relaxation, this unusual stress relaxation behaviour is encountered.

Furthermore, long-term service below 525°C also leads to low ductility and low impact energies. This is also shown by the brittle fracture surface: smooth grain boundary facets without any evidence of deformation.

Therefore, it is concluded that the brittle fracture of the bolts is caused by LRO. Another item is that the (Al + Ti)-content of the $1^1/_2$ in. bolts is the highest of all. The $1^1/_2$ in. bolts ruptured after 9 months of service, which is very fast compared to other reported cases of brittle failure by LRO [1]. Other phenomena may have contributed to the brittle failure:

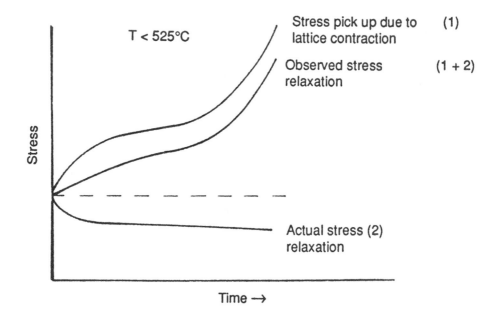

Fig. 5 *Schematic representation of the unusual stress relaxation behaviour of Alloy 80A below 525°C, due to lattice contraction by LRO [1].*

- very coarse grain size (ASTM 1–1$^1/_2$, 200–250 μm),
- thread application method (notch sensitive material)
- cooling rate after solution annealing is high (oil quenched),
- the antimony (Sb) found on the fracture surface.

It is known that an uneven heat distribution can be present during start-up of the turbine. This leads to an additional thermal loading of the bolts; this is also demonstrated by the observation that most bolts failed at the turbine side of the valve chest. Therefore, the final conclusion is that this bolting failure is caused by the combined actions of embrittlement by LRO and additional thermal loading.

5. Material Requisition for Alloy 80A Bolting Material

From literature it is known that embrittlement by LRO can be decreased by lowering the γ'-content or by coarsening the γ'-phase. The first can be achieved by lowering the (Al + Ti)-content (see Fig. 6); the second by the cooling rate after solution annealing.

The (Al + Ti)-content has been limited to 4.0 wt% maximum, which corresponds to approximately 7 at.%. This is above the range of (Al + Ti)-contents investigated by Marucco and Nath [2] (cf. Fig. 6). The content has been limited to 4 wt% in order ro maintain the high strength of Alloy 80A. The cooling rate after solution annealing using air cooling is considered sufficient for dimensions up to 3 in.; above that cooling in oil is preferred.

The coarse grain size in the ruptured 1$^1/_2$ in. bolt is obviously caused by the lack of cold stretching before solution annealing. A minimum of 10% cold stretching is absolutely necessary for a fine grain size (see Fig. 7). This cold deformation introduces enough nuclei for re-crystallisation.

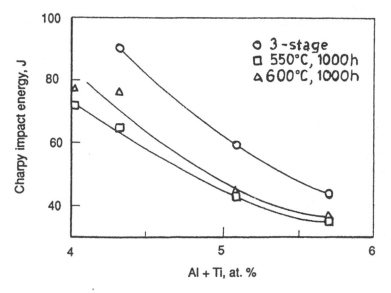

Fig. 6 Impact energy vs (Al + Ti)-content [2].

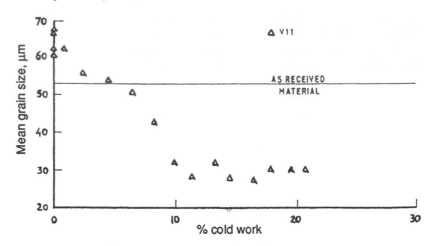

Fig. 7 *Effect of cold work on grain size of alloy 80A after solution treatment showing that 10% cold work is necessary for small grain sizes [1].*

Bolt manufacture will be performed using the following procedure:

– extrusion of bar at 1125–1150°C and water quench
– cold stretching (10% minimum)
– solution annealing: 1080°C / 8 h / air cool
– thread cutting using single point cutting
– ageing : 850°C / 24 h / AC + 700°C / 16 h / AC.

With these heat treatments a grain size of ASTM $4^{1}/_{2}$–5 minimum (75 μm max.) can be achieved and guaranteed. Thread cutting is preferred instead of rolling, because rolling may introduce grain growth and high hardness at the threading surfaces. When thread cutting is performed in a high-quality workshop, then the surface roughness is similar to that of rolled threading. No solution annealing is performed after thread cutting in order to avoid grain growth.

Knowing that phosphorus segregations did not cause the bolt failure, it is obvious that the phosphorus content must be maintained as low as possible. For application below 550°C a P-content of 30 ppm maximum will be allowed; above 550°C the content must be below 20 ppm.

The presence of antimony (Sb) on the fracture surface lead to the question of the maximum allowable content of trace elements (such as Sn, Bi, Pb, Zn). It is possible that these elements are present in ferro-chrome, which is used for Cr-addition. Therefore, the Fe-content is limited to 1.0% maximum in order to minimise the use of ferro-chrome, and thus the presence of trace elements.

A Charpy-V impact energy of 27 J minimum is required for new material. When the above mentioned requirements (fine-grained material, less and coarser γ′-precipitates) are fullfilled, then this impact energy will be met.

6. Conclusions

The brittle bolt failure occurred by long-range ordering (LRO) and additional thermal loading during start-up of the turbine. Other parameters influencing the observed intergranular fast fracture are:

- very large grain size
- application of thread rolling after heat treatment
- antimony (Sb) found on the fracture surface
- oil quenching after solution annealing.

The material specification for Alloy 80A bolting material has been revised. This includes (Ti + Al)-content, heat treatments, and thread application. The B16-bolting used for replacement after the failure will be replaced by Alloy 80A stud bolts at the first available occasion.

7. Acknowledgements

The authors acknowledge DSM's permission to publish this information. K. Brown, P. Williams and D. Busby of Inco Alloys (Hereford) and I. Lennox of Prosper Engineering (Irvine) are thanked for their helpful discussions. The help from our collegues at Mechanical Plant Services is appreciated.

References

1. B. Nath, 'An assessment of nickel-based alloys for high-temperature bolting apllications', CEC-report EUR-13802 (1991).
2. A. Marucco and B. Nath, 'Effects of composition on ordering behaviour and rupture properties of Alloy 80A', *COST Conf. on High Temperature Materials for Power Engineering*, Liège, 1990, p.535–544.
3. S. Beech *et al.*, 'An assesment of Alloy 80A as a high-temperature bolting material for advanced steam conditions', *Int. Conf. on Advances in Materials Technology for Fossil Power Plants*, Chicago, 1987, p.403–410.
4. B. Reppich, 'Negatieves Kriechen', *Z. Metall.*, 1984, **75**, 193.
5. K. H. Mayer and H. König, 'Operational characteristics of highly creep resistant Nimonic 80A bolts', see Ref. 3, p.395–402.
6. K. Brown, P. Williams and D. Busby, private communications (Inco, Hereford).
7. I. Lennox, private communications (Prosper Engineering, Irvine).

22

A Ni–25Mo–8Cr Alloy for High Temperature Fastener Applications

S. K. SRIVASTAVA

Haynes International, Inc., Kokomo, IN 46902, USA

ABSTRACT

HAYNES® 242TM alloy (Nom. Comp. Ni–25Mo–8Cr) is a high-strength, low thermal expansion alloy. The alloy was originally developed for gas turbine applications such as seal rings and has also been specified for gas turbine fasteners. The paper will review the strengthening mechanisms and thermal stability of the alloy and the effect of cold work on mechanical properties. In addition, the paper will discuss environmental considerations relevant to the fastener application.

1. Introduction

High performance fasteners require very high strength and corrosion resistance when cold worked. For high-temperature application it is imperative that the materials possess adequate strength, thermal stability as well as oxidation resistance. The NiMo–Cr system is the basis for HASTELLOY® series of alloys such as HASTELLOY C22TM alloy, HASTELLOY C-276 alloy, etc. These alloys are well known for their corrosion resistance, but the length of a heat-treatment to produce strengthening (200–500 h) is too long to be economical. HAYNES 242 alloy, based on the Ni–Mo–Cr system was originally developed for gas-turbine application requiring a combination of high strength and low-thermal expansion for service up to about 700°C. The alloy has already been specified for various wrought gas-turbine components such as seal rings, casings and containment structures.

In addition to the high strength and low thermal expansion, it exhibits good oxidation resistance, good thermal stability and good fatigue resistance. These are the same characteristics that also favor the alloy for fastener application in and around the hot component. A major engine manufacturer in the U.S. already has specified 242 alloy for fasteners. Traditionally, aerospace fasteners have been made from alloys A-286, 718, MP35N, etc. Alloy A-286 and alloy 718 are γ' and γ'' precipitation strengthened alloy, respectively. MP35N derives its strengthening from a deformation induced martensitic type fcc–hcp transformation. Nominal compositions of the alloys studied in this paper are given in Table 1. For fastener application, materials are generally used in cold-worked and direct-aged condition. The use of 242 alloy as a fastener is relatively new. The objectives of this paper are, therefore, to discuss metallurgy of the alloy (strength and stability), effect of cold work on the mechanical properties, and, finally, the environmental properties relevant to fastener applications.

Table 1. *Nominal compositions of alloys*

Material	Composition, wt %											
	Ni	Fe	Cr	Mo	Cb	Ti	Al	Si	Mn	C	B	Co
HAYNES ® 242™ alloy	Bal.	2*	8	25	–	–	0.5*	0.8*	0.8*	0.03*	0.006*	2.5*
MP35N® alloy	35	1	20	10	–	0.8	–	–	–	0.005	–	Bal.
Alloy 718	Bal.	19	18	3	5	1	0.6	0.35*	0.35*	0.05	0.004	1*
Alloy X-750	Bal.	7	15.5	–	1	2.5	0.8	0.35*	0.35*	0.08*	–	1*
Alloy A-286	26	Bal.	15	1.25	–	2.15	0.2	0.4				

HAYNES is a registered trademark of Haynes International, Inc.
242 is a trademark of Haynes International, Inc.
MP35N is a registered trademark of SPS Technologies, Inc.

2. Metallurgy of 242 Alloy

In the annealed condition, 242 alloy consists of gamma-phase matrix, with a disordered fcc structure, M_6C carbides and some μ phase. When it is aged for relatively short times between 595–705°C (1100–1300°F), the alloy undergoes long-range-order strengthening. The strengthening response results in the formation of small ordered domains, *ca.* 10 nm in size, of a $Ni_2(Mo, Cr)$ phase with crystal structure similar to Pt_2Mo [4]. Figure 1 shows transmission electron micrographs of the material in the heat treated condition. The dark-field image shows a large volume fraction of homogeneously distributed particles. The ellipsoidal or disk-shaped morphology of the ordered microdomains promotes strengthening without adversely affecting tensile ductility. The closely spaced particles contribute to a substantial increase in the strength. A typical set of tensile properties in the standard heat-treated condition, i.e. annealed + aged 650°C/24 h, is shown in the column 1 of Table 2.

A measure of thermal stability of the alloy can be obtained by determining the residual ductility or toughness after exposure at various service temperatures. Table 2 (columns 2 and 3) shows tensile properties after 4000 and 16 000 h exposure at 650°C (1200°F). It is important to note that on prolonged exposure, 242 alloy does not overage, in contrast to gamma prime-strengthened alloys. The strength is stable after prolonged exposure (it does not fall off), and it is accompanied by retention of good ductility. Table 3 shows toughness data after exposure up to 16 000 h at 650°C (1200°F). The high residual toughness values for 242 alloy are believed to derive from the stable ordering transformation. Furthermore, it implies that no embrittling phases, such as sigma, formed during this ageing process.

3. Effect of Cold Work

The effect of prior cold work upon the nature of the long-range order strengthening mechanism in 242 alloy was studied by Rothman and Srivastava [5]. Mill produced 3.2 mm (0.125

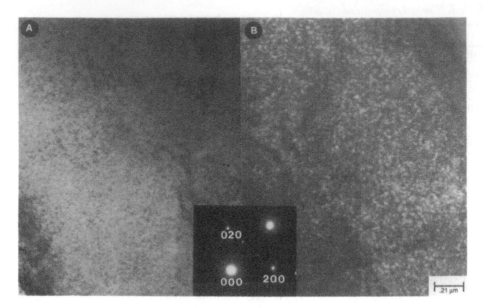

Fig. 1 TEM micrographs of 242 alloy, annealed plus aged condition (650°C/24 h); A: Bright field; B: Dark field using 1/3 (220); Inset SADP [001] zone axis.

Table 2. Tensile properties and thermal stability of 242 alloy (plate data)

Test Temp., °C (°F)	Ageing Time @ 650°C, h		
RT	24*	4000	16 000
0.2% Y.S., Ksi (MPa)	110 (159)	122 (841)	125 (862)
U.T.S., Ksi (MPa)	179 (1234)	196 (1352)	193 (1331)
%El	39	25	21
%RA	44	37	39
540 (1000)			
0.2% Y.S., Ksi (MPa)	70 (483)	106 (731)	103 (710)
U.T.S., Ksi	145 (1000)	163 (1124)	164 (1131)
%El	47	27	22
%RA	41	33	37
650°C (1200°F)			
0.2% Y.S., Ksi (MPa)	76 (524)	98 (676)	96 (662)
U.T.S., Ksi	142 (979)	159 (1097)	153 (1055)
%El	43	17	18
%RA	36	19	26

* Standard heat treatment.

Each data being averages of two test results.

Table 3. *Thermal stability of 242 alloy*

Material Condition	Charpy Impact Toughness	
	ft-lb	J
Annealed	99.0	134
Heat Treated*	66.0	90
650°C/1000 h	41.0	56
650°C/4000 h	31.3	42
650°C/8000 h	26.0	35
650°V/16 000 h	26.5	36

* 650°C (1200°F)/24 h.

in.) sheet was subjected to 0, 20 and 40% cold reduction in the laboratory followed by aging at 650°C (1200°F) for 24 h. Subsequently, hardness measurements were taken. Room-temperature tensile tests were performed on transverse samples for each condition. Results are shown in Table 4. Without any prior cold work, standard aging treatment caused the 0.2% yield strength to double, the hardness to increase from Rc23 to Rc38, and the tensile ductility to drop from 46 to 32%.

In contrast, the hardening response (i.e. change in strength and hardness) for material in the cold-worked condition appears to be considerably less. For example, when aged, cold-worked material showed an increase of only 235–260 MPa (34 to 38 Ksi) in the yield strength, compared to the yield strength in the as cold worked condition. The slight increase in the tensile elongation data suggests some recovery of the cold work concurrent with the aging. An examination of the 40% cold-worked and aged material by electron microscopy revealed ample presence of residual substructure together with the ordered domains [5].

Table 4. *Room temperature tensile and hardness properties for 242 alloy sheet product as influenced by cold work**

Condition	Ultimate Tensile Strength		0.2% Yield Strength		Elongation (%)	Hardness (Rockwell C)
	Ksi	MPa	Ksi	MPA		
M.A. + 0% C.W.	137.6	950	65.3	450	46.7	23.3
M.A. + 20% C.W.	169.6	1170	138.5	955	19.6	39.0
M.A. + 40% C.W.	217.6	1500	181.3	1250	7.5	44.2
M.A. + 0% C.W. + Age	192.0	1325	130.4	900	32.1	38.0
M.A. + 20% C.W. + Age	209.5	1445	173.0	1195	20.8	46.4
M.A. + 40% C.W. + Age	244.7	1685	219.7	1515	10.9	50.1

* Average of 2 or more tests per condition.

M.A. = Mill Anneal, C.W. = Cold Work, Age = 1200°F (650°C)/24 h/AC.

In order to obtain some base-line comparison with the traditional fastener alloys, similar cold rolling and direct aging were performed for alloy 718. Tensile results for cold worked and aged material are shown in Fig. 2, together with those for 242 alloy and results of Raring *et al.* [6] for A-286. The yield strength and ultimate tensile strength for 242 alloy are comparable to those for alloy 718 and significantly better than A-286. Of note is the fact that tensile elongation for 242 alloy is significantly better than that for either alloy. This may have positive ramifications for fatigue resistance, crack growth rate and critical flaw size considerations. The elevated-temperature tensile properties of 40% cold-reduced and direct aged 242 alloy and alloy 718 sheet materials are given in Table 5. The results show that alloy 718 is somewhat stronger than 242 alloy; however the strength advantage is accompanied by considerable loss of ductility.

For creep rupture experiments, mill-annealed 242 alloy bar of about 25 mm (0.98 in.) diameter was reduced in diameter to 19.8 mm (0.78 in.) by cold working on a pilger mill. This amounted to about 37% RA. The hardness of the material in the cold worked, and cold-worked plus direct aged conditions were Rc 40.6 and Rc 48.5, respectively. Optical micrographs of the material in two conditions are shown in Fig. 3 (p.290). The effect of cold work is easily observed in Fig. 3(a). Aging results in some precipitation of secondary phases and

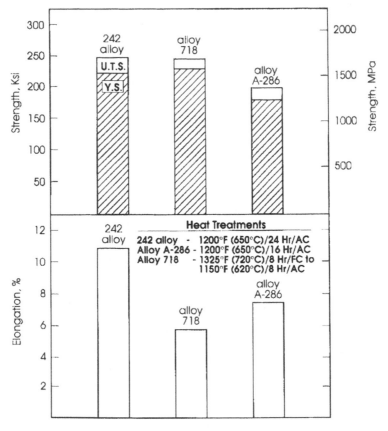

Fig. 2 *Room temperature tensile properties of 40% cold-reduced and direct-aged sheet.*

Table 5. *Elevated temperature tensile properties for 40% cold-reduced and direct-aged sheet of 242 alloy and alloy 718* [1][2]

Test Temperature		Material	Ultimate Tensile Strength		0.2% Yield Strength		Elongation
°F	°C		Ksi	Mpa	Ksi	MPa	%
70	20	242 Alloy	245	1690	220	1515	10.9
		Alloy 718	243	1675	229	1580	5.6
1000	540	242 Alloy	208	1435	180	1240	13.7
		Alloy 718*	214	1475	200	1380	4.1
1100	595	242 Alloy	203	1400	175	1205	11.0
		Alloy 718	208	1435	194	1340	4.1
1200	650	242 Alloy	195	1345	160	1105	8.0
		Alloy 718	203	1400	183	1260	4.4
1300	705	242 Alloy	184	1270	134	925	11.5
		Alloy 718	190	1310	161	1110	14.1

* Single test results.

[1] Two or more tests per condition.

[2] 242 alloy aged 1200°F (650°C)/24 h/AC.

possibly some recovery. The LRO-microdomains are not observable by optical microscopy. For an initial definition of the material, tensile properties of the cold-worked and cold-worked plus direct-aged materials are given in Table 6. The results are very much in line with the results shown in Table 5 for 40% cold reduced 242 alloy sheet material. From creep-rupture tests at various temperatures and employing Larson–Miller extrapolations, 100 and 1000 h rupture strengths were calculated. The approximate rupture strengths for the cold worked + direct-aged 242 alloy, along with those for the standard heat-treated materials are presented in Table 7. The data for standard heat-treated material were obtained from previous investigations [2, 7]. The 1000 h data are plotted in Fig. 4. It is clearly seen that over the temperature range studied, cold working significantly increased the rupture strength of 242 alloy.

4. Environmental Considerations

In addition to the high strength and thermal stability, 242 alloy possesses excellent oxidation resistance. The oxidation resistance for the alloy has been reported by Srivastava [7] and Barnes and Srivastava [8]. Based on this data, it is believed the alloy does not require any protective coating for service up to 815°C (1500°F). Frequently, fasteners are reported to fail due to stress corrosion cracking and hydrogen embrittlement [9]. These two parameters are critical performance determinants for an alloy. Stress corrosion tests were carried out using a modified ASTM G-36 procedure. Tests were run using standard U-bend specimens stressed and immersed in 154°C (310°F) boiling 45% $MgCl_2$. Results of triplicate tests, for

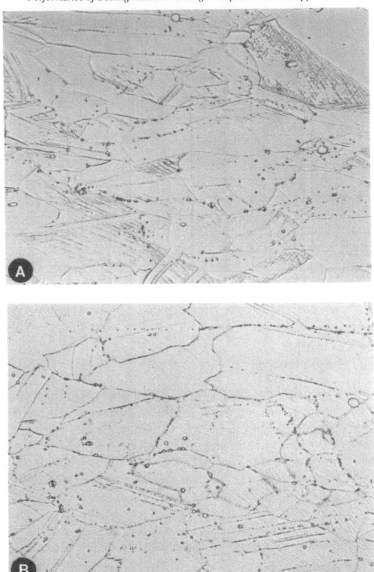

Fig. 3 *Optical micrograph of 242 alloy; A: 37% cold worked; B: 37% cold worked plus aged 650°C/24 h.*

242 alloy subjected to standard heat-treatment, showed no cracking after 1008 h of exposure. Similar results were also obtained when welded plus aged specimens were tested. However tests performed on alloy 718, using less severe C-shaped specimens, showed cracking in just 24 h. Also, immersion corrosion tests were performed in boiling seawater adjusted to a pH of 2 for 242 alloy, aged at 650°C (1200°F) for 100 h, and 30% cold worked followed by the similar aging treatment. Duplicate tests were run in both the condition for 72 h; no discernible corrosion was observed in either case.

Table 6. Tensile properties of a 37% cold-worked bar

As Cold-Worked				
Test Temp. (°C(°F))	0.2% Y.S. (Ksi (MPa))	UTS (Ksi(MPa))	El/ (%)	R.A. (%)
RT 195 (1345)	200 (1379)	8.6	41.2	
Cold Worked + 650°C/24 h				
RT 204 (1407)	239 (1648)	11.6	32.0	
430 (800)	173 (1193)	207 (1428)	11.1	29.6
540 (1000)	171 (1179)	204 (1407)	11.0	28.7
650 (1200)	161 (1110)	193 (1331)	11.0	30.0
705 (1300)	152 (1048)	182 (1255)	13.8	37.3
760 (1400)	124 (855)	158 (1090)	26.2	60.0

Each data point is the average of two results.

Table 7. Comparison of stress rupture strengths.
Standard heat treated material* vs 37% cold-worked plus direct-aged material**

	Approximate Stress to Rupture in Time Shown							
	Heat Treated				Cold Worked + Aged			
	100 h		1000 h		100 h		1000 h	
Temp. °C (°F)	Ksi	MPa	Ksi	MPa	Ksi	MPa	Ksi	MPa
540 (1000)	140	965	120	825	–	–	184	1269
595 (1100)	110	760	93	640	160	1103	130	897
650 (120)	90	620	75	515	120	828	90	621
705 (1300)	69	475	35	240	78	538	48	331
760 (1400)	29	220	17	115	43	297	27	186
815 (1500)	16	110	11	76	–	–	–	–

*Annealed + Aged 650°C/24 h.

** Aged 650°C/24 h.

In order to determine susceptibility of 242 alloy to hydrogen embrittlement, standard notched room-temperature tensile tests were performed in pressurized hydrogen and air. The results show that 242 alloy is superior to alloy 718, alloy X-750 and MP35N in resisting hydrogen embrittlement. Notched tensile strength ratios for these alloys were reported by Franklin [10] and Harris and Van Wanderham [11] and are compared to 242 alloy in Table 8.

5. Summary and Conclusions

The paper briefly reviewed strength and stability of HAYNES 242 alloy. The alloy achieves its very high strength when aged at 650°C (1200°F)/24 h. The strengthening is achieved

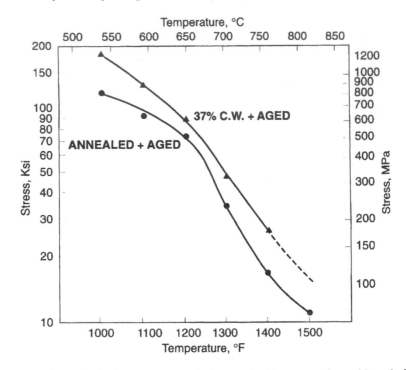

Fig. 4 *Comparison of 1000 h rupture strengths for standard heat treated vs cold-worked plus direct-aged material.*

Table 8. *Notched tensile strength ratios for pressurised hydrogen vs air for various alloys*

Material	Hydrogen Pressure		Kt	Ratio of Notched Tensile Strength Hydrogen/Air
	Psi	Mpa		
242 alloy	5000	34	8.0	0.74
Alloy 718	5000	34	8.0	0.53
MP35N	10 000	69	6.3	0.50
Alloy 718	10 000	69	8.0	0.46
Alloy X-750	7000	48	6.3	0.26

through the formation of long-range order microdomains, *ca.* 10 nm in size, of a $Ni_2(Mo,Cr)$ phase. This strengthening is achieved with only a moderate loss of ductility. In prolonged aging at 650°C (1200°F) up to 16 000 h, the strength appeared to be stable after an initial increase. No loss of strength was observed due of overaging, in contrast to the traditional strengthened fastener alloys. In addition, the strength after prolonged exposure was accompanied by excellent residual ductility and toughness.

It was found that 242 alloy subjected to cold work and direct aging, is capable of developing strength properties comparable with those of cold-worked plus direct-aged alloy 718, and superior to those of cold-worked and direct-aged A-286. The 242 alloy particularly excels in ductility; its tensile ductility was double that of alloy 718 and 1.5 times that of A-286. Furthermore, a substantial increase in rupture lives was noted as the result of cold work plus direct aging. This is important since it indicates that in longtime testing, the effect of cold work was not altogether mitigated by recovery.

The alloy was investigated for critical environmental characteristics. Results of both stress corrosion cracking in boiling 45% $MgCl_2$ and hydrogen embrittlement resistance indicate superior performance relative to alloy 718 and some other fastener materials.

To date, only limited data are available on cold-worked plus direct-aged 242 alloy. However, based on these, in conjunction with other characteristics such as lesser susceptibility to thermal fatigue due to its low-thermal expansion properties, it is believed 242 alloy fasteners should find increasing success in high-temperature applications.

References

1. M. F. Rothman and S. K. Srivastava, 'A New Long-Range-Order-Strengthened Superalloy', World Aerospace Profile, Sterling Publications. London, pp.121–124 (1989).

2. S. K. Srivastava and M. F. Rothman, 'An Advanced Ni–Mo–Cr Alloy for Gas Turbines', in *High Temperature Materials for Power Engineering*, Conf. Proc., Kluwer, Dordedt, pp. 1357–1366 (1990).

3. S. K. Srivastava, 'A Low-Thermal Expansion, High-Strength Ni–Mo–Cr Alloy for Gas Turbines', in *Superalloys '92*. TMS, Warrendale, PA, 1992, pp. 227–236.

4. H. M. Tawancy, 'Order Strengthening in a Nickel-base Superalloy (HASTELLOY alloy S)', *Met. Trans.*, 1980, **11A**, 1764–1765.

5. M. F. Rothman and S. K. Srivastava, 'Effect of Cold Work and Aging Upon the Properties of a Ni–Mo–Cr Fastener Alloy'. *IGTI Conf.*, Paper 91-GT-14, ASME, New York.

6. R. H. Raring, J. W. Freeman, J. W. Schultz and H. R. Voorhees, 'Progress Report of the NASA Special Committee on Mat. Res. for Supersonic transports', NASA Technical Note D-1798, Washington, D.C., 1963, pp. 131–137.

7. S. K. Srivastava, 'Mechanical Properties, Oxidation Resistance and Their Interaction for Two Gas-Turbine Seal Ring Alloys'. *IGTI Conf.*, Paper 92-GT-341, ASME, New York.

8. S. K. Srivastava and J. J. Barnes, 'Oxidation of Gas Turbine Seal Ring Alloys and its Effect on Mechanical Properties', in *Superalloys '92*. TMS, Warrendale, PA 1992, pp. 825–834.

9. Thomas A. Roach, 'Aerospace High Performance Fasteners Resist Stress Corrosion Cracking', *Mat. Perform.*, 1984, (9), 42–45.

10. D. B. Franklin, 'Corrosion of Space Boosters and Space Satellites', *Metals Handbook*, 9th Edn, Vol. 13, ASM, Metals Park, Ohio, 1987, pp. 1101–1104.

11. J. A. Harris, Jr and M. C. Van Wanderham, 'Properties of Materials in High Pressure Hydrogen at Cryogenic, Room and Elevated Temperatures', NASA CR-124394, Marshall Space Flight Center, AL, 1973, p. VIII-2.

Influence of γ' Volume Fraction on Ordering and Rupture Properties of Alloy 80A Bolting Material

A. MARUCCO and B. NATH*

CNR-ITM, Via Induno 10, 20092 Cinisello Balsamo (Milan), Italy
*National Power PLC, Swindon, Wilts SN5 6PB, UK

ABSTRACT

Alloy 80A bolting material operating at $T < 550°C$ undergoes atomic ordering based on Ni_2Cr superlattice, resulting in lattice contraction which exceeds in magnitude very limited creep occurring in this γ'-hardened alloy and leads to higher stress on the bolt than the design intent. The magnitude of stress increase could be minimised by reducing either the degree of order or the creep strength, so as to offset some of the lattice contraction by creep strain. Both objectives were attained by reducing the γ' volume fraction from 20% to 10–15%, which also reduces rupture strength and susceptibility to embrittlement. These changes in composition do not affect ductility and Young's modulus adversely and SCC resistance in acidic environment remains similar to that of the standard alloy.

1. Introduction

Alloy 80A (Ni–20Cr–2.4Al–1.4Ti) is a high strength material with excellent mechanical properties at elevated temperatures (–800°C). The alloy has been widely employed as bolting material in steam turbines operating below 580°C for up to 130 000 h service and is also considered as a primary candidate for bolting applications in advanced steam cycle power plants. The ability to maintain mechanical properties during long service derives mainly from the presence of 20% volume fraction of the intermetallic phase γ', $Ni_3(Al,Ti)$. Although Alloy 80A bolting material has been reliable in service, ~0.4% of bolt population has failed much earlier than the expected rupture life. With the exception of some early failures by creep due to incorrect heat treatment/installation others have been attributed mainly to SCC and intergranular fast fracture [1]. The alloy, which does not creep significantly below 600°C, also undergoes atomic ordering based on Ni_2Cr superlattice below 550°C [2]. Atomic density changes on ordering and a lattice contraction occurs in the matrix phase of this γ'-hardened alloy, leading to higher stress on the bolt than the design intent [3]. Ex-service material also exhibits very low Charpy impact energy (5–17 J) and the combination of high stresses acting on a brittle material can lead to failure of the fasteners. Although grain boundary P-segregation is considered the main cause of embrittlement [1], some very low P casts of Alloy 80A became brittle on prolonged exposure at $T < 550°C$ and this is attributed to ordering [4]. SCC occurs in acidic environment by H-cracking: clearly, high stresses and embrittlement can exacerbate this phenomenon. Thus, ordering has two consequences: (i) high stresses increasing the risk of fast fracture and (ii) embrittlement.

The magnitude of stress increase due to ordering could be minimised either by reducing the creep strength of the alloy, so as to offset some of the lattice contraction by creep strain, or by reducing the degree of order. Both these objectives can be attained by decreasing the (Ti + Al) content of the alloy and hence the volume fraction of the strengthening phase from nominal 20% to 10–15%. This would be achieved by relatively minor changes in the nominal composition [5] and therefore extensive validation data may not be necessary. Moreover, these changes are unlikely to alter the coefficient of thermal expansion of the alloy, which is an important consideration for the application as bolting material. The influence of variation of γ' volume fraction on magnitude and kinetics of lattice contraction and stress rupture strength was determined in four low (Al + Ti) versions of Alloy 80A, which were also evaluated for mechanical properties of interest in steam turbine bolting material such as tensile properties, susceptibility to embrittlement and SCC resistance.

2. Materials and Methods

Four experimental melts of Alloy 80A were produced. Chemical compositions are compared to Alloy 80A (average cast) in the table. Samples were treated in standard 3-stage condition for bolting application (1080°C, 8 h, AC + 850°C, 24 h, AC + 700°C, 16 h, AC) before testing or ageing at 450 to 600°C, interrupted by water quench after different ageing times for up to 30 000 h. Lattice parameter variations were measured by X-ray diffraction and then correlated to different degrees of order [6].

Tensile tests were conducted at 20 to 700°C at constant loading rate of 3.25 MPa s^{-1} [6] and stress rupture tests (plain and notch bars) in multispecimen creep machine at 550°C and stress levels of 500 to 800 MPa [7]. The embrittlement susceptibility of alloys either 3-stage treated or aged for 1000 h at 550–600°C has been determined by monitoring Charpy impact energy [4]. The SCC resistance was evaluated by comparing properties obtained from tensile tests at constant strain rate of 1.2×10^{-6} s^{-1}, at 90°C, at rest potential, in acidic environment of 4% H_2SO_4 and control tests in silicone oil [4].

3. Experimental Results and Discussion

3.1. Ordering Behaviour

Examination of ex-service material revealed that most of failed bolts had contracted during service. This has been attributed to atomic ordering of the matrix, essentially a solid solution of Cr in Ni [2, 8, 9]. Ordering kinetics decrease on departure from stoichiometric Ni_2Cr (Ni–33 at.% Cr) composition and hence the ordering transformation of Alloy 80A matrix, containing ~26 at.% Cr, is very sluggish. Alloy 80A undergoes short-range order (SRO) at $T < 550$°C, characterised by an electrical resistivity increment of up to 5% due to scattering of electrons by SRO nuclei and a limited lattice contraction of 0.02–0.03%. Long-range order (LRO) forms after very long ageing below 525°C, with resistivity reduction and larger lattice contractions (Fig. 1). In constant-strain stress-relaxation tests the strain of Alloy 80A increases from original value of ~0.15 to ~0.26% and the stress from ~300 to > 600 MPa [3]. Although this increase in stress can be relieved by slackening and retightening bolts during

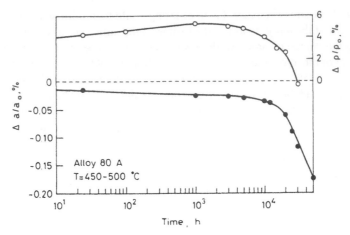

Fig. 1 *Order-induced resistivity and lattice variation in Alloy 80A upon ageing at 450–500°C.*

service, it would be preferable that bolting material could guarantee reliable performance for over 30 000 h service without any intermediate retightening. The material contraction leads to high stress on the bolt, which can be partly mitigated by creep of the ferritic flange material, but the stress on the bolt can still increase by up to 140 MPa [10]. One possible means of minimising the stress increase is to reduce the creep strength of the alloy, so as to partially offset lattice contraction by creep strain. To reduce the strength of this γ'-hardened alloy, the interparticle spacing must be increased either by coarsening γ' precipitates by modifying the heat treatment [7] or by reducing the γ' volume fraction.

Lower (Ti + Al) concentrations reduce the γ' volume fraction and increase the Ni:Cr ratio of the matrix since fewer Ni atoms are tied up as γ'. The estimated γ' volume fraction is thus ~15% in Alloys N80B and N80E and ~10% in Alloys N80C and N80D, while Cr in the matrix is ~25 at.% in Alloys N80B and N80E and ~24% in Alloys N80C and N80D. The degree of order of the alloy, which primarily depends on Cr content [11], decreases. As a consequence, the low (Al + Ti) versions of Alloy 80A show smaller lattice contractions than the standard alloy, particularly after ageing for 10 000 h or more when LRO is expected to form (Fig. 2). In particular, the maximum contraction measured after 30 000 h ageing at 450–500°C in alloys with 10% of γ' volume fraction was ~0.04%, while in alloys with 15% of γ' volume fraction was ~0.06%, i.e. well lower than the ~0.11% of the standard alloy. Ordering kinetics also decrease with reducing the γ' volume fraction, as observed from comparison of lattice contractions in Alloys N80B and N80C during ageing at different temperatures (Fig. 3).

The dimensional stability of the material aged below 525°C thus increases with decreasing the γ' volume fraction, meeting one of the objectives of developing the new alloy compositions.

3.2. Mechanical Properties

Tensile properties of modified alloys were determined and compared to Alloy 80A [12]. Both ultimate tensile strength, R_m, and 0.2% proof stress, $R_{p0.2}$ decreased with increasing

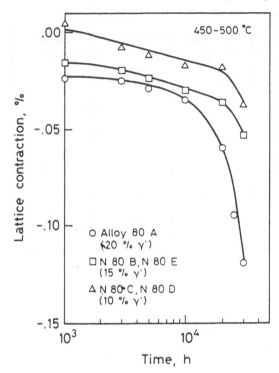

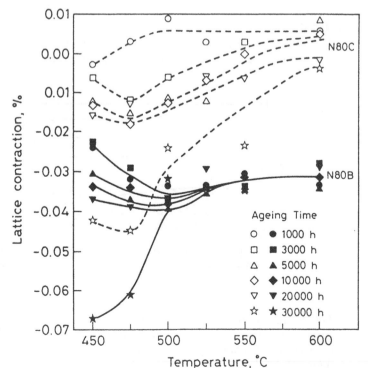

Fig. 2 *Lattice contractions upon ageing at 450–500°C in Alloy 80A and its modified versions.*

Fig. 3 *Ordering kinetics in Alloys N80B (~15% γ' vol. fraction) and N80C (~10% γ' vol. fraction).*

Table 1. *Chemical compositions of alloys (wt%)*

	Alloy 80A	Alloy 80B	Alloy 80D	Alloy 80D	Alloy 80E
C	0.08	0.11	0.09	0.12	0.09
Si	0.27	0.01	0.01	–	–
Mn	0.14	0.01	0.01	–	–
Al	1.41	1.24	1.15	0.96	1.18
Cr	19.50	19.57	19.68	19.25	19.53
Fe	0.81	0.04	0.04	0.03	0.04
Ti	2.40	2.16	1.67	1.72	2.10
Ni	75.32	76.90	77.40	77.80	77.00
P	<50 pmm	~20 ppm	<50 ppm	~25 ppm	~25 pmm
Ni_3 (Al, Ti)	17.14*	13.12	5.77	7.21	12.48

Measured in cast V_3 [4].

temperature as in Alloy 80A (Fig. 4(a)), however, at $T \leq 600°C$, strength was clearly dependent on γ' volume fraction, whereas at 700°C no dependence on γ' volume fraction was observed (Fig. 4(b)). Changes in composition did not affect Young's modulus adversely, which is an important point for bolting application, as this parameter lay within ± 10% data band of Alloy 80A (Fig. 4(c)). The ductility values of the modified alloys were also comparable with those of the standard alloy (Fig. 4(d)). SEM fractography revealed transgranular fracture mode [6].

In addition to changes in material density due to ordering, embrittlement is often associated with bolt failures. LRO contributes to embrittlement of Alloy 80A after prolonged ex-

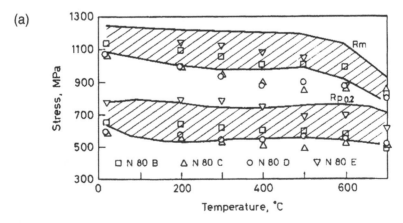

Fig. 4 *((a), above) Ultimate tensile strength and 0.2% proof stress of modified alloys vs temperature compared to Alloy 80A data band [12]; (opposite): (b) ultimate tensile strength vs (Al + Ti) concentration; (c) Young's modulus and (d) ductility of modified alloys compared to Alloy 80A data band [12].*

(b)

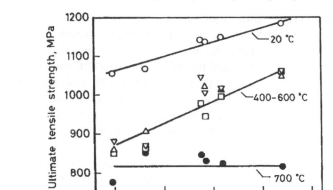

(c)

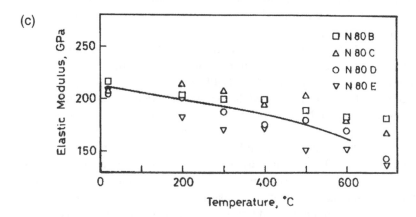

(d)

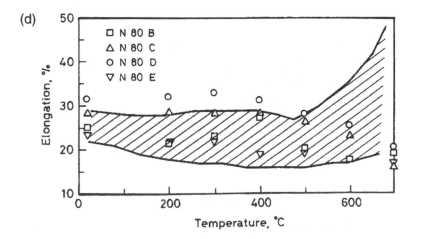

posure below 550°C, but the alloy is also susceptible to embrittlement after shorter treatments, i.e. 1000 h at 550–600°C [4]. Kinetics and magnitude of embrittlement differ possibly due to variation of P content in different casts [1]. Charpy impact energies of the modified alloys either 3-stage treated or aged for 1000 h at 550–600°C were significantly higher than in Alloy 80A (Fig. 5), leading to the conclusion that the resistance to embrittlement increases by reducing the γ′ volume fraction. SEM fractography revealed transgranular fracture mode in 3-stage treated alloys (Fig. 6(a)) changing to transgranular/ intergranular mode due to ageing (Fig. 6(b)), with a small increase in the proportion of brittle intergranular facets at 600°C [6].

A considerable number of failures in Alloy 80A bolts have been attributed to SCC. The SCC resistance of the modified alloys were evaluated by constant strain rate tensile tests in acid. Control tests were also performed in silicone oil to isolate the environmental effects. The main crack in failed bolts often did not form at position of highest stress concentration, but multiple cracks were observed on fracture surface, mainly due to S contamination [1, 4]. Ultimate tensile strength, time to rupture and reduction in area were measured and the latter was used as a measure of the SCC resistance. Results show that the SCC resistance of the alloy is not affected by reducing the γ′ volume fraction from 20% to 10–15% (Fig. 7, overleaf). Microstructural examination revealed intergranular fracture mode in samples tested in acid, with multiple intergranular cracks on the gauge length (Fig. 8, overleaf), while samples tested in oil were observed to fracture in transgranular manner [6].

Stress rupture strengths of alloys N80B-N80E were below the corresponding values of the standard alloy (Fig. 9(a), p.303), while the ductility values were similar (Fig. 9 (b)). Moreover Alloys N80B and N80E with 15% of γ′ volume fraction are stronger than Alloys N80C and N80D with only 10% of γ′ volume fraction. The alloy N80D was notch strengthening like many casts of Alloy 80A, while other alloys were notch weakening like some casts of Alloy 80A [4, 6]. The reasons why some casts of Alloy 80A are notch strengthening and others are notch weakening are not understood. SEM fractography revealed typical mixed transgranular/intergranular fracture mode; in particular, plain samples showed transgranular fracture at the outer circumference and brittle intergranular fracture in the central region of

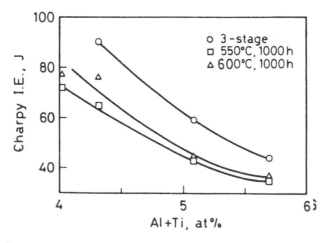

Fig. 5 *Charpy impact energy vs (Al + Ti) concentration.*

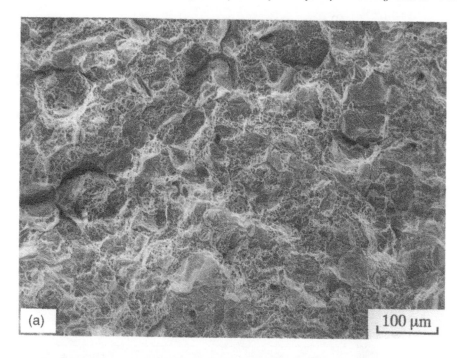

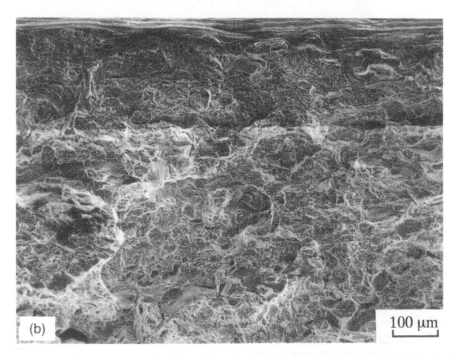

Fig. 6 *Fracture surfaces after impact test of (a) 3-stage treated N80C sample, and (b) N80B sample aged at 600°C for 1000 h.*

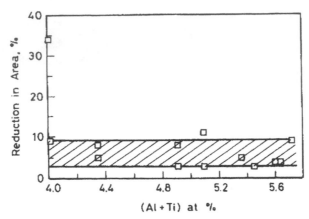

Fig. 7 *SCC resistance vs (Al + Ti) concentration.*

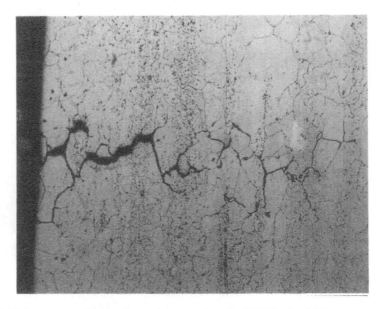

Fig. 8 *Multiple intergranular cracks on the gauge length of N80C after SCC test.*

the fracture surface (Fig.10 (a), p.304), whereas in notch samples fractures are intergranular in the outer zone and transgranular in the inner one (Fig. 10 (b)). Cracks growing perpendicularly to the stress direction were observed in notch samples. The alloy N80D did not exhibit different behaviour and hence the notch strengthening mechanism of this alloy can not be ascribed to a different fracture mode.

The second aim of the work, to decrease strength in Alloy 80A by reducing the γ' volume fraction, was thus also attained and the relationship between strength and γ' volume fraction is consistent with data in the literature [13].

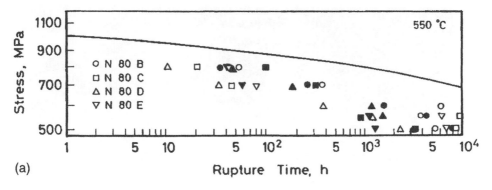

(a)

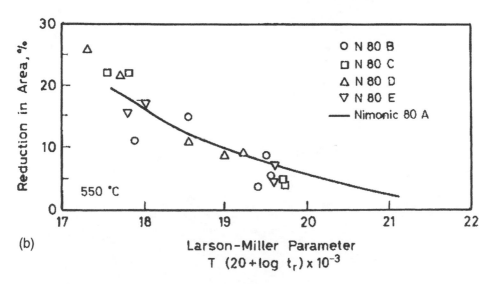

(b)

Fig. 9 *(a) Plain (open symbols) and notch (solid symbols) bar rupture strength, and (b) ductility of modified alloys as compared to Alloy 80A.*

4. Conclusions

The (Ti + Al) content of Alloy 80A has been reduced to weaken the material and to increase the Ni:Cr ratio in the matrix phase. Both these were expected to minimise the potential for the time dependent increase in stress on fasteners operating at 550°C and below. The results show that:

• Low (Ti + Al) melts of Alloy 80A exhibit smaller lattice contraction, and hence lower degree of order, particularly after ageing for over 10 000 h, when LRO is expected to form;

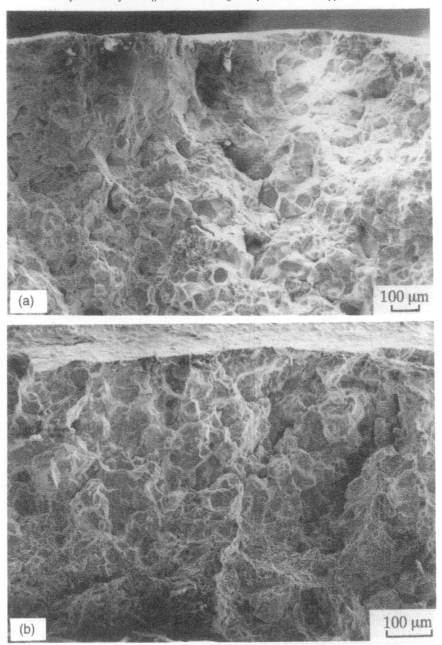

Fig. 10 *Fracture surfaces after creep rupture tests at 800 MPa at 550°C in (a) plain N80E, and (b) notch N80C samples.*

- Low (Ti + Al) melts of Alloy 80A are weaker than the standard alloy, consistently with the main rationale of this work, which intended to reduce the stress increase by weakening the alloy;

- The modified alloys exhibit higher impact energy and slower rate of embrittlement, while the SCC resistance in acidic environment remains similar to that of Alloy 80A.

5. Acknowledgements

The authors are indebted to Ing. K. Loser for stress rupture tests data and fruitful cooperation during Programme COST 505, to Dr J. D. Parker and to Mr G. Carcano and E. Signorelli for valuable help in the experimental work.

References

1. S. M. Beech, S. R. Holdsworth, H. G. Mellor, D. A. Miller and B. Nath, in *Proc. Int. Conf. on Advances in Materials Technology for Fossil Power Plants*, Chicago, ASM/EPRI, 1987, 403–410.
2. E. Metcalfe and B. Nath, in *Proc.Conf. on Phase Transformations*, York, UK, Inst. of Metals Conf. Series. 3, 2, Vol. II, 1979, 50–53.
3. E. Metcalfe, B. Nath and A. Wickens, *Mat. Sci. Engng*, 1984, **67**, 157–162.
4. B. Nath, COST 505-Project UK21, Final Report, 1990.
5. T. B. Gibbons and B. E. Hopkins, *Met. Sci. J.*, 1971, **5**, 233–243.
6. A. Marucco, COST 505-Project I8 Final Report, 1989.
7. K. Löser and J. Granacher, COST 505-Project D11 Final Report, 1989.
8. A. Marucco and B. Nath, in *Proc. 5th Int. Conf. on Mechanical Behaviour of Materials, Beijing, China* (Eds M. G. Yan *et al.*), Pergamon Press, Oxford, UK, Vol. 2, 1987, 949–956.
9. A. Marucco and B. Nath, *J. Mat. Sci.*, 1988, **23**, 2107–2114.
10. K. H. Mayer and H. König, in *Proc. Int. Conf. on Advances in Materials Technology for Fossil Power Plants*, ASM/EPRI, 1987, 395–402.
11. A. Marucco and B. Nath, in *Proc. Int. Conf. on Phase Transformations '87*, Cambridge, UK, (Ed. G. W. Lorimer), The Metals Soc., London, 1988, 588–591.
12. Wiggin and Co. Ltd. Publication 3663, 1975.
13. R. F. Decker, in *Proc. Symp. Steel Strenghtening Mechanisms*, Climax Molibdenum Co, Zurich, CH, 1969, 1–24.

24

Recent Developments in Alloy 80A for High Temperature Bolting Applications

B. NATH*, K.H. MAYER†, S. BEECH‡ and R. VANSTONE§

*National Power PLC, Swindon, UK
†MAN Energie GmbH, Frankenstraße 150, D-8500 Nürnberg 44, Germany
‡IRD, Fossway, Newcastle-upon-Tyne, UK
§GEC-ALSTHOM Large Steam Turbines, Rugby, UK

ABSTRACT

European experience with Alloy 80A bolts in turbine components is very good and failures are relatively infrequent. However, increased integrity of bolted joints is required for greater safety, for longer overhaul periods of turbines, and for use at up to 600°C. Most of the fasteners which have failed have done so either by stress corrosion cracking (SCC) or by inter-granular fast fracture. Therefore, European collaborative research projects, COST 505 and COST 501-Round II, were aimed at improving the SCC resistance and the toughness of the alloy without undue adverse effects on relaxation and rupture properties at up to 600°C. This has been attained by modifying the composition and the heat treatment of the material. This paper reviews the results of the programmes.

The main conclusions are as follows:

• Alloy 80A is suitable for bolting at up to 600°C. However, engineering design must accommodate a mismatch in the coefficient of thermal expansion of Alloy 80A bolts and that of the flange made in 9–12Cr steels.

• Significant improvements in the toughness of the material and in its resistance to SCC are attainable by limiting phosphorus content to below 20 ppm.

• A modified heat treatment is highly effective in improving the toughness and the resistance to SCC of Alloy 80A, without undue reduction in relaxation and rupture strengths.

1. Introduction

At more advanced steam temperatures of 600°C, low alloy ferritic steels are unsuitable for use either as the flange or as the bolting material because of their inadequate creep properties. Higher strength 9–12Cr steel flanges are required for such applications. Either a matching 9–12Cr steel or Alloy 80A might be suitable for bolting at these temperatures except that the advanced martensitic steels are not well characterised for this purpose and Alloy 80A, which has sufficient relaxation strength, may be deficient in other respects. For example, some casts of the material become very brittle at these temperatures; the material is susceptible to SCC; and its coefficient of thermal expansion is significantly larger than those of the 9–12Cr steels.

In a European collaborative programme, COST 501 Round II Work Package 3, high strength 9–12Cr martensitic steels and Alloy 80A have been evaluated as potential bolting materials for use at 600°C. The stress relaxation, stress-rupture, notch sensitivity, embrittlement and SCC behaviours of a controlled purity cast of Alloy 80A are summarised in this paper.

2. Material

Based on the embrittlement data on 16 commercial casts [1], a detailed specification has been developed for Alloy 80A. The feasibility of producing the controlled purity composition has been established. A 100 mm dia. bar was fabricated by Firth Rixson Plc, Glossop Superalloys (UK), using a VIM/ESR route. The chemical composition of the cast No. E5662 is shown in Table 1.

One half of the material (labelled V_{12}) had been heat treated through the standard 3-stages, viz.,

1080°C, 8 h, air cool to room temperature;
850°C, 24 h, air cool to room temperature;
700°C, 16 h, air cool to room temperature.

The other half of the stock was in an 'as-forged' state. Some of the 'as-forged' material (labelled V_{12N}) was treated as follows:

1080°C, 8 h, furnace cool at 2°C min^{-1} to 850°C;
hold at 850°C, 24 h, air cool to room temperature;
700°C, 16 h, air cool to room temperature.

Table 1. Specification for the controlled purity material

Parameter	Specification	Cast E5662
C	0.06 ± 0.02	0.072–0.073
Cr	19.5 ± 0.5	19.6
Al	1.4 ± 0.15	1.35–1.38
Ti	2.4 ± 0.15	2.5–2.54
Mn	0.02–0.1	0.06
Si	<0.1	<0.01
Fe	<0.2	0.06–0.09
B	20–30 ppm	20–30 ppm
P	<20 ppm	<20 ppm
S	<30 ppm	5 ppm
Grain Size	ASTM 4 max.	ASSTM 5–7
Room Temperature Charpy Impact Energy After 3-stage Treatment	>40 J	45.5 J

The intention of the modified treatment was to coarsen precipitates which, in a different cast, V_{11}, led to a marked increase in the resistance to SCC [2]. The grain size after the treatment was comparable with that of the 3-stage material.

3.1. Uniaxial Stress Relaxation

Stresses necessary to maintain a constant strain of 0.15% in uniaxial samples were monitored at 550 and 600°C. At 600°C, the initial stress in the 3-stage material, V_{12}, relaxed from 283 MPa to 250 MPa in 10 000 h (Fig. 1). The corresponding relaxation at 550°C was from 259 MPa to under 219 MPa (Fig. 2). (The difference in the initial stress is considered to arise from different loading rates used by two laboratories.) The magnitude of stress relaxation is comparable with published values [3].

The relaxation was higher after the modified treatment. For example, after around 5000 h the stress decreased from 291 MPa to 232 MPa at 600°C (Fig. 1) and from 243 MPa to 205 MPa at 550°C (Fig. 2). The larger magnitude of stress relaxation, compared with the 3-stage treatment, was due to enhanced creep after the modified treatment. The decrease in strength was the intended consequence of coarsening precipitates.

3.2. Stress Relaxation in Model Bolted Assemblies

The relaxation of both the fastener and the flange contribute to the stress relaxation of a bolted joint. In dissimilar metal joints (e.g. Alloy 80A fastener and 9–12Cr steel flange), the differential thermal expansion also plays a significant role in the behaviour of a joint [4]. Thus the relaxation strength of the joint can differ from that of the bolt material measured in uniaxial stress relaxation tests.

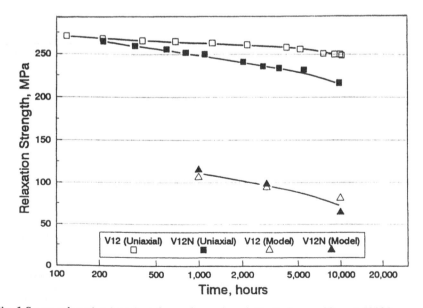

Fig. 1 *Stress relaxation in uniaxial samples and model bolted assemblies at 600°C.*

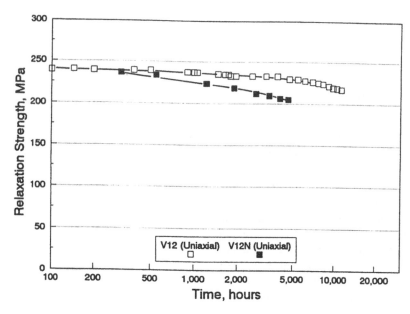

Fig. 2 *Uniaxial stress relaxation at 550°C.*

Therefore, models of a bolted assembly were used to monitor stress relaxation of joints at 540, 570 and 600°C. Alloy 80A bolts were tightened at room temperature to an initial strain of 0.2 or 0.25% using Alloy 80A nuts and the modified 9CrMo flange (Fig. 3). At test temperatures, these would correspond approximately to 0.1 and 0.15% initial strain respectively — typical of tightening strains used on fasteners.

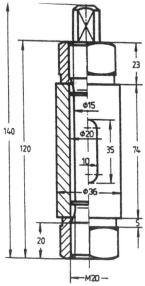

Fig. 3 *Model bolted assembly.*

, the initial stress decreased from 180 MPa to 81 MPa in 10
gnitude of stress relaxation is much greater in bolted assem-
 and 4) because of the plastic deformation of the thread and
 creep of the flange. Relaxation was more limited as the test
d 540°C because of deformation.

000 h relaxation at different temperatures are compared *in*
gram includes the relaxation behaviour of a high strength
lange was used for the TAF bolts which were tightened to
e [5]. This prestrain did not change on heating to the test
nperature prestrain on an Alloy 80A bolt decreased to ap-
ause of differential expansion. The differential thermal ex-
 into account in the design of Alloy 80A fasteners on high
ified 9CrMo).

ure prestrain from 0.2 to 0.25% is one way of accommodat-
 would correspond to 0.1 and 0.15% strain respectively, at
sses were higher in bolts tightened to the higher strain level
tion strengths at all three temperatures (Fig. 4). Enhanced
d to greater relaxation at 600°C than at lower temperatures

of Alloy 80A offers a significant advantage over the ad-
F steel) (Fig. 4). However, advanced 9–12Cr steels may be

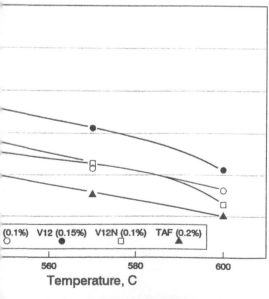

(0.1%)   V12 (0.15%)   V12N (0.1%)   TAF (0.2%)
 ○   ●   □   ▲

560   580   600

Temperature, C

n strength after 10 000 h at 540–600°C.

potential alternatives to Alloy 80A fasteners for bolting high alloy steel flanges at temperatures below 550°C. It should be noted that the current practice is to use either low alloy steel or Alloy 80A fasteners at these temperatures and the flange material is usually a low alloy steel.

3.2.2. Modified Treatment Material V_{12N}

The modified heat treatment had a marginal effect on the stress relaxation of bolted assemblies at 600°C but not at lower temperatures (Figs. 1 and 4).

3.3. Embrittlement Tests

The embrittlement of V_{12} and V_{12N} has been monitored as a function of ageing times of up to 10 000 h at temperatures of 450–650°C.

3.3.1. 3-Stage Material, V_{12}

The Charpy impact energy (C_v) of the 3-stage treated material was 44.5 J which is high within the databand [6] for Alloy 80A. On ageing it decreased gradually (Fig. 5) but remained at the top of the databand even after prolonged ageing to up to 10 000 h (Fig. 6). The observed embrittlement behaviour is comparable to the cast V_3, typical of a 'slow kinetics' material [2, 6]. The minimum C_v of the aged V_{12} was 26 J after 10 000 h at 650°C.

Scanning electron microscopy revealed narrow regions of trans-granular fracture, just below the notch root, the proportion decreasing with increasing embrittlement. The remaining fracture surface was predominantly inter-granular with dimples on the grain boundary

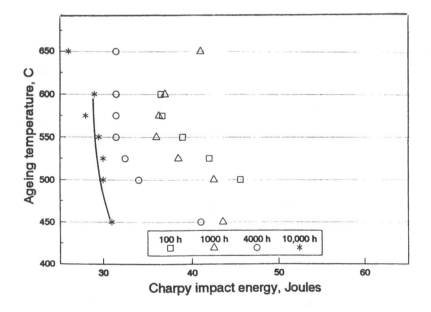

Fig. 5 *Embrittlement kinetics in 3-stage treated cast V_{12}.*

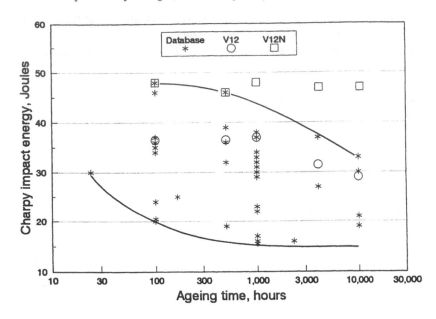

Fig. 6 *The impact energy of the controlled purity material lies at the top of the databand for Alloy 80A at 600°C.*

facets. Previous studies have shown that grain boundary phosphorus segregation make Alloy 80A brittle on ageing at 450–650°C [2, 6].

Thus, the improved material with low phosphorus concentration and fine grain size has met one of the objectives of the research programme — C_v in excess of 20 J after prolonged ageing.

3.3.2. Modified Treatment Material V_{12N}

The impact energy of V_{12N} decreased from approximately 55 to around 44 J within 500 h ageing 600°C (Figs 6 and 7). It did not decrease on further ageing for up to 10 000 h at 480–650°C. The results demonstrate the significant beneficial effect of the modified heat treatment on the embrittlement behaviour of Alloy 80A. It is believed that the coarsening of precipitates is responsible for the improvement.

3.4. SCC Tests : 3-Stage Material, V_{12}

Tensile tests have been conducted under freely corroding conditions in 4v/o H_2SO_4 at a constant strain rate of 1.2×10^{-6} s^{-1} and a temperature of 90°C to evaluate stress corrosion cracking [6]. Control tests were in an 'inert' environment of silicone oil and at the same strain rate and temperature. The % reduction in area in acid is used as a measure of the resistance to SCC. The data show that the ductility of V_{12} in acid is relatively high in comparison with commercial grades of the material, thus showing greater resistance to SCC (Fig. 8). The fracture in acid was predominantly inter-granular although initiation may have occurred at the surface in a trans-granular mode.

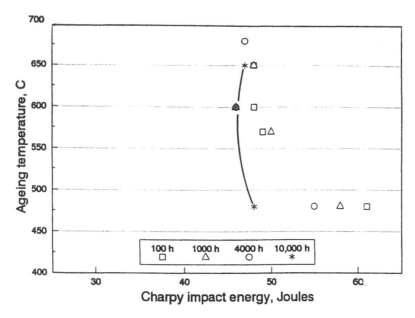

***Fig. 7** The magnitude of embrittlement is significantly lower after the modified heat treatment.*

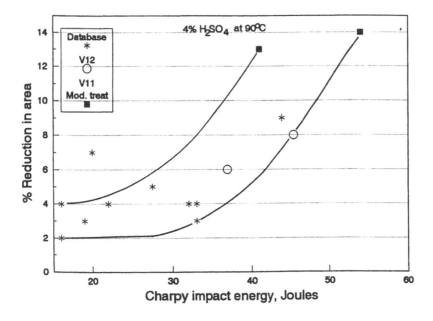

***Fig. 8** SCC resistance of the controlled purity material is high in comparison with the database for Alloy 80A.*

impact energy and the resistance to SCC (Fig. 8) suggests
ve a high resistance to SCC. Indeed, data on another cast of
resistance to SCC after identical modified heat treatments

3.5. Stress Rupture Tests

out on plain bar and notch bar samples at 550 and 600°C.
ed heat treatment materials have been tested.
-stage material at both 600 and 550°C lay at the top of the
9–11). The material exhibited a notch strengthening behav-
, notch weakening was observed in short term tests at 550°C
at this may change to notch strengthening in the long term
d life in excess of 15 000 h) (Fig. 10). The reasons for notch
ng are not understood although Alloy 80A is known to ex-

ed marginally after the modified treatment, as expected, but
(Figs. 9–11). Further, the material was notch strengthening

COST 501-II WP3 project was to evaluate candidate bolting
ious work in COST 505 and other programmes had identi-
roving the behaviour of Alloy 80A further. These were (i) a
m and (ii) a modified heat treatment to coarsen precipitates.
ed for the 600°C application.

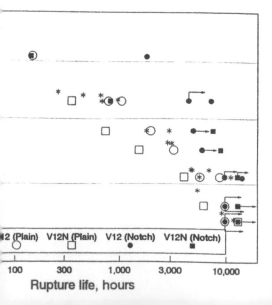

Rupture life, hours

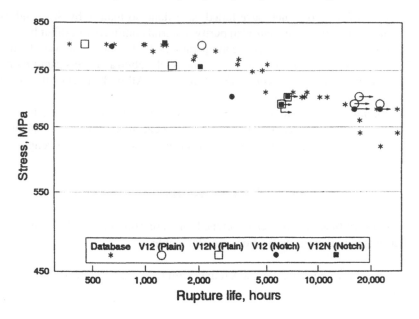

Fig. 10 Stress rupture strength at 550°C.

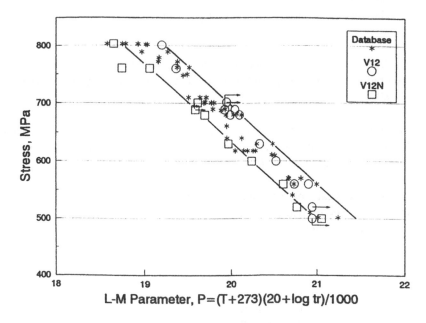

Fig. 11 The rupture strength of the 3-stage treated material lies at the top of the databand at 550–600°C.

Few Alloy 80A fasteners which have failed have done so mainly by SCC and by fast fracture in a brittle material. The controlled purity material continues to exhibit high impact energy after prolonged exposure to service temperatures. It also has a relatively high resistance to SCC. Other properties of the material, e.g. stress relaxation and stress-rupture strengths compare favourably with data on conventional grades of Alloy 80A. Thus the controlled purity material offers an advantage.

The embrittlement behaviour of the material improves, and the SCC resistance is expected to increase very significantly after a modified heat treatment. It is considered that these outweigh a relatively small decrease in relaxation and stress-rupture strengths.

5. Conclusions

1. A detailed specification has been developed for a controlled purity Alloy 80A. Materials fabricated to these specifications are obtainable.

2. The relaxation strength of Alloy 80A remains high at 600°C and is not affected by the improved purity of the material.

3. A comparison of residual stresses in different bolted assemblies shows that Alloy 80A is superior to advanced 9–12Cr steels, for application at 600°C but relatively high initial tightening strains may have to be used to maintain leaktight joints.

4. The controlled purity material exhibits a 'slow kinetics' of embrittlement and retains relatively high impact energy through prolonged exposure at service temperatures.

5. The SCC resistance of the controlled purity material is similar to that of the conventional grade casts with comparable compositions and lies at the top of the databand for Alloy 80A.

6. The rupture strength of the controlled purity material is high within the databand for Alloy 80A. In general, the material is notch strengthening except for tests of less than say, 15 000 h duration at 550°C.

7. A modified heat treatment markedly improves the resistance of Alloy 80A to embrittlement and probably to SCC, without significant adverse effects on the relaxation and stress rupture strengths.

References

1. B. C. Edwards, D. A. Miller and B. Nath, 'The long-term embrittlement characteristics of Alloy 80A', reported by Beech *et al.*, in *Proc. Conf. on Advances in Materials Technology for Fossil Power Plants*, Chicago, 1987, ASM/EPRI, 403–410.

2. K. Löser, J. Granacher and K. H. Kloos, 'Behaviour of selected heat resistant nickel-base alloys for steam turbine bolting, Final report on Project COST 505-D11', Report No. BMFT 03 K0507 7, Institut für Wekstoffkunde, Darmstadt, 1989.

3. E. Metcalfe, B. Nath and A. Wickens, *Mat. Sci. Eng.*, 1984, **67**, 157–162.

4. R. H. Mayer and H. Koenig, 'High temperature bolting of steam turbines for improved coal-fired power plants. Research Project 1403-15', EPRI, 1991.

5. K. H. Mayer and H. König, 'Operational characteristics of 10–12%CrMoV bolt steels for steam turbines', this volume, pp.150–162.

6. B. Nath, 'An assessment of nickel-based alloys for high temperature bolting applications', 1991, Commission of the European Communities Publ. EUR 13802.

25

High Strength, High Temperature Bolting Alloys for Turbine Applications

J. NEWNHAM and S. BUZOLITS*

SPS Technologies Ltd, Leicester, UK
*SPS Technologies Inc, Jenkintown, PA, USA

ABSTRACT

A sequence of tests is described which is normally used to determine whether an existing alloy will have capability as a bolting material. The range of commonly used high temperature bolt alloys is presented in charts which allow their main properties to be differentiated. The most frequently used bolt materials for aircraft gas turbine applications are reviewed. The Multiphase alloy MP159 is described, this alloy has the highest strength and creep resistance of any fastener material up to about 600°C. A new material, Aerex 350, has been developed which was based upon MP159 but with adjusted chemistry to give the possibility of application temperatures up to 760°C. Some of the properties of this material are presented.

1. Introduction

The main requirements for fasteners used in high temperature applications are the ability to survive for the required lifetime under the prevailing conditions of stress and temperature, and a compatibility with the joint materials and the application environment. Most of the materials used for bolting are derived from alloys which have been originally developed for other applications. For example, at the higher range of temperatures, materials which have been developed for turbine disc or blade applications are commonly used for fasteners.
There are major benefits in finding materials for bolting which are already used in other applications. These are price, availability, and a broad knowledge of alloy behaviour and properties. However, not all alloys with properties suitable for other applications will be capable of being used for high temperature bolts and nuts. Once having identified an alloy with high temperature properties which are of interest, there is a sequence of tests which is usually followed to determine whether the alloy is suitable as a bolt material.

2. Evaluation of Suitability as a Bolting Material

The steps in the evaluation procedure usually followed by bolt manufacturers are shown in Fig. 1. Initial testing is on standard specimens as defined for example by ASTM E8, for room temperature tensile, or ASTM E21 for elevated temperature tensile strength. These are machined test pieces with a waisted central section. Factors such as the relatively smooth variation of strength with temperature, Yield/Ultimate ratio, and the presence of any ductility troughs would determine whether to discontinue the evaluation.

A. SPECIMEN TESTING

B. THREADED STUD TESTING

C. BOLT TESTING

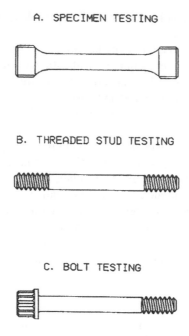

Fig. 1 *Progression of test samples used for the evaluation of bolting materials.*

If tensile testing is satisfactory, a notched tensile test usually follows, to ASTM E602. This is to simulate a screw thread on the part, and to look for notch sensitivity, which usually precludes the use of the material for bolts. Other properties such as creep strength, shear strength and some physical properties may also be evaluated on test specimens.

The next step is to manufacture threaded test samples as shown in Fig. 1(b). The threads would normally be rolled, and often rolled after final heat treatment if fatigue is likely to be an issue. For most high temperature materials, this means that thread rolling procedures have to be developed. The main objectives of these tests is to determine whether the 'notches' provided by the threads have a substantial detrimental effect on the properties already known from the specimen testing, and whether the thread rolling procedures have degraded the properties.

The third stage is to manufacture a headed sample, in other words a bolt. The objective here is to determine whether there are major problems in forging the head, either by cold or hot working, whether substantial metallurgical damage is done by forging the head, and whether the properties are maintained in the headed sample.

Alloys which survive such preliminary testing are then put through a comprehensive programme of mechanical tests. Obviously, not all alloys which have good performance in other turbine applications will be capable of being used as bolts. Some are too notch sensitive and some cannot be headed without degradation. For some materials, the process technique may not lend itself to bolt manufacturing, for example isothermal forging. So there is a limited number of bolting materials available for the more demanding applications involving high temperature and applied stress.

3. Alloys used in Aircraft Engine Applications

Commonly used bolting alloys in aircraft engines are classified according to maximum application temperature and room temperature tensile strength in Table 1. This also shows some airframe bolting materials such as titanium alloys, which are not commonly used in engine applications. For H11, Aermet 100 and alloy steels, it is assumed that these metals would be coated in order to prevent corrosion, and the limiting use temperature is that of the coating, not the alloy.

This table shows many alloys which were originally developed for other applications. A286 and Alloy 718 were developed as turbine disc materials, while the original applications for Alloy 80A and Waspaloy were as turbine blade materials. In the former application, tensile strength and low cycle fatigue are important, while for blade materials, creep resistance is required [1]. The maximum use temperature may vary according to the precise application but the table gives a good general picture of some of the alloys available and their capabilities. Room temperature strength does not really give information at the application temperature, but the relative strengths are often maintained at the maximum service temperature.

There have been some significant trends in gas turbine bolting materials over the last several years. Some materials are now less frequently used than they were in the past, as more suitable alloys have become available. For example Alloy 80A is not frequently specified as a bolting material in new applications. Its replacement is usually Alloy 718, which has a higher strength at both room and application temperatures. Jethete M152 has also been displaced because although it is relatively low cost, its maximum operating temperature is quite low, and it suffers from stress corrosion in some environments. FV 535 is a replacement in some applications. Another commonly used material is A286 for temperatures up to about 600°C, but its high expansion coefficient limits its use. Waspaloy is almost the only commonly used alloy for temperatures up to about 700°C. The most widely used engine bolt materials both in the USA and Europe are A286, Alloy 718 and Waspaloy.

Another useful classification for high temperature bolting is given in Table 2. This shows application maximum temperature against coefficient of linear expansion. In most joints where thermal cycling occurs, the matching of expansion of bolt and joint is important. In situations where the bolt has a higher coefficient of expansion than the joint, e.g. A286 bolt in a titanium alloy joint, as the joint heats up the clamp load will be reduced significantly. For this particular application, there is currently some alloy research activity to develop low expansion bolting materials. Incoloy 909 is used in some applications, but is not suitable universally because of limited oxidation resistance.

Some of the alloys which have originally been developed for high temperature use can be strengthened by cold working prior to aging. These include both A286 and Alloy 718, which have found widespread use at the higher strength but at lower maximum temperatures. In Tables 1 and 2, these alloys are suffixed 'CW'.

The relative costs of the materials is important, though for critical applications, the possibility of an engine failure cannot be tolerated, and if the application requires very high strength at high temperatures, then even the high priced materials come into consideration. Table 3 shows some approximate relative raw material prices using low alloy steel as a base line.

Table 1. High temperature fastener alloys classified by maximum use temperature and room temperature strength

Room Temperature Strength MPa	Maximum Use Temperature, °C							
	<235	<315	<425	<450	<550	<650	<730	<760
1800	Aermet 100	MP35N				MP159		
1550	H11 PH13–8MoCW		Alloy 718 CW					
1250	Alloy Steels	PH13–8Mo		A286 CW AMS 6304		Alloy 718		
1210						Incoloy 909	Waspaloy	
1100		410 SS Ti–6Al–4V		17–4PH Jethete M152		A286		
1000					FV535	Alloy 80A		
900		304 SS CW						

Note: CW = Cold worked before heat treatment.

Table 2. High temperature fastener alloys classified by expansion coefficient and maximum use temperature

Maximum Temperature °C	Expansion Coefficient, μm m °C							
	<10.0		10.0–12.4		12.5–15.0		>15.0	
	Alloy	Strength	Alloy	Strength	Alloy	Strength	Alloy	Strength
<235			PH13 – 8Mo CW	1550	Aermet 100	1800		
					H11	1550		
<315	Ti-6Al-4V	1100	PH13 – 8Mo	1250	Alloy Steels	1250	304 SS CW	900
					MP35N	1800		
<425	17–4PH		Jethete M152	1100				
<450					Alloy 718 Cw	1550	A286 CW	1350
					AMS 6304	1250		
<550	FV535	1000						
<650	Incoloy 909	1210			MP 159	1800	A286	1100
					Alloy 718	1250		
					Alloy 80A	1000		
<760					Waspaloy	1210		

Note: Strength is ultimate tensile strength at room temperature in MPa.

Table 3. *Approximate relative costs of fastener alloy raw materials*

Candidate materials	Cost
Low alloy steels	1
AMS 6304	2
H11	5
A286	8
FV535	17
Alloy 80A	18
Alloy 718	18
Incoloy 909	24
Waspaloy	33
Multiphase MP159	60

4. 'Multiphase' Alloy MP159

The 'Multiphase' materials, MP35N and MP159, though they have outstanding properties as shown by Table 1, are based predominantly upon cobalt, chromium and nickel, and have very high raw material prices, as shown in Table 3. Some of the reason for the high price is also because of relatively low usage.

The original Multiphase alloy MP35N was developed in the 1960s and had outstanding strength and corrosion resistance, but limited temperature capability. MP159 resulted in 1973 from development efforts to increase the resistance of Multiphase alloys to high temperatures, and since its introduction, has been used in critical applications in gas turbine engines.

The nominal chemistry of AEREX 350 is as follows:

Ni	Co	Cr	Mo	Ti	Al	Nb	Fe
Bal.	35%	19	7	3	$^1/_4$	$^1/_2$	8

MP159 exhibits the highest strength and creep resistance of any fastener material up to *ca.* 600°C. Bolts fabricated from this material have the following mechanical property capabilities:

- 1795 MPa (260 kpsi) room temperature tensile strength;

- 910 MPa (132 kpsi) room temperature shear strength;

- 1415 MPa (205 kpsi) 650°C tensile strength. Virtual immunity to general corrosion, crevice corrosion, stress corrosion and hydrogen embrittlement.

MP159 gains its strength from a combination of cold work and age hardening. Cold deformation promotes a martensitic transformation in which thin platelets of hexagonal close packed phase form in a face centred cubic matrix. Aging causes the precipitation of the gamma prime intermetallic Ni (TiAl), which gives the alloy its heat resistance. Because cold working is necessary for full strengthening, there is a limit to the diameter of material which can achieve full strength. In bolts this limit is usually set at 1.5 in. dia., after which the strength capability reduces quite rapidly.

5. Development of a High Strength High Temperature Material

Examination of Table 1 shows a gap in the region above 650°C and 1210 MPa strength. A new alloy, AEREX 350, has been developed [2] with the objective of filling this gap. This alloy blends the characteristics of the Multiphase alloys and nickel based superalloys. The composition of the new alloy was based upon the 'Multiphase' type of material, that is, cobalt, chromium and nickel. This was to take advantage of their known properties of low notch sensitivity, high strength and creep resistance, and excellent corrosion resistance. An additional objective was to optimise the high temperature properties by selection of hardener elements (Al, Ti, Nb, Ta, and W) and heat treatment cycle, to give a gamma prime intermetallic precipitate of suitable size, stability and lattice mismatch for creep and stress rupture resistance to be developed. Consideration was also given to the thermal expansion coefficient, which was required to be similar to that of conventional nickel based superalloys, so that joint integrity would be maintained throughout the application temperature range. The new alloy has therefore been developed specifically for bolting, rather than borrowed from existing turbine disc or blade applications.

6. Properties of the New Alloy AEREX 350

The nominal chemistry of AEREX 350 is as follows:

Ni	Co	Cr	Mo	Ti	Al	Nb	Ta	W
Bal.	25%	17	3	2	1	1	4	2

The material was initially evaluated from an experimental heat of *ca.* 250 pounds weight. A second, larger heat of 1500 lb was then processed, and finally a 6000 lb heat, which for this material is likely to be typical of a production melt, was produced for evaluation. The properties of specimens and bolts produced from the larger heat were little different from, or often better than, those from smaller heats. In each case, bolts were readily forged and manufactured into complex configurations using conventional equipment.

6.1. Tensile Strength

Results of room and elevated temperature tensile tests conducted on bolts of $^3/_8$ and $^1/_2$ in. UNJF thread sizes, and then averaged, are shown in Fig. 2. Both sets of samples had threads

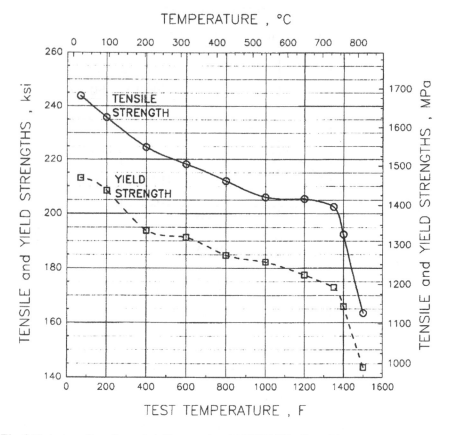

Fig. 2 *Variation of tensile and yield strengths of AEREX 350 alloy with temperature.*

rolled after aging heat treatment. In all tests, the failures were in the threads, so that the heads of the bolts did not degrade the tensile performance. The tensile results show a reasonable amount of separation between the yield and the ultimate stresses across the entire range. This shows that there is no ductility trough or embrittlement occurring which might cause service problems. In fact, compared with smooth specimen tensile tests, the results for bolts were about 4% higher, which indicates good notch toughness.

6.2. Impact Properties

Additional information on the notch toughness of the alloy was obtained by conducting Charpy impact tests on notched samples. The method used was according to ASTM method E23. The tests were done on a Tinius Olsen impact machine, at temperatures between –200 and 375°C. The results are given in Fig. 3. Some established high temperature alloys were also tested for comparison purposes. The nominal strength level of these materials was as follows:

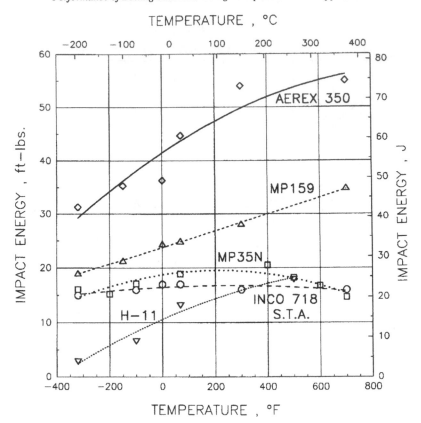

Fig. 3 *Charpy impact data on notched samples, comparing AEREX 350 behaviour with that of other high temperature materials.*

MP159	1800 MPa
Alloy 718	1250 MPa
MP35N	1800 MPa
H 11	1800 MPa.

Figure 3 shows that although the peak temperatures were modest, the new alloy had a high degree of toughness compared with the established materials. Note that this toughness extended down to cryogenic temperatures.

6.3. Creep Strength

Creep strength was determined on samples which were 0.252 in. in diameter, according to ASTM method E139. The tests were run on a Satec Model M-3 dead weight test machine with a 16:1 lever arm ratio. The parts were held at temperature for 45 min–1 h before loading and then loaded in five equal steps.

Table 4. *Stress in MPa to give 0.2% creep in 1000 h*

	650°C	705°C
AEREX 350	607	345
MP 159	455	138
Waspaloy	407	207
Alloy 718	380	117

The times to reach 0.1 and 0.2% creep were measured for several temperature/stress combinations. Again, a comparison was made with other alloys, but in this case, reference data was used. The calculated stresses in MPa to give 0.2% creep in 1000 h are given in Table 4 above. Data was also converted to allow a comparative graph to be drawn, using the Larson–Miller parameter. The result, given in Fig. 4, demonstrates the increased temperature capability of the new alloy. Stress rupture tests on bolts were also conducted, and a similar set of curves resulted.

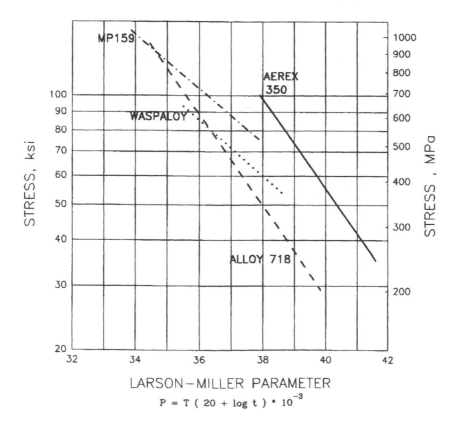

Fig. 4 *0.2% creep properties of AEREX 350 compared with other high temperature materials.*

6.4. Thermal Expansion Coefficient

The thermal expansion coefficient was determined to ASTM E228 on 0.375 in. dia. $\times$ 2 in. long blanks. The measurement range was from room temperature to 730°C, and two runs were made on each of two samples. The results are compared with published data for three other high temperature alloys in Table 5 below.

Table 5. *Thermal expansion coefficient of AEREX 350 compared with other high temperature materials*

Temperature range, °C	AEREX 350	MP 159	Waspaloy	Alloy 718
20–425	13.5	14.6	13.9	14.2
20–540	13.9	14.9	13.9	14.4
20–650	14.4	15.1	14.9	15.1
20–705	14.8	16.0		15.5

Numerous tests of various types have been run in order to evaluate this new alloy. On the evidence to date, AEREX 350 appears to be capable of service temperatures up to around 750°C, with very high strength and corrosion resistance. Since its coefficient of expansion is similar to that of many nickel-based superalloys, it is likely to find service in critical applications clamping superalloy components. At the present time, no other bolting material has comparable potential for such applications.

References

1. D. Driver, *Metals and Materials*, June 1985, 345–354.
2. G. L. Erickson, European Patent Application No. 93113435.7, 'Nickel–Cobalt Based Alloys', 9 March 1994.

SESSION 6

Life Management

Chairman: DJ Gooch
National Power PLC, Swindon, UK

26

Review of HT Bolt Tightening, Removal and Replacement Procedures

G. T. JONES

ERA Technology, Leatherhead, UK

1. Introduction

High temperature bolting used in turbine and pipework applications is an expensive and very important structural component. Economic life management of these items requires an understanding of their service function, their failure mechanisms and the effects of maintenance actions. This paper sets out to briefly review these aspects in an attempt to set the scene for best practice covering both mechanical and metallurgical considerations.

2. Stresses in Bolted Joints
2.1. Elastic Considerations

Before considering bolt tightening methods it will be useful to consider some of the important aspects of bolt/joint design. An excellent guideline on this subject for low temperature applications is provided by Ref. [1].

Disregarding temperature effects the clamping force produced by a bolt in a flange joint is a function of the elastic resilience of the bolt and the flange. Taking the simplest model of a bolt tightened against a cylindrical sleeve with a diameter similar to the nut size (Fig. 1), where the force in the bolt is identical but opposite to the force in the sleeve, the force–displacement diagram is directly related to the respective cross-sectional areas and elastic modulii (Fig. 2). Thus, if the areas are Ab and As and the elastic modulii are Eb and Es for the bolt and sleeve respectively and their length is L, to a first approximation the clamping force F is:

$$F = \frac{\delta b \; Eb \; Ab}{L} = \frac{\delta b \; Es \; As}{L} \tag{1}$$

where δb and δs are the bolt and sleeve displacements.

The sum of the displacements in the joint are given by

$$\delta b + \delta s = FL\left\{\frac{1}{Eb \; Ab} + \frac{1}{Es \; As}\right\} \tag{2}$$

If the sleeve in this model is now made larger than the nut (i.e. tending towards the situation in a flange joint) the compressive stress within it is no longer uniform but is distrib-

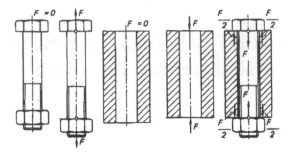

Fig. 1 *Simple bolt and sleeve model.*

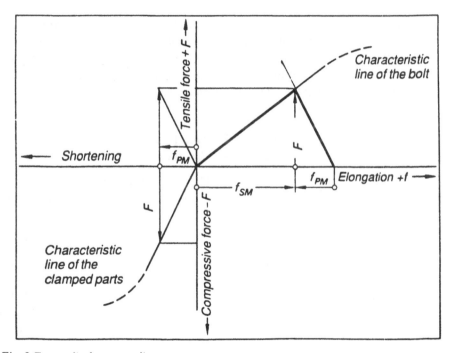

Fig. 2 *Force–displacement diagram.*

uted approximately as shown in Fig. 3. For this situation an effective sleeve area Ase may be used in these equations as given by:

$$Ase = \tfrac{\pi}{4}(dn^2 - db^2) + \tfrac{\pi}{8}dn(D - dn)\left[(x+1)^2 - 1\right]$$

where

$$x = \left[\frac{Ldn}{D^2}\right]^{1/3}$$

with the variables as shown in Fig. 3.

Thus, the effective sleeve area becomes proportionally less as its diameter increases but approaches a constant value when D is greater than $dn + L$, as shown in Fig. 4.

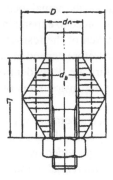

Fig. 3 *Compressed regions in simple bolted joint.*

This example demonstrates that even for the simplest of cases — an axisymetrical model with no external loading, or temperature effects — calculation of the tightening required to produce a desired average joint pressure is not straight forward. In reality, the resilience of the bolt itself cannot be simply calculated; it being made up of the combination of the resiliences provided by the shank and the two threaded regions, including that of the nut (or head, as the case may be). The resilience of the threaded region is particularly difficult to

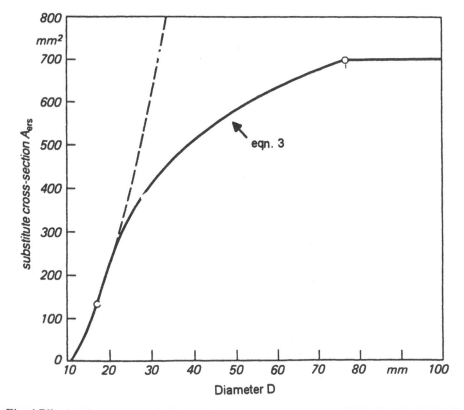

Fig. 4 *Effective sleeve area for M10 bolted joint.*

calculate; it being critically dependent on the actual local bolt thread geometry in relation to that of its female counterpart which is forced to expand in a radial direction in reaction to the axial forces. Account must also be taken of the resilience of any washers or gaskets, and sleeves. The combined resilience is therefore dependent upon the combined elastic response the local frictional forces, which in turn are dependent on the local design and tolerances, and also, at temperature, the relative expansion rates of each component having different coefficients of thermal expansion.

In operation at temperature an axial temperature variation usually exists along the length of a bolt and this will not usually correspond with the equivalent temperature variation through the joint section. The working pressure provided by the bolt is thereby critically dependent on the overall temperature expansion differential of the two components, and this is not easy to calculate.

The other major problem in calculating the tightening loads (in the elastic domain) required to ensure both a leak-free joint and the integrity of the bolt is the resolution of the non-axisymmetric imposed forces. In most bolt connected flanges, bending stresses in the bolts are induced, due either to the working pressure, or to the local temperature gradients.

In a flanged pressurised cylinder or a capped pipe design (e.g. Fig. 5), the working pressure produces bending stresses that are relatively easy to estimate. The additional stresses due to thermal gradients, however, are less readily established, they being very dependent upon the local heat transfer conditions dictated primarily by the condition/quality of the insulation.

In summary, therefore, even in global elastic terms, it is not easy to establish the original tightening load required to maintain joint efficiency under the working load. When global and local time dependent deformation is taken into consideration rigorous design methods become even more difficult.

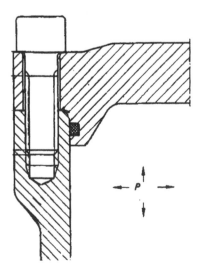

Fig. 5 *Bolted cylinder cap.*

2.2. Time Dependent Operational Effects

The most important time–temperature phenomenon affecting bolt tightness is stress relaxation, during which elastic strains are converted to permanent deformation with a consequent loss of load, which eventually has to be restored by retightening (Fig. 6). Again, because of local stress and temperature inhomogeneity, and elastic follow up, the actual joint tightness as a function of time is difficult to calculate. In a typical turbine flange design this is complicated further by the fact that adjacent bolts will be relaxing at different rates. A secondary time-temperature effect on bolt tightness is oxide forrnation leading to oxide jacking which generally results in progressive tightening and is therefore opposite in effect to stress relaxation.

Nimonic 80A bolting material undergoes ordering reactions at temperatures of less than 550°C resulting in lattice contraction that tends to increase the service stress with time and these effects must be taken into account when setting the preload.

2.3. The Practical Result

In view of the complex nature of the stresses generated in a bolted joint during tightening and the way that they change with time it is not surprising that general rules for tightening have emerged mainly based on past experience. For many years it has been common practice to tighten bolts to a specified cold elastic overall strain over the 'active' bolt length, typically 0.15% for HT bolts in turbine flanges when the flange and bolt materials are similar and to specify a maximum period between retightenings of about 30 000 h.

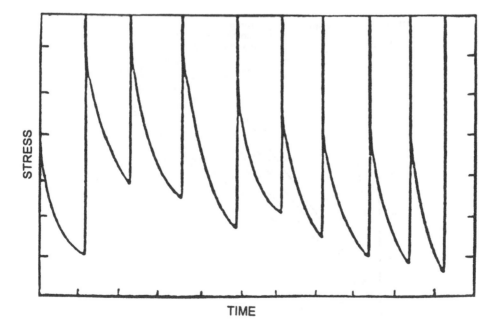

Fig. 6 *Stress relaxation for a series of retightenings.*

3. HT Bolt Tightening Procedures

All HT bolting standards call for 'controlled' tightening procedures. Uncontrolled techniques such as flogging and impact torque tightening are discouraged but controlled torque tightening is usually allowable for small bolts.

Acceptable procedures for ensuring an accurate preload are pre-heating; the use of hydraulic bolt tensioning devices; or by the use of special nuts.

3.1. Torque Wrenching

Torque wrenching is quick a cheap, but does not provide good control of the tightening forces. The fact that up to 90% of the applied torque may be necessary simply to overcome friction effects, and that normal friction coefficient variations can cause a 50% variation in clamp load (Fig. 7) are the main reasons why this method provides poor control.

Torque wrenches range from hand tools to hydraulically operated tools that can give very precise applied torque. They should not be used in critical applications, however, unless accompanied with accurate pre-load measurement.

3.2. Bolt Heating

Bolt heating is the method most commonly employed for establishing a controlled pre-tightening bolt displacement. In order to achieve the designed cold tightening force, or strain in the bolt, the initial expansion required is clearly a function of design with many factors referred to in Section 1 playing a part.

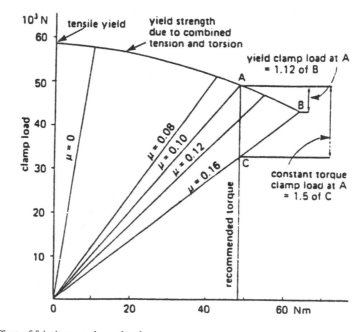

Fig. 7 *Effect of friction on clamp load.*

The heating of bolts for both assembly and removal must be well controlled. For this purpose bolts to be tightened in this manner usually have central through or blind holes used both to insert a heating element and as a mean of accurately measuring bolt expansion for tightening control and for measuring in-service permanent extension. Heating is performed using either gas torches or electrical resistance probes, the latter being preferred because they provide better control. Measurements are made using either customised extensometers or by ultrasonic methods.

The advantages of the heating method are that it is well established, reliable and comparatively cheap. The disadvantages are the potential safety hazards, the potential for local overheating (particularly with gas), but most importantly the time penalty of having to heat and then cool prior to checking each individual fastener.

3.3. Hydraulic Tensioning

Hydraulic tensioning devises which overcome some of the disadvantages of bolt heating have been available for many years. They generally operate on the piston and cylinder principle; the piston being threaded to the bolt which is stretched, with its nut loosely in place, by hydraulic pressure. At a predetermined pressure the nut is hand-tightened and the pressure released. Figure 8 shows a typical device.

This method is quick and accurate but is expensive. Each tensioner is dedicated to a particularly thread size and unless the bolts/nuts were originally designed to suit its application may be limited by the length of available thread and the spacing between the bolts. In

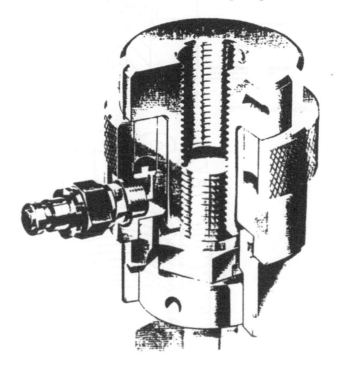

***Fig. 8** Typical hydraulic tensioning device.*

situations where there are considerable cost benefits to be gained by reduced down-time this method offers the distinct advantage of pre-loading all bolts on a flange simultaneously by use of multiple devises and a common hydraulic pump.

3.4. Special Nuts

There are basically two types of special nut designed for HT bolting; one is hydraulically loaded in a manner similar to the above, and the other uses a system of inset jacking bolts.

Early designs of hydraulic nuts were limited to low temperature applications but the use of plastically deformed metallic seals has allowed high temperature application. A typical design is shown in Fig. 9. As with the above, although these nuts are expensive, they offer the advantage of multiple concurrent loading. Designs of jack-bolt loaded nuts, or 'torque nuts' as they are commonly referred to, have been developed over a number of years and are now a sophisticated combination of engineering and material selection. The basis principle of the design is shown in Fig. 10. Some advantageous features that advanced designs possess are:

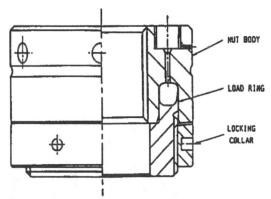

Fig. 9 Hydraulic nut design.

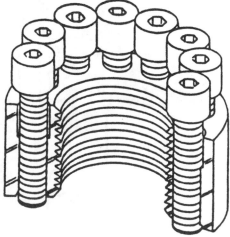

Fig. 10 Jack-bolt loaded nut design.

(1) the ability of the nut to deform radially in a controlled manner that distributes the load along the thread length;

(2) the ability to use a nut/jackbolt/washer materials combination to mitigate bolt loosening due to expansion differentials; and

(3) the ability to make full use of the nut/jack bolt resilience thereby reducing the need for long bolts with sleeves or compressible washers.

Both types of special nut are relatively easy to remove and this alone can make them an attractive alternative to conventional designs.

4. HT Bolt Removal

The disassembly of HT nuts and bolts during overhauls can cause major disturbances to maintenance schedules if seizures or severe galling are encountered.

Clearly, care during assembly and the use of high temperature anti-seize compounds and lubricants reduces the risk of seizure. For austenitic stainless and Nickel based alloys, however, which are susceptible to stress-corrosion cracking, compounds containing sulphur and chlorides must be avoided.

Prescaling can also reduce the occurrence of seizures by inhibiting the formation of an interfacial oxide. Treatments found to be beneficial for two of the major categories of nut and bolt materials are:

• Low alloy (or CrMoV range) – 1 h at 550°C;

• Austenitic stainless (12%Cr) – 1 h at 650°C.

Incidences of gauling can be reduced by ensuring that nuts are made from an alloy grade lower (in hardness and strength) than the grade used for the bolts.

Experience suggests that if a nut or stud is found to be difficult to remove there is nothing to be gained by continued flogging, and the maintenance engineer should be armed with a battery of alternative procedures at his disposal. For stud removal these may include:

(i) Pressurised penetrating oil may be applied to studs with existing through-holes or with holes drilled for this purpose.

(ii) Pneumatic impact and hydraulic wrenches.

(iii) Explosive shock treatment can be used to break the oxide film in the stud thread by passing a high energy stress wave along its axis.

(iv) Shock heating can have a similar effect but the heating rate required frequently precludes subsequent reuse.

(v)   If other methods have failed or, it is simply expedient to do so, destruction of a stud is achieved by drilling and collapsing the thread.

Since nuts are more easy to replace, the destructive route is usually the most cost effective.

5. HT Bolt Replacement

The service life of HT bolts can be influenced by a number of failure mechanisms which must be considered for condition assessment and replacement criteria. Basically, bolted joints must remain leak-free between overhauls implying that the bolts must not fail or accumulate excessive strain between re-tightenings. Material selection is based primarily on the creep rupture and relaxation properties required for the design temperature. Premature bolt failures, however, frequently occur for reasons other than creep rupture. The most important of these are listed below.

5.1. Brittle Fracture

Brittle fracture in HT bolts is usually the result of temper embrittlement induced by service at temperature but is frequently initiated by other mechanisms such as low cycle fatigue. Both the CrMoV alloy range and Nimonic 80A are susceptible. The 12% Cr steels are largely imrnune but tend to have rather low fracture toughness anyway. Brittle fracture is most likely to occur when the bolts are stressed in a cold condition, ie. on retightening, but when influenced by other factors may occur during normal shutdowns.

5.2. Low Cycle Fatigue

Cyclic stresses produced by startup and shutdown cycles in a properly designed bolted joint should not be sufficient to cause fatigue crack initiation. Unfortunately design does not always take account of the cyclic stresses induced by local thermal gradients and expansion mismatches. Poor insulation is also a frequent cause. Secondary cyclic stresses often cause localised bending or shear stresses in bolts leading to crack initiation near to the first engaged thread in normal bolt designs.

5.3. Stress Corrosion Cracking (SCC)

Both the austenitic and nickel based bolting alloys are susceptible to SCC in the presence of aggressive species such as sulphates, sulphides and chlorides. It is also known that once cracks have been initiated they can propagate by SCC in relatively pure condensing steam.

5.4. High Cycle Fatigue

High cycle fatigue in HT bolts is rather rare and is mainly confined to joints that can suffer from vibration induced stresses, e.g. pipework flange joints.

5.5. Condition Assessment

Condition assessment techniques that may be used to trigger appropriately time bolt replacement should take into account all of the potential failure mechanisms.

Non-destructive techniques that address some of these problems are:

(1) Elongation measurement for creep strain evolution.

(2) Crack detection by dye penetrant and ultrasonics (particularly the thread roots).

(3) Hardness tests to determine 'effective' temperature gradients, and also to position bolts in the plain and notched creep rupture scatter band.

(4) Replication for determination of localised (thread root) damage (creep cavitation).

(5) Compositional and microstructural examination for embrittlement susceptibility.

Destructive techniques that can be used to refine replacement decisions include hot tensile, accelerated creep, and toughness (usually impact) testing.

The author knows of no national, utility or OEM based HT bolt replacement criteria that utilises all of these techniques for condition assessment.

6. Replacement Criteria and Best Practice

Recommended bolt replacement criteria usually rely on some, but sometimes none, of the condition assessment techniques described above. Most apply a maximum measured strain criterion (usually between 0.5 and 1.0% — which approximately corresponds to a thread root ductility of the order of 5%). Some, however, simply recommend replacement (irrespective of condition) at a certain number of operating hours, whilst others are based simply on a creep life fraction calculation with or without an allowance for retightenings.

A number of manufacturers base their replacement criteria on the attainment of a certain hardness value irrespective of the initial installed value.

Probably the most logical and comprehensive guideline for bolt condition monitoring and replacement criteria is that published by VGB Techniache Vereinigung dev Grosskraftwerksbetreiber EV. (VGB-R505 Me). It utilises non-destructive testing techniques (1)–(3) above together with selective destructive testing. Replacement recommendations are based on strain evolution, crack detection, hardness and toughness evaluation.

The VGB guideline would form a sound basis for an internationally agreed best practice development.

Areas in which further work would be useful for best practice development include:

(1) Thin film and hard replication of bolt threads;

(2) Non-destructive evaluation of embrittlement; and

(3) Compositional and hardness calibration of creep life.

7. Conclusions

HT bolting design, maintenance, and life management as reviewed in this paper is a mature technology but it does not enjoy universal application.

There is an opportunity, perhaps starting as a result of this conference, to develop an industry wide international best practice for bolt life management, which would eliminate inconsistences and promote economical performance.

8. Acknowledgement

This paper is published by permission of ERA Technology Ltd.

Reference

1. Systematic calculation of high duty bolted joints, VDI Guidelines VDl 2230, Part 1, 1986 (English Trans. 1988).

27

Damage Assessment of Service Stressed Nimonic 80A Steam Turbine Bolts

K. H. MAYER and H. KÖNIG

MAN Energie, Nürnberg, Germany

ABSTRACT

Since the middle of the 1960s high-temperature-resistant bolts of the nickel-based alloy Nimonic 80A have been successfully used by European turbine builders for highly stressed bolted joints of HP turbine casings and HP valve bodies. This material is used on account of its high relaxation strength and its thermal expansion behaviour which is almost identical with that of the low-alloy CrMoV steels generally used for turbine casings and valve bodies operating at steam admission temperatures of up to 565°C.

The bolts have set up a good operating record. The failure rate across Europe is as low as 0.37%. The most frequent cause of failure is recorded to be stress corrosion cracking in combination with mechanical overloading and negative creep.

COST Programme 505 provided the opportunity of a joint investigation in the form of destructive tests on an M 140 bolt which had been used at 480°C over a period of roughly 105 000 operating hours, to check the relaxation, creep, toughness and corrosion behaviour. The results of the investigation showed that long-term stressing reduces fracture toughness, relaxation strength and resistance to stress corrosion cracking. Contraction due to disordering effects was at 0.05–0.154%. By regeneration annealing at 700°C it was found that properties roughly tallied with those of the new bolt.

1. Introduction

Highly stressed bolts made of the nickel-chromium-base alloy Nimonic 80A have been successfully used in the HP section of steam turbines for about 30 years [e.g. 1–3]. A survey made under the EPRI project of the results obtained in service with high heat-resistant bolts made of super alloys in steam turbines has revealed a very low failure rate of 0.37% for Nimonic 80A bolts [2].

The survey covered 20 291 bolts in 231 turbines with steam parameters of 566°C/240 bar maximum and service lives of up to about 130 000 h. The bolts were used in HP inner casings, valves, h.p. piping flanges, gland carriers and strainer covers. The most frequent type of damage was stress corrosion. Some bolts also failed due to high strain fatigue, creep embrittlement and long-term embrittlement. These causes can be countered by:

• Elimination of sulphur and lead-bearing lubricants;

• Design measures to eliminate the formation of stagnating steam condensate;

- Melting of Nimonic 80A with the lowest possible content of trace elements to prevent long-term embrittlement and reduction of stress corrosion susceptibility. In particular, it is important to maintain a P content of 0.002% maximum [4];

- Provision of thermo-elastic bolted joints with low stress concentration, and

- Selecting and strictly maintaining a pre-load matched to the bolt temperature.

On the basis of this experience acquired during long-term turbine operation with a large number of turbines from different manufacturers, the nickel alloy — Nimonic 80A — is confirmed also as a prospective candidate for bolts of improved coal-fired power plants having steam parameters of 600°C and about 300 bar.

Both in respect of the use of Nimonic 80A in these new plants and an evaluation of the bolts that have been in service since the sixties, the question arises as to how to assess the residual life of bolts subjected to long-term operational stressing.

As early as 1966, Buchan *et al.* [5] suggested that Nimonic 80A bolts be replaced upon a permanent creep deformation of about 0.40% having been reached. Their recommendation was based on the results of cyclic retightening tests with a relatively short cycle time on specimens subjected to two-stage heat treatment (8 h 1080°C/air +16 h 700°C/air).

On the strength of supplementary tests with relaxation specimens subjected to three-stage heat treatment, the Nimonic Alloys company [6] later recommended the replacement of the bolts after about 1% creep deformation.

However, the information accumulated over the past 25 years from laboratory tests, failure analyses and overhaul inspections showed that it is also necessary to take into consideration additional criteria in assessing the residual life of long-term stressed Nimonic 80A bolts.

Apart from the relaxation strength, the following operational properties are of specific interest:

- Fracture toughness;

- Stress corrosion behaviour, and

- Creep ductility behaviour.

Another aspect is the possibility of regeneration by additional heat treatment. In order to determine these properties, a large horizontal joint bolt (M140 × 930 mm) of a super high-pressure casing (see Fig. 1) whose bolting temperature was about 480°C maximum was removed after about 105 000 h operation and studied under the European joint project COST 505 [7]. This test project was especially suitable because extensive results were available from relaxation and long-term embrittlement tests on specimens of the initial material of this bolt, i.e. from tests made in the sixties and early seventies.

2. Material Data of Virgin Condition

The bolt had been made in 1966 from a 1500 kg ingot of an electric-arc-furnace melt which was vacuum-arc remelted.

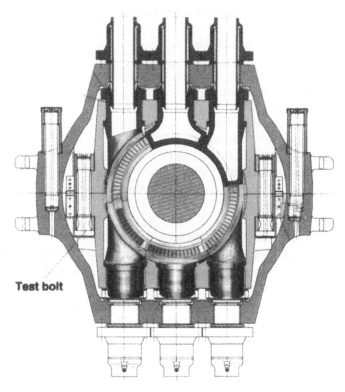

Fig. 1 HP casing, steam admission end.

The forging had a maximum diameter of 200 mm. Heat treatment was applied in three stages: 8 h 1080°C/air + 24 h 850°C/air + 16 h 700°C/air.

The check analysis of the bolt revealed the following chemical composition (data in% by weight):

C	Si	Mn	P	S	Cr	Ti	Al	Fe	Co	Ni
0.026	0.47	0.11	0.002	0.001	20.6	2.3	1.3	1.0	0.018	bal.

The low P and S contents suggest that melting was under high-purity conditions. As can be seen from the mechanical properties in the virgin condition indicated below, a high ductility and toughness were verified during the acceptance tests:

0.2-Limit	Tensile Strength	Strain*	Reduct. of Area	Impact Energy[†]
MPa	MPa	%	%	J
618.0	1062.4	26	24	62.5–57.7
612.1	1062.4	29	28	61.1

*Lo = 5 do.   [†]DVM Specimen.

3. Test Programme

The test programme included the following tests:

- Determination of the contraction condition (negative creep;
- Tensile and Charpy-V-notch tests;
- Fracture tests at RT and 480°C;
- Relaxation tests 480°C and 540°C;
- Creep rupture tests at 500°C and 600°C;
- Creep crack growth tests at 500°C and 600°C;
- Stress corrosion test in Na2S203 at 90°C;
- Auger investigation to determine grain boundary segregation; and
- Microstructure investigation.

The orientation of the specimens was generally selected so as to assess the property profile of the complete bolt. In order to be able to study the effect of a regenerative heat treatment, some of the properties were determined after a 700°C heat treatment and after a completely new (3-stage) heat treatment.

Part of the tests were carried out jointly with the partners under the COST 505 project. An overview of some of the results obtained is given below.

4. Results of Tests
4.1. Contraction Test

Nimonic 80A, like many other nickel–chromium alloys is subject to an order transformation in the temperature range beneath about 550°C [6]. In new condition after the heat treatment the matrix reflects an irregular distribution of nickel and chromium atoms. Due to exposure at temperatures in the range between 450 and 550°C there is an occurrence of short-term close and long-term remote orders on the basis of the Ni_2Cr superstructure. These order transformations produce an increase in strength, a loss in ductility and a contraction in volume of up to 0.16% [4] which in the literature is frequently termed 'negative creep'. The degree of magnitude essentially depends on the time, temperature and on the external stress.

For the purpose of checking the contraction existing after 105 000 h operation at a maximum component temperature of 480°C, specimens taken from various zones of the bolts were examined by means of X-ray microstructure measurements and by mechanically measuring the length of about 90 mm long specimens before and after heat treatments. In the case of the X-ray microstructure examinations, contraction was determined in such a manner that specimens of about 4 mm thickness and a diameter of 20 mm were separated into two halves, one half being brought into a disordered state by applying a completely new 3-stage heat treatment. The contraction was determined from the difference of the lattice spacing of the ordered condition and the new heat-treated condition.

The values determined ranged between 0.07 and 0.13% on average — depending on specimen orientation (Fig. 2). A maximum value determined reached 0.154% and, consequently, the same level that had been established by the same test method in [4] in a bolt that failed after about 50 000 h (0.16%).

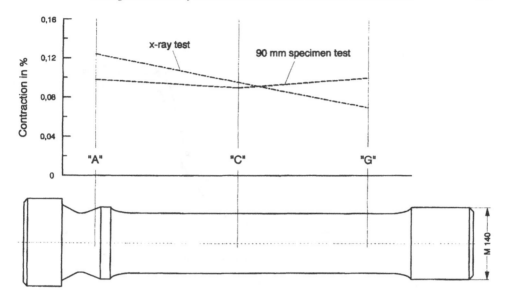

Fig. 2 Contraction vs position along the bolt.

In the case of the mechanical determination of contraction of the above-cited 90 mm long specimens, contraction due to operational stressing was determined by comparing the specimen length before and after a two hour heat treatment in a temperature range of 450 to 650°C. Figure 3 shows the results obtained. Order transformation under these conditions occurred in the temperature range from 525–575°C. Compared to the X-ray microstructure measurement, the mechanical determination of contraction provided lower values:

Position	Mechanical test	X-ray test
A	0.097%	0.13%
B	0.085%	0.10%
C	0.091%	0.067%

Considering that in the case of the X-ray microstructure test, contraction is determined in the micro range, global mechanical measurement of the 90 mm long specimens is probably more representitive of the effective contraction of the 930 mm long bolt.

4.2. Tension and Impact Energy Test

Figure 4 shows the results obtained at room temperature. The results in the virgin condition have been marked on the ordinate for the purpose of a comparison.

The results are indicative of the following global changes in the mechanical properties due to operational stresses:

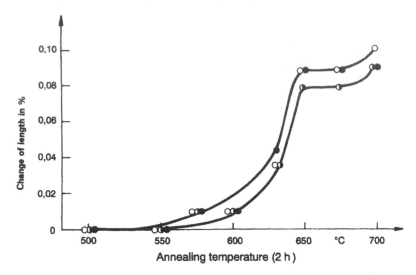

Fig. 3 *Change of specimen length of 105 000 h service stressed Nimonic 80A bolt material due to a disorder annealing treatment.*

- an increase of 21–32% in yield strength (0.2 limit) and a distinct decrease in ultimate elongation of 45– 63% and in reduction of area of 17–40%.

As a result of a completely new heat treatment, the 0.2 creep limit has been reduced from *ca.* 830 MPa to *ca.* 700 MPa. The 0.2 limit of *ca.* 615 MPa determined in the virgin condition cannot be reached by the 3-stage heat treatment.

Due to the new heat treatment, ultimate elongation and reduction of area have been distinctly increased from values of *ca.* 12–21% and 16–25% respectively. Surprisingly, tensile strength, too, has increased from values around 1115 MPa to about 1170 MPa.

The three-hour heat treatment at 700°C has also resulted in a decrease in the 0.2 limit to values around 690 MPa and a distinct increase in ultimate elongation and reduction of area.

The impact energy determined on the Charpy-V-notch specimens is also shown in Fig. 4. In detail, the following values were established:

Condition	Impact Energy (J)
• 105 000 h service	9–16
• 105 000 h service + 3 stage heat treatment	33–36
• 105 000 h service + 3 h 700°C/air	38–43

The low toughness after 105 000 h service points to a marked reduction in toughness due to operational stressing.

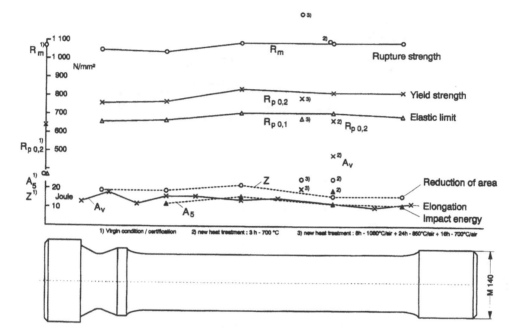

Fig. 4 *Mechanical properties at room temperature after 105 000 service hours and new heat treatment.*

Regeneration heat treatments make it possible again to obtain toughness values of the order obtained with bolts in the virgin condition using modern melting technology.

4.3. Fracture Toughness Investigation

The abovementioned lower impact energy after 105 000 h service at 480°C brings up the question of safety against embrittlement in the case of Nimonic 80A bolts at installation and lower starting temperatures. Testing of fracture toughness was carried out by means of 20 mm thick CT specimens at 20 and 480°C. In spite of the lower impact energy of 9 to 16 J at RT, it was not possible to establish a valid K_{IC} value. The measured K_{IQ} values were as follows:

Test Temperature	K_{IQ} in N/mm$^{3/2}$
20°C	2338
480°C	2863

Figure 5 provides a comparison of these values with K_{IQ} values determined earlier with 20 and 60 mm CT specimens for materials that had not, or only for a short time, been subjected to service stressing [10].

In agreement with the lower impact energy at RT, fracture toughness after 105 000 h

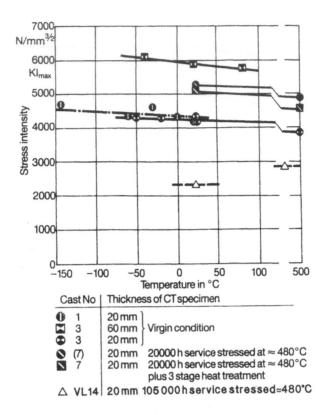

Cast No	Thickness of CT specimen	
◑ 1	20 mm	⎫
⬚ 3	60 mm	⎬ Virgin condition
⬤ 3	20 mm	⎭
◔ (7)	20 mm	20000 h service stressed at ≈ 480°C
◣ 7	20 mm	20000 h service stressed at ≈ 480°C plus 3 stage heat treatment
△ VL14	20 mm	105 000 h service stressed ≈480°C

Fig. 5 Fracture toughness of Nimonic 80A.

service has been reduced by about 50% compared to the initial condition. At 480°C, the decrease in fracture toughness is only half as much.

For an assessment, the results are compared in Fig. 6 with the fracture toughnesses (K_{IC} and K_{Imax}) obtained with ferritic bolt steels with similar specimen tests [11]. As can be seen, fracture toughness of the long-time service-stressed bolt material Nimonic 80A is roughly equal to the fracture toughness of the high-heat-resistant ferritic steel X 19 CrMoVNbN 11. 1 (DIN 17 240) in the virgin condition.

4.4. Relaxation Tests

The tests were made by means of a small bolted joint model [12] because this closely simulates the actual service stressing, and extensive results were available from a similar programme providing for reference tests on non-service-stressed Nimonic 80A melts [10].

Matching the super-high-pressure inner casing, the flange material consists of a 1% CrMoV steel and the nut of Nimonic 80A. The determination of the initial stresses and the residual stresses was carried out by means of electric strain gauges applied to the bolt shank.

Figure 7 shows the results obtained in the 480°C test for an initial strain of 0.18% as a function of testing time, compared to the behaviour in the virgin condition of various Nimonic

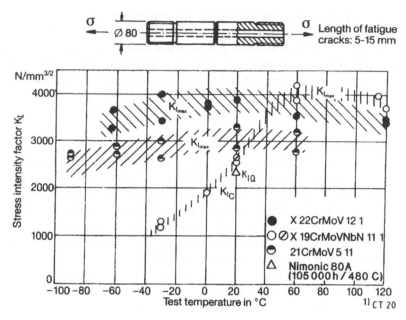

Fig. 6 *Toughness of heat-resistant bolt materials as a function of test temperature.*

80A melts. The elastic residual strain was determined after exposure of the test models which were removed from the laboratory furnace after 3000, 10 000 and 20 000 h.

The results found are in the lower range of the scatter band for the virgin condition. A 3 h regeneration heat treatment at 700°C resulted in a slight increase in relaxation strength.

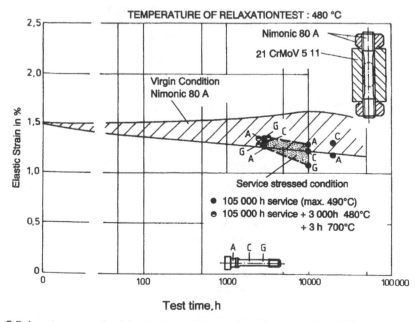

Fig. 7 *Relaxation strength of the virgin condition and service stressed condition.*

The results obtained in the 540°C tests (Fig. 8) were in full agreement with this. A noteworthy fact in this situation is that the relaxation strength determined earlier in the virgin condition of the melt of the service bolt is at the upper limit of the scatter band of the virgin condition, and the relaxation strength of the 700°C regeneration heat treatment, which roughly attains the mean value of the scatter band, no longer matches the originally existing relaxation strength.

4.5. Creep Rupture Test

A special object of the investigations was to check the creep rupture deformation behaviour. Based on the service temperature of the bolt of 480°C, a test series was carried out at the standard temperature of 500°C and another test series using a time-lapse technique at 600°C. The results of the tests extending over up to about 16 000 h are shown in Fig. 9. In the case of both test temperatures, the strength of the smooth and notched creep specimens is within the scatter band of the virgin condition of the material Nimonic 80A (DIN 17 240).

In the 500°C tests, the ductility is still satisfactorily high and the notched specimens (α_K = 4–5) have the same or a higher ultimate strength compared to the smooth specimens. In the case of the 600°C test, this applies only to test times up to about 3000 h. Thereafter, the time to rupture of the notched specimens is shorter and the ultimate strain and reduction of area on rupture are very low at values of 1.9 and 2.5% respectively.

4.6. Stress Corrosion Test

As mentioned above stress corrosion has in the past been the most frequent cause of failure

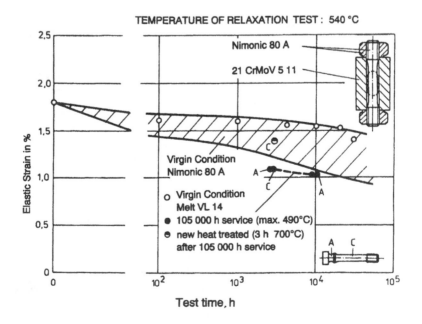

Fig. 8 Relaxation strength of the virgin condition and service stressed condition.

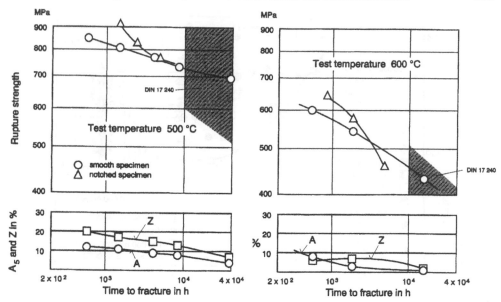

Fig. 9 *Creep rupture behaviour of Nimonic 80A at 500 and 600°C after 105 000 h service at about 480°C (bolt of HP inner casting).*

of steam turbine bolts made of Nimonic 80A, and there were indications in more or less most of the cases that sulphur-bearing steam contaminations play an important role in crack initiation [1, 2]. This susceptibility of Nimonic 80A in sulphur-bearing solutions was also established in stress corrosion tests [1, 10].

Figure 10 shows a summary of the results obtained by the authors to date. The shortest test times were obtained with smooth tensile specimens stressed below the 0.2 limit of Nimonic 80A in sulphur-bearing media.

The test results established with specimens of the 105 000 h service-stressed bolts and specimens after a 3 h regeneration heat treatment at 700°C are also included in Fig. 10.

The shortest test time of all specimens examined to date was recorded after the 105 000 h service-stressed state and a test time of 15.3 h at a test stress of 111% of the 0.2 limit.

As a result of the regeneration heat treatment at 700°C, the corrosion resistance has obviously distinctly improved again. After about 530 h at a stress level of 120% of the 0.2 limit, the specimen was still completely free from cracks.

4.7. Auger-Electronic Spectroscopic (AES) Investigation

Three different bolt zones have been investigated, viz. in the service-stressed state, after the 3 h regeneration heat treatment at 700°C, and after a completely new 3-stage treatment.

All specimens tested failed at −100°C, failure being mostly intercrystalline.

In the service stressed state, considerable enrichment of boron was found to exist at the grain boundaries, also enrichment of silicon, carbon and titanium.

As a result of the completely new heat treatment, boron and carbon were substantially reduced at the grain boundaries. Also, less titanium was found but, on the other hand, there was an enrichment in chromium.

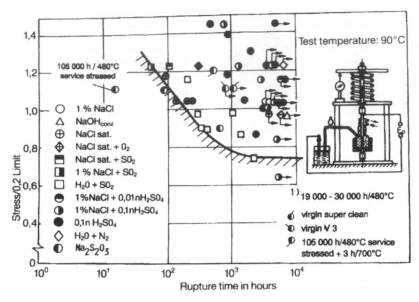

Fig. 10 Stress corrosion behaviour of service stressed Nimonic 80A bolt material.

The 700°C heat treatment reduced the silicon, boron, titanium and nickel proportions and increased the chromium and carbon proportions. No phosphorus was found in any of the specimens tested at the grain boundaries. The reason is possibly in the lower phosphorus proportion of 20 ppm.

5. Summary and Conclusions

Highly stressed bolts in the HP sections of steam turbines made of the nickel–chromium–base Nimonic 80A alloy have been most successfully used for about 30 years with only a low failure rate.

However, there is still little experience available for an assessment of residual life expectancy after long-time service stressing. Evaluation of creep deformation alone is not sufficient because, on the one hand, ordering changes during long-time service below 550°C tend to cause material contraction as well as a decrease in ductility and, on the other hand, grain boundary segregations are liable to cause susceptibility to corrosion and a reduction in toughness.

With a view to gaining experience and a better understanding, a large horizontal joint bolt of a super-high-pressure casing, which had logged about 105 000 h service with a component temperature of about 480°C maximum, was removed and subjected to extensive testing under the European joint programme COST 505. The tests confirmed that the long-term service stressing in the case of the Nimonic 80A material tends to result in a distinct decrease in ductility toughness. On the other hand, no significant reduction occurs in relaxation strength and creep ductility in 500°C creep tests.

The global contraction due to order processes is of the order of 0.10%. There is, however, no distinctly noticeable increase in bolt preload due to plastic and creep deformation of the lower strength flange or casing materials which have a lower strength. A most unfavourable effect consists in a distinct sensibilisation to stress corrosion in sulphur-bearing media under service stressing. This can be obviated by discontinuing the use of sulphur-bearing lubricants and by physical design measures to avoid stagnating steam condensate.

Another important result of the investigation is that a 3-hour regeneration heat treatment at 700°C enables the properties of the material to be substantially restored to the level in the virgin condition.

6. Acknowledgements

The authors are grateful to their colleagues and partners in the COST 505 programme for their contributions and many helpful discussions during the course of the work. Special thanks go to the Institut für Werkstoffkunde of the Darmstadt Technical University for the conduct of the creep and stress corrosion tests, to MPA Stuttgart for the rupture toughness tests and the Max-Planck-Institut for the performance of the Auger tests.

References

1. S. M. Beech, S. R. Holdsworth, H. G. Mellor, D. A. Miller and B. Nath, 'An Assessment of Alloy 80A as a High Temperature Bolting Material for Advanced Steam Conditions', *Int. Conf. on Advances in Materials Technology for Fossil Power Plants*, 1–3 Sept. 1987, Chicago, IL, USA.
2. K. H. Mayer and H. König, 'High Temperature Bolting of Steam Turbines for Improved Coal-Fired Power Plants', *2nd Int. Conf. on Improved Coal-Fired Power Plants*, 2–4 November, 1988, Palo Alto, CA, USA.
3. S. M. Beech, 'Keynote Paper: 25 Years Experience with Nickel Bolting Alloys', this volume, pp. 259–270.
4. B. Nath, K. H. Mayer, S. M. Beech, R. W. Vanestone, 'Recent Developments in Alloy 80A for High Temperature Bolting Application', this volume, pp.306–317.
5. I. Buchan, B. Kent and M. Kirkei, 'Stress Relaxation Properties of some Nickel–Chrome Alloys for Steam Plants', *The Engineer*, 1966.
6. Letter of Nimonic Alloys company (today Inco, Hereford dated 26th Sept., 1969 to MAN, Nuremberg).
7. K. H. Mayer, 'Damage Assessment of Service Stressed Nimonic 80A Steam Turbine Bolt', Final Report of COST 505, D 29 Project of MAN U13/89, 28th Feb. 1989.
8. E. Metcalfe and B. Nath, 'Some Effects of the Ordering Transformation on Nimonic 80A on Stress Relaxation Behaviour', *Mat. Sci. Engng*, 1984, **67**, 157–162.
9. B. Nath, 'An assessment of nickel-based alloys for high temperature bolting applications', 1991 Commission of the European Communities Publ. EUR 13802.
10. K.H. Mayer and H. König, 'Operational Characteristics of Highly Creep Resistant Nimonic 80A Bolts for Steam Turbines', *Int. Conf. on Advances in Materials Technology for Fossil Power Plants*, 1–3 Sept. 1987, Chicago, IL, USA.
11. K. Wellinger, A. Erker, K. H. Mayer and R. Schäfer, 'Sprödbruchuntersuchungen an warmfesten Schraubenwerkstoffen', *VGB Kraftwerkstechnik*, 1975, **7**, 455–466.
12. H. König and K. H. Mayer, 'Testing of the relaxation strength of bolted joints', this volume, pp. 54–69.

28

A Mechanistic Basis for Long-Term Creep Life Prediction of a High Temperature Bolting Steel

P. F. APLIN and J. M. BREAR

ERA Technology Ltd, Leatherhead, Surrey, UK

ABSTRACT

The very high costs and long timescales associated with the direct measurement of long-term creep properties necessitates the use of extrapolation procedures for life prediction of bolting steels under service conditions. The present paper addresses the development of a mechanistic basis for performing such extrapolations using the Arrhenius-Norton creep law. Evidence is presented for a 1CrMoVTiB bolting steel to demonstrate that high and variable minimum creep rate stress exponents in the Arrhenius-Norton creep law are a natural consequence of the combined effect of primary and tertiary creep processes. It is shown that incorporation of these effects leads to a rupture life stress exponent value of around 4 which is consistent with dislocation creep theory and with very similar values obtained for a $^1/_2$CrMoV and Type 316 steel.

1. Introduction

The very high costs and long timescales associated with the direct measurement of long-term creep properties necessitates the use of extrapolation procedures for life prediction of bolting steels under service conditions. One approach has been based on the use of an Arrhenius type Norton creep law of the form:

$$\frac{1}{t_r} \propto \dot{\epsilon}_s \propto \sigma^{n'} \exp\left[-Qc' / RT\right] \tag{1}$$

for describing the dependence of rupture life t_r and secondary creep rate $\dot{\epsilon}_s$ on stress σ and temperature T. However, this approach often leads to high and variable stress exponent n' and activation energy Qc' values (e.g. [1]), that cannot be interpreted easily in terms of creep mechanisms and thus can lead to considerable inaccuracies in extrapolated creep lives. As a result in recent years considerable research effort has been directed to rationalising these high observed stress exponents and activation energies.

Recently there has been an increasing realisation that 'secondary' creep is unlikely to be a separate stage of the creep process but instead is associated with a minimum in creep rate resulting from a balance between primary creep processes. that lead to a decay in creep rate. and tertiary creep processes, that lead to an acceleration in creep rate [2].

In an earlier paper [3], it has been proposed for some creep resistant steels (and other materials) that high minimum creep rate stress exponents and activation energies in the

Arrhenius-Norton creep law are a natural consequence of the combined effect of primary and tertiary creep processes. Furthermore. it is shown that incorporation ol these effects leads to stress dependencies and activation energies for creep that are consistent with dislocation creep theory. ie stress exponents of around 4 and activation energies for creep equal to those for self-diffusion.

The present paper examines the applicability of this approach to a 1CrMoVTi (Durehete 1055) bolting steel.

2. Basis of the New Approach

For situations where normal three-stage primary-, secondary- and tertiary-creep curves are observed, it is always found that the lower is the stress, then the longer is the time (and in most cases the lower is the strain) to reach the minimum creep rate. On the basis that this minimum arises from a balance between primary and tertiary creep processes, the implication of this observation is that the stress dependence of the minimum creep rate will always be greater than the true stress dependency of the creep rate as measured at some constant time or strain in the undamaged condition (i.e. where tertiary creep effects are negligible). The situation is illustrated schematically in Fig. 1. (It is assumed that tertiary creep effects are negligible in the primary creep regime.) A similar implication applies regarding the activation energy for creep.

High observed stress exponent and activation energy values for minimum creep rate can then be explained on the basis that they are not a true measure of the stress and temperature dependence of creep. If this explanation is correct, it should be possible to rationalise high values of rupture life stress exponent observed in 1CrMoVTi steel (Fig. 2) by including

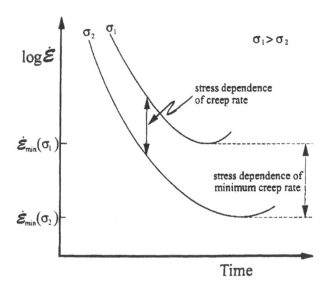

Fig. 1 Schematic creep rate behaviour.

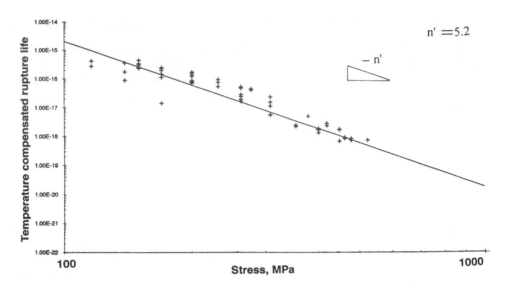

Fig. 2 *Temperature compensated rupture life vs stress.*

primary and tertiary creep terms in the Arrhenius-Norton creep law. The results of such an exercise are described in the next section.

3. Application to 1CrMoVTi Bolting Steel

The investigation was performed on a multi-cast data set of 1CrMoVTiB (Durehete1055) steel in the stress range 116–525 MPa and the temperature range 475–600°C. The data were obtained from ERA's Project 2021: Creep of Steel which is a large collaborative applied research programme performed on behalf of alloy producers. plant manufacturers and utilities into the high temperature mechanical behaviour of steels and alloys for power and other high temperature plant.

The particular creep law used in the present work was based on the following Andrade primary creep modified Kachanov-Rabotnov formulation [4]:

$$\dot{\epsilon} \propto \left(\frac{\sigma}{1 - \omega} \right)^n \bullet t^{\frac{1}{\mu} - 1} \bullet \exp\left[-Q_c / RT \right] \tag{2}$$

where $\dot{\epsilon}$ is the creep rate, t is time, ω is the Kachanov-Rabotnov damage term characterising tertiary creep and μ is a term characterising primary creep. (Further details are given in the Appendix.) It is important to emphasise that this formulation is considered to have no special significance and other formulations may be equally applicable for examining the present hypothesis. Indeed a simpler formulation (see Appendix) was applied under some conditions and similar results were obtained. It is noteworthy that both formulations yield the same relationships between minimum creep rate stress exponent n' and true stress exponent n and between minimum creep rate activation energy $Q_{c'}$ and true activation energy Q_c, i.e. $n = n'/\mu$ and $Q_c = Q_{c'} / \mu$ respectively.

The values of the parameters n' and μ obtained from eqns (1), (A8) respectively were 5.2 and 1.27 (Figs. 2 and 3) giving $n = 5.2/1.27 = 4.09$. This is in excellent agreement with the value expected from creep theory, i.e. around 4 and with values of 3.98 and 3.94 respectively, obtained for $^1/_2$CrMoV and AISI Type 316 steel using the same approach [3]. These results provide strong evidence for 1CrMoVTiB steel that high observed stress exponents in the Arrhenius-Norton creep law are a natural consequence of the combined effect of primary and tertiary creep processes.

4. Conclusions

The minimum creep rate and rupture life behaviour of a 1CrMoVTiB bolting steel (Durehete 1055) has been investigated and the following conclusions have been drawn:

1. The high value of stress exponent in the Arrhenius-Norton creep law is a natural consequence of the combined effect of primary and tertiary creep processes.

2. Incorporation of primary and tertiary creep effects into this creep law leads to a stress exponent value of around 4 which is consistent with dislocation creep theory and with very similar values obtained for a $^1/_2$CrMoV and Type 316 steel.

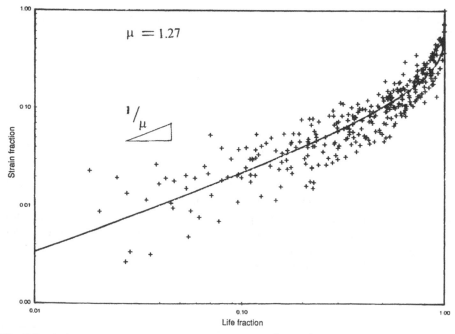

Fig. 3 Strain fraction vs life fraction.

5. Acknowledgements

This paper is published with the permission of ERA Technology Ltd. The sponsors of ERA's Project 2021 are acknowledged for the supply of data used in validating the relationships addressed in the present work.

References

1. D. D'angelo, S. Ragazzoni and V. Regis, 'Room and high temperature creep behaviour of an AISI 316 steel plate for fast reactor structures', in *Proc. Conf. 'Structural Mechanics in Reactor Technology Conference SMIRT-7*, Chicago IL, Aug. 1983, 495–506.
2. R. W. Evans, J. D. Parker and B. Wilshire, *Recent Advances in Creep and Fracture of Engineering Materials and Structures* (Eds B. Wilshire and D. R. J. Owen), Pineridge Press. Swansea, 1982, 135.
3. J. M. Brear and P. F. Aplin, 'Rationalisation of high minimum creep rate stress exponents and activation energies in the Norton creep law through incorporation of primary and tertiary creep effects', in *Proc. 4th Int. Conf. on Creep and Fracture of Engineering Materials and Structures*, The Institute of Materials, 1993.
4. J. M. Brear, P. F. Aplin and R. Timmins, 'The effect of primary creep on the Kachanov-Rabotnov model — results on $^1/_2$CrMo, 1 CrMo and Type 316 steel', in *Proc. Conf. on Behaviour of Defects at High Temperatures*, 30 Mar.–3 Apr. 1992, University of Sheffield, UK.

Appendix
Creep Laws

The creep law used in the present work is based on the following Andrade primary-creep-modified Kachanov-Rabnotnov formulation:

$$\begin{cases} \dot{\epsilon} = C\sigma^n / (1-\omega)^n \bullet t^{1/\mu-1} \\ \dot{\omega} = D\sigma^\upsilon / (1-\omega)^\eta \bullet t^{1/\mu-1} \end{cases} \qquad \begin{matrix}(A1)\\(A2)\end{matrix}$$

$$1 - (t/t_r)^{\frac{1}{\mu}} = (1 - \epsilon/\epsilon_r)^\Lambda = (1-\omega)^{n\Lambda/(\Lambda-1)} \qquad (A3)$$

where Λ is the ratio of rupture ductility to minimum creep rate/rupture life product and C, D are temperature dependent reference creep and damage accumulation rates.

Though written here in a form appropriate to time hardening (following Andrade), an equivalent strain hardening formulation can be used — it is phenomenologically indistinguishable at constant load and temperature.

Integrating the above equations as they stand gives [3]:

$$\dot{\epsilon}_{min} \propto C\sigma_o^n (D\sigma_o^\upsilon)^{\mu-1} \qquad (A4)$$

Hence, assuming that $n \sim \upsilon$, the true stress exponent, n, and activation energy, Q_c, are simply related to the minimum creep rate stress exponent, n', and activation energy Q_c' as:

$$n = n'/\mu \tag{A5}$$
$$Q_c = Q_c'/\mu \tag{A6}$$

This describes the situation for a test where the minimum creep represents a genuine balance between primary hardening and tertiary damage processes — i.e. it is a reflection of true materials behaviour.

Other possibilities exist, within the same general formulation — thus in the absence of damage but allowing large strains to give a Hoff type tertiary and rupture:

$$\left.\begin{matrix} \dot{\epsilon} = C\sigma^n \bullet t^{\frac{1}{\mu}-1} \\ \sigma = \sigma_o(1+\epsilon) \end{matrix}\right\} \tag{A7} \\ \tag{A8}$$

which integrate to give:

$$\dot{\epsilon}_{min} \propto (\mu C\sigma_o{}^n)^\mu \tag{A9}$$

which ajgain leads to relationships given above, i.e. eqns (A5). (A6), and which was also used in the present work. This describes a test in which the minimum creep rate is the balance between primarv hardening and gross thinning of the specimen — it is thus a mixed materials/structural description.

A third case, arising from this second, is where the minimum creep rate occurs when primary hardening is offset by local necking at the Considère strain

$$\epsilon_I = (\mu - 1)/n\mu \tag{A10}$$

Again, this leads to the relation eqns (A4), (A5), in this case as a result of a balance between hardening and instability.

Derivation of Primary Creep Parameter

As discussed in [A1], μ may be obtained by rigorous fitting of creep curve data to eqn (A3), using an appropriate optimisation routine. A more rapid method, which is usually of sufficient accuracy for practical purposes, involves the linearisation of eqn (A3).

$$\Lambda(\epsilon/\epsilon_r) \approx (t/t_r)^{\frac{1}{\mu}} \tag{A11}$$

Reference

A1 A. M. Othman and D. R. Hayhurst, 'Multi-axial creep rupture of a model structure using a two parameter material model', *Int. J. Mech. Sci.*, 1990, **1**, (32), 35–48.

29

Physically-Based Modelling of Stress-Relaxation in Superalloys and Ferritic Steels

S. OSGERBY and B. F. DYSON*

Division of Materials Metrology, National Physical Laboratory, Teddington, Middlesex, UK

ABSTRACT

A physics-based constitutive law for high temperature deformation and fracture is presented. The model has been validated previously for creep, using data for $1Cr^1/_2Mo$ steel and the parameters derived from these data have been used to predict trends in the stress-relaxation behaviour of engineering alloys. Sensitivity analyses have identified the critical model parameters to be those related to primary creep and thermal coarsening of precipitate particles. Softening caused by strain-controlled processes have little effect on stress-relaxation behaviour. The implications for practical stress-relaxation testing have also been considered. The model predictions indicate that, although the time taken to load a test to the control-strain affects the initial stress in a test, subsequent stress values are little affected by the loading sequence. The ability of stress-relaxation tests to predict creep rates for design has been investigated computationally and found to be unreliable due to uncertainty in the precise nature of the creep rate obtained by this method.

Nomenclature

A dot over any symbol signifies the derivative with respect to time.

B material parameter in sinh law
C coefficient for material softening due to dislocation multiplication
d grain size
E Young's Modulus
f damage strength parameter for internal oxidation
H internal variable describing material hardening during primary creep
$H*$ saturation value of H
h' coefficient for material hardening during primary creep
K' coefficient for material softening due to precipitate coursening
k' coefficient for material softening due to cavitation
K_p parabolic corrosion kinetic coefficient
l cavity spacing
N cavity density
n stress exponent for power law creep
P precipitate particle spacing
P_i initial precipitate particle spacing

---

*now at Imperial College of Science, Technology and Medicine, Dept of Materials, London SW7 2BP, UK.

R testpiece radius
r cavity radius
S internal variable describing material softening
S_c internal variable describing material softening due to corrosion
T temperature
t time
x depth of corrosion product

α exponent for damage due to oxide scale spallation
$\dot{\varepsilon}$ creep strain rate
$\dot{\varepsilon}_o$ material parameter in sinh or power law
ε_u uniaxial creep ductility
κ coefficient for damage due to oxide scale spallation
ρ mobile dislocation density
ρ_i initial mobile dislocation density
σ applied stress
σ_i initial applied stress
σ_o material strength parameter in sinh law
ω internal variable describing material softening and fracture due to creep cavitation

1. Introduction

Stress-relaxation behaviour of engineering alloys is important to engineers who design or operate bolted structures at high temperature. Current structures are designed for intervals of 30 000 h between overhauls and there are numerous data available for durations up to this time-scale. However it is predicted that service intervals will increase to 50 000 h in future installations and there is a need for validated extrapolation procedures in lieu of available data. Current extrapolation techniques are largely empirical in nature, e.g. Larson-Miller, and have only limited value as design tools when the material is thermally unstable. A physics-based constitutive law however can provide a sound basis for extrapolation to service conditions. This paper presents a framework for a physics-based description of high temperature deformation; extends the model to predict stress-relaxation behaviour; analyses the implications for practical stress-relaxation testing; and assesses the ability of the stress-relaxation test to obtain creep rates for design.

2. Constitutive Model for High Temperature Deformation and Fracture

Traditionally creep rates have been represented as a function of stress and temperature alone. The most common representation is the Norton creep law

$$\dot{\varepsilon} = \dot{\varepsilon}_o \left(\frac{\sigma}{\sigma_o} \right)^n \tag{1}$$

although the Nadai creep law

$$\dot{\varepsilon} = \dot{\varepsilon}_o \sinh(B\sigma) \tag{2}$$

and its derivatives were also very popular prior to the 1950s, particularly in the USA. It is now realised that these creep laws alone are inadequate to describe high temperature behaviour and mathematical representations of the entire curve have been developed. The basis of these representations range from empirical [1] through a generalised creep mechanism [2] to those based upon specific mechanisms of deformation and fracture [3, 4].

Physics-based continuum damage mechanics (CDM) have been used to describe creep behaviour in both power-law [3] and sinh law [4] formalisms. In each case, a state-variable damage parameter is defined for each micromechanism of deformation and fracture. The evolution rate of each state-variable is also defined together with its influence on strain-rate. A summary of the damage state variables that have been identified and defined to date is shown in Table 1 using the sinh formulation: a similar table has previously been constructed for the power law formalism [3]. Although at first sight the list of potential variables looks daunting, it must be emphasised that for any one material only a limited number of these micromechanisms will be active. Thus, in most cases, equation sets will be limited to two or three damage state-variables and one of these will frequently dominate.

Physics based CDM has been used to analyse and predict creep behaviour in Ni-base superalloys [5] using an equation set that can be rewritten as:

$$\dot{\varepsilon} = \dot{\varepsilon}_o (1 - H)(1 + S)\left[\frac{\sigma}{\sigma_o (1 - \omega)}\right]^n$$

$$\dot{H} = \frac{h'\dot{\varepsilon}}{\sigma}\left(\frac{1 - \dfrac{H}{H*}}{1 - H}\right) \tag{3}$$

$$\dot{S} = C\dot{\varepsilon}$$

$$\dot{\omega} = k'\dot{\varepsilon}$$

For ferritic steels an alternative equation set based upon the sinh law formalism has been developed and used successfully to analyse creep behaviour [6] in $1Cr^1/_2Mo$ steel, viz.

$$\dot{\varepsilon} = \dot{\varepsilon}_o \sinh\left[\frac{\sigma(1 - H)}{\sigma_o(1 - S)}\right]$$

$$\dot{H} = \frac{h'\dot{\varepsilon}}{\sigma}\left(1 - \frac{H}{H*}\right) \tag{4}$$

$$\dot{S} = \frac{K'}{3}(1 - S)^4$$

In both equation sets, stress evolves according to $\sigma' = \sigma\varepsilon'$ when creep tests are conducted under constant load conditions.

Table 1. *Creep damage categories, mechanisms, parameters, rates and effects on uniaxial strain rates using a hyberbolic sine law material model*

Creep Damage Category	Damage Mechanism	Damage Parameter S, ω	Damage Rate $\dot{S}$, $\dot{\omega}$	Strain Rate $\dot{\varepsilon}$
Strain-induced $\dot{\omega} = \dot{\omega}(\sigma,T,\omega)$ or $\dot{S} = \dot{S}(\sigma,T,S)$	Uniform straining under constant load	$S_1 = 1 - \dfrac{\sigma_i}{\sigma}$	$\dot{S}_1 = (1-S_1)\dot{\varepsilon}$	$\dot{\varepsilon}_o \sinh\left[\dfrac{\sigma_i}{\sigma_o(1-S_1)}\right]$
	Creep-constrained cavity growth	$\omega_1 = \dfrac{\pi d^2 N}{4}$ $\omega_2 = \left(\dfrac{r}{l}\right)^2$	$\omega_1 = 0$ $\omega_2 = \dfrac{d}{2l\omega_2^{1/2}}\dot{\varepsilon}$	$\dot{\varepsilon}_o \sinh\left[\dfrac{\sigma_i}{\sigma_o(1-\omega_1)}\right]$
	Continuous cavity nucleation with creep-constraint	$\omega_3 = \dfrac{\pi d^2 N}{4}$	$\omega_3 = \dfrac{1}{3\varepsilon_u}\dot{\varepsilon}$	$\dot{\varepsilon}_o \sinh\left[\dfrac{\sigma_i}{\sigma_o(1-\omega3)}\right]$
	Multiplication of the dislocation substructure	$S_2 = 1 - \dfrac{\rho_i}{\rho}$	$\dot{S}_2 = C(1-S_2)^2\dot{\varepsilon}$	$\dot{\varepsilon}_o(1-S_2)^{-1}\sinh\left[\dfrac{\sigma}{\sigma_o}\right]$
Thermally-induced $\dot{S} = \dot{S}(T,S)$	Particle-coarsening	$S_3 = 1 - \dfrac{P_i}{P}$	$\dot{S}_3 = \dfrac{K'}{3}(1-S_3)^4$	$\dot{\varepsilon}_o \sinh\left[\dfrac{\sigma_i}{\sigma_o(1-S_3)}\right]$
Environmentally-induced $\dot{S} = \dot{S}(\sigma,T,S,R)$ or $\dot{S} = \dot{S}(T,S,R)$	Fracture of corrosion product	$S_4 = \dfrac{2x}{R}$	$\dot{S}_4 = \dfrac{1}{R}\left(\dfrac{2K_p\dot{\varepsilon}}{\varepsilon^*}\right)^{1/2}$	$\dot{\varepsilon}_o \sinh\left[\dfrac{\sigma_i}{\sigma_o(1-S_4)}\right]$
	Internal oxidation	$S_5 = \dfrac{2x}{R}$	$\dot{S}_5 = \dfrac{2K}{R^2 S_5}$	$\dot{\varepsilon}_o \sinh\left[\dfrac{\sigma_i}{\sigma_o(1-S_5)}\right]$

3. Modelling Stress-Relaxation Testing

During stress-relaxation testing, stress continually evolves as elastic strain is converted to inelastic strain. Since the total strain rate is zero, stress evolution is expressed by

$$\dot{\sigma} = -E\dot{\varepsilon} \qquad (5)$$

This equation is be substituted for $\sigma' = \sigma\varepsilon'$ in equation sets (3) and (4) to provide a description of stress-relaxation behaviour in ferritic steels and superalloys. Modelling stress-relaxation from creep data has been promoted theoretically in the literature [7, 8], but experiments are rare although some evidence to justify the approach has been provided for constant strain rate tests [9].

In the following sections, unless otherwise specified, the model predictions have been generated using the sinh-law formalism (Equation Set (4)), with the model parameters defined in Table 2, at a temperature of 550°C. Unless stated otherwise, plastic strain is assumed not to accumulate on loading to a strain level of 0.015%, which is one of the common loading levels for tests performed to generate industrial data.

3.1. Sensitivity Analysis for Model Predictions

The low levels of inelastic strain accumulation inherent to stress-relaxation testing implies intuitively that the parameters describing primary creep behaviour will be one of the major influences on predicted stress-relaxation behaviour. Analysis of the creep behaviour of a $1Cr^{1}/_{2}Mo$ steel [6] indicated that a variation in the primary creep parameter, h', of a factor of four is possible within a single batch of material and Fig. 1 shows the predicted influence of this variation on stress-relaxation behaviour. The effect is largest in the early stages of relaxation and the curves converge as relaxation progresses. Conversely any variation in the saturation parameter H^* has little effect on the early stages of relaxation but begins to exert a modest influence as the residual stress falls to less than 40% of the initial value (Fig. 2).

Values for Young's Modulus measured during a creep test procedure can show a scatter of up to ±15% about the mean in a series of tests on a single batch of material [10]. The value taken for modulus has a direct influence both on the initial stress when loading to a total strain and on the stress evolution during relaxation. This is demonstrated in Fig. 3 where a deviation of ±10% in modulus has been input into the model simulation; it can be seen that although the higher Youngs modulus results in a higher initial stress, there is a more rapid reduction of stress with time. The overall effect therefore is minor.

Due to the low level of plastic strain associated with stress-relaxation, any tertiary creep mechanisms that evolve with strain will have only minor effects. However material softening due to particle coarsening, which is dependent upon time and temperature alone, may

Table 2. *Intrinsic model material parameters for a $1Cr^{1}/_{2}Mo$ ferritic steel*

$\dot{\varepsilon}_o$ (h^{-1})	σ_o (Mpa)	h' (Mpa)	H^*	K' (h^{-1})
$\dfrac{1.4 \times 10^{9}}{\sigma_o} \exp\left[\dfrac{31\ 000}{T}\right]$	$8 \times 10^{-3} \exp\left[\dfrac{6000}{T}\right]$	10^{5}	0.4	$1.4 \times 10^{12}\ \sigma_o^{3}\ \exp-\left[\dfrac{36\ 000}{T}\right]$

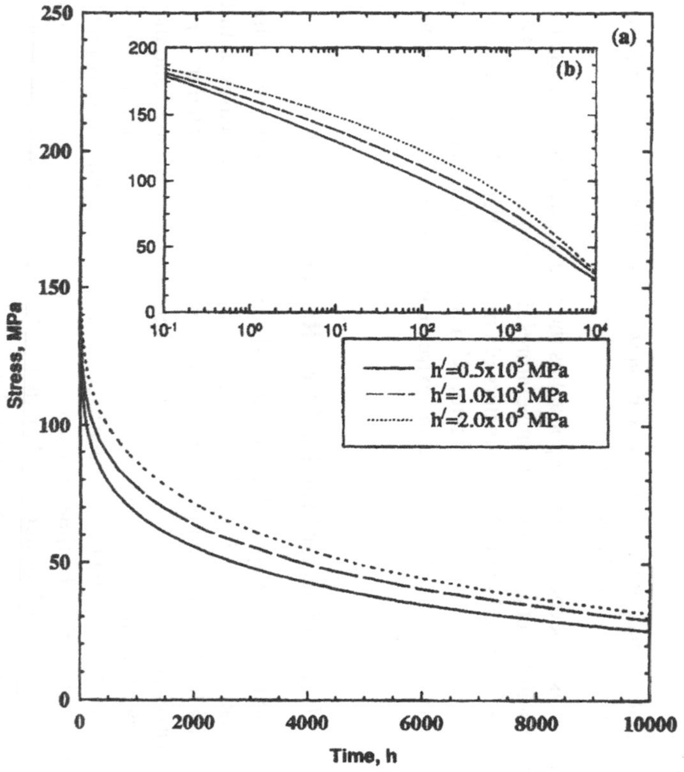

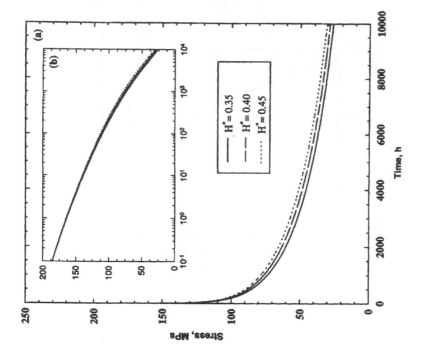

Fig. 1 The predicted influence of the primary creep coefficient, h′, on stressrelaxation behaviour in ferritic steels.(a) Linear time axis (b) Log time axis.

Fig. 2 The predicted influence of the primary creep coefficient, H* on stressrelaxation behaviour in ferritic steels.(a) Linear time axis (b) Log time axis.

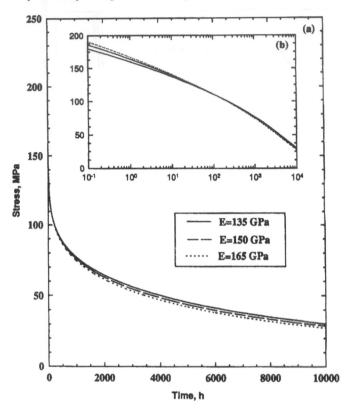

Fig. 3 *The predicted influence of Young's Modulus, E, on stress-relaxation behaviour in ferritic steels. (a) Linear time axis; (b) Log time axis.*

have significant effects. It becomes more important as the coarsening kinetics increase and hence will become significant at higher test temperatures. Figure 4 illustrates this: the middle value taken for K′ is based upon the parameters obtained for $1Cr^1/_2Mo$ ferritic steel at 550°C [6]; in the other two curves on this figure, K′ has been decreased to zero and increased by an order of magnitude to represent the extreme limits taken by this parameter at a particular temperature.

3.2. Implications for Practical Stress-Relaxation Testing

In laboratory tests, loads are not applied instantaneously and some inelastic creep strain is invariably accumulated during loading. The magnitude of this strain increases with time taken to achieve full load for a given total strain. This behaviour is illustrated in Fig. 5(a) using data from Durehete 1055 at 550°C. Four levels of total strain were used and the percentage difference between measured and elastically-calculated stress plotted as a function of loading time. In Fig. 5(b), the model has been used to simulate this behaviour by inputting a constant loading rate to give the time to reach the total strain. There is of course only a qualitative similarity between data and model prediction because no attempt has been made to optimise the $1Cr^1/_2Mo$ steel model parameters for Durehete 1055.

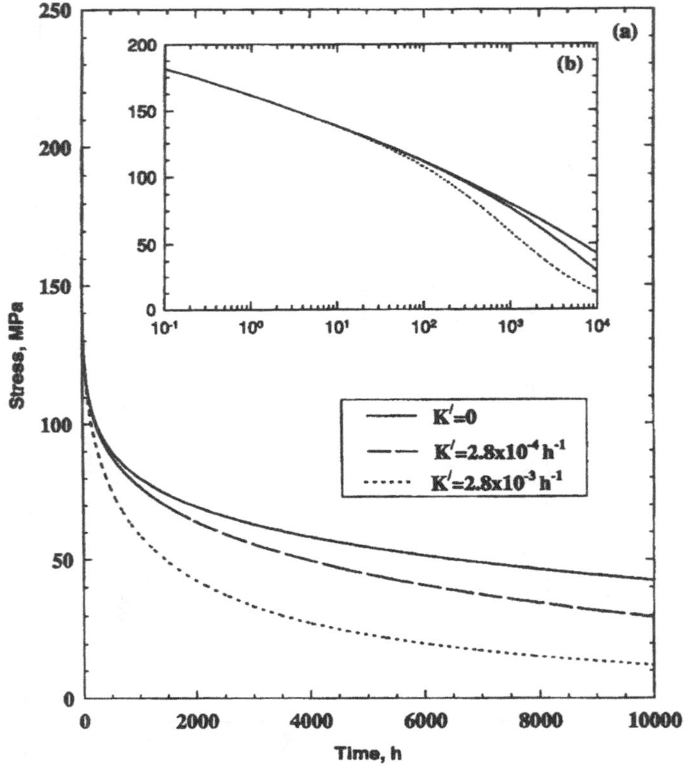

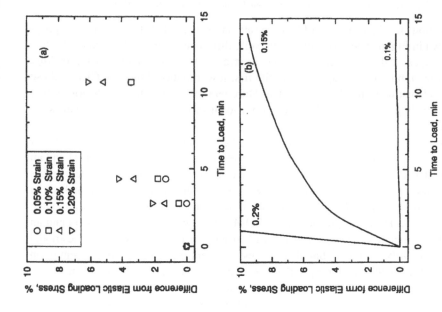

Fig. 5 *Comparison of model predictions with data for the error in initial stress due to loading time. (Data courtesy of A. Strang, GEC ALSTHOM Turbine Generators.) (a) Data; (b) Model predictions.*

Fig. 4 *The predicted influence of precipitate coarsening on stress-relaxation behaviour in ferritic steels. (a) Linear time axis; (b) Log time axis.*

The implications for the influence of loading rate on the extended stress-relaxation properties are shown in Fig. 6. There is a very noticeable difference in predicted behaviour during the early stages of the test (see insert) due to stress-relaxation effects during loading, illustrated in Fig. 5, but these disappear after the first few minutes of the test. Clearly these calculations underestimate the likely effect in Durehete 1055, but the conclusion is that differences in loading rate will not have any significant influence on long-term stress-relaxation behaviour.

The repeatability of high temperature testing has been a subject of interest during recent times. A reference material with certified creep properties has been established [11] and an analysis of the uncertainties in creep properties due to uncertainties in testing practices has been undertaken [12] utilising a steady state creep law. Application of such an analysis to stress-relaxation behaviour is less straightforward due to the absence of any steady state behaviour in this test. However physics-based CDM can be applied to this problem and an example of the influence of one parameter is illustrated in Fig. 7. Several of the model parameters, particularly ε_o and K', but also to a lesser extent E and σ_o, are dependent upon the test temperature. Using the appropriate temperature dependencies for these parameters (see Table 2) the influence of an error in test temperature of 3°C has been calculated for a test temperature of 550°C. It can be seen that the influence of this error in the test tempera-

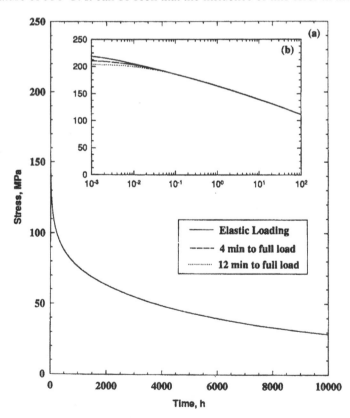

Fig. 6 *The predicted influence of loading rate on stress-relaxation behaviour in ferritic steels. (a) Linear time axis; (b) Log time axis.*

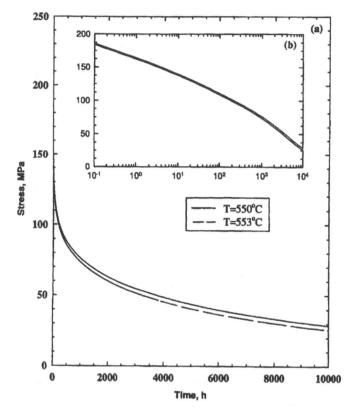

Fig. 7 *The predicted influence of a small temperature difference on stress-relaxation behaviour in ferritic steels. (a) Linear time axis; (b) Log time axis.*

ture (–5% in remaining stress) is much less than the corresponding error in strain rates during creep testing (~20–25%) for an identical temperature uncertainty [12]. This result, which is a consequence of non-linear creep behaviour, indicates that stress-relaxation is much less sensitive to temperature fluctuations than creep tests performed at the same temperature.

There is a growing body of opinion in the literature that constant strain rate [13] or stress-relaxation tests [14] can be used to generate creep rate data for design purposes. The case for constant strain rate tests was refuted in an earlier publication by the authors [9] but the case for stress-relaxation testing persists. Figure 8 shows calculated creep rates as a function of stress at 550°C for creep and stress-relaxation testing. Curve (a) shows the predicted minimum strain rates from a series of constant-load creep tests whilst curve (b) shows creep rates calculated from predictions of stress-relaxation behaviour at the same temperature creep rate is calculated as $-\sigma/E$ since the total strain rate is zero and so creep rate balances the rate at which the testpiece contracts elastically due to the reduction in applied stress. At high stresses, there is a difference of more than an order of magnitude between the two sets of predictions and, although the curves become closer at lower stresses, there is still a factor of two between the two predictions at the lowest stress simulated. This discrepancy can be explained by consideration of the nature of the creep rate measured by stress-relaxation tests. If it is assumed that loading is purely elastic, the creep rate at the start of the stress-

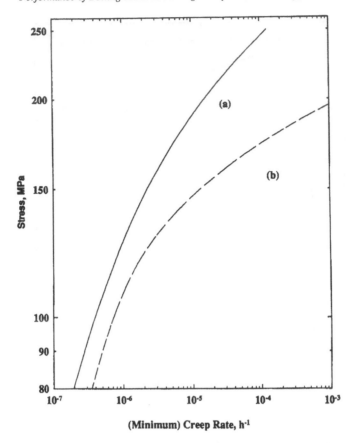

Fig. 8 *Model predictions for creep rates as a function of stress obtained from (a) creep simulation and (b) stress-relaxation simulation.*

relaxation test is equivalent to the initial creep rate in a conventional creep test, then as plastic strain accumulates, the equivalent creep rate moves from the initial rate towards the minimum rate in a creep test. Only if the accumulated plastic strain in the stress-relaxation test is sufficient for the minimum creep rate to be achieved will the two procedures give the same result.

Even in this case, the stress dependency of creep properties generated by the stress-relaxation test would be incorrect. It must therefore be concluded that there is no quick experimental route to generating long-term creep data and that the future lies in combining targeted testing with materials behaviour modelling along the lines explored in this paper.

4. Concluding Remarks

A physics based continuum damage mechanics approach to predicting stress-relaxation properties has been presented. Using values for the model parameters derived to explain the creep properties of a $1Cr^1/_2Mo$ steel, it has been shown that the most important model parameters are those controlling primary creep and material softening by thermal coarsening of particles. Tertiary creep parameters that evolve with strain have only a minor influence on stress-relaxation behaviour.

The model has been used to predict the influence of loading rate and uncertainty in test temperature on stress-relaxation behaviour; both experimental parameters were shown to have relatively minor and short-term effects on the predicted behaviour, indicating that the stress-relaxation test can be very forgiving of certain testing procedures.

The recently revived concept that stress-relaxation tests can be used to generate minimum creep rate data for design purposes has been refuted on the basis of a self consistent comparison of model predictions for both types of test.

References

1. A. Graham and K. F. A. Walles, *J. Iron Steel Inst.*, 1955, **119**, 105–120.
2. R. W. Evans, J. D. Parker and B. Wilshire, in *Recent Advances in Creep and Fracture of Engineering Materials and Structures*, Pineridge Press, 1982, 135–144.
3. B. F. Dyson, *Rev. Phys. Appl.*, 1988, **23**, 605–613.
4. B. F. Dyson, in *Aspects of High Temperature Deformation and Fracture in Crystalline Materials* (Eds Y. Hosoi *et al.*), The Japan Inst. Metals, 1993, 657–666.
5. J. Ion, A. Barbosa, M. F. Ashby, B. F. Dyson and M. McLean, 'The Modelling of Creep for Engineering Design – I', NPL Report DMA(A)115, 1986.
6. B. F. Dyson and S. Osgerby, 'Modelling and Analysis of Creep Deformation and Fracture in a $1Cr^1/_2Mo$ Ferritic Steel', NPL Report DMM(A)116, 1993.
7. A. K. Ghosh, *Acta Metall.*, 1980, **28**, 1443–1465.
8. R. N. Ghosh and M. McLean, *Acta Metall.*, 1992, **40**, 3075–3084.
9. S. Osgerby and B. F. Dyson, in *Creep and Fracture of Engineering Materials and Structures* (Eds B. Wilshire and R. W. Evans), The Institute of Metals, 1993, 53–61.
10. G. D. Dean, M. S. Loveday, P. M. Cooper, B. E. Read and B. Roebuck, in *Materials Metrology and Standards for Structural Performance* (Eds B. F. Dyson, M. S. Loveday and M. G. Gee), Chapman and Hall, 1995, 150–209.
11. D. Gould and M. S. Loveday, in *Harmonisation of Testing Practice for High Temperature Materials* (Eds M. S. Loveday and T. B. Gibbons), Elsevier Applied Science, London, 1992, 85–109.
12. M. S. Loveday and R. Morrell, in *Mechanical Testing of Engineering Ceramics at High Temperatures* (Eds B. F. Dyson, R. D. Lohr and R. Morrell), Elsevier Applied Science, London, 1989, 11–30.
13. M. Steen and M. DeWitte, in *Creep and Fracture of Engineering Materials and Structures* (Eds B. Wilshire and R. W. Evans), The Institute of Metals, London, 1987, 773–788.
14. D. A. Woodford, D. R. Van Steele and D. Stiles, in *Creep and Practure of Engineering Materials and Structures* (Eds B. Wilshire and R. W. Evans), The Institute of Metals, 1993, 603–612.

30

Development of Ultrasonic Bolt Axial Force Inspection System for Turbine Bolts in a Thermal Power Plant

S. KAWAKAMI, H. HAYAKAWA, H. WATANABE, Y. OGURA*,
T. TAKISHITA* and Y. SUZUKI*

Research Laboratory, Kyushu Electric power Co. Inc., 2-147 Shiobaru, Minami-ku, Fukuoka 815, Japan
*FA Factory, Hitachi Construction Machinery Co. Ltd, 650 Kandatsu-machi, Tsuchiura, Ibaraki 300, Japan

ABSTRACT

The only definitive method for controlling turbine bolt axial force is the measurement of the elastic elongation of bolts with gauges after fastening. This tends to result in an imbalanced bolt fastening force and might affect the remnant life of the bolts. For this reason, acoustoelastic theory was applied to the measurement of bolt axial force and an ultrasonic inspection system was developed. This takes advantage of ultrasonic time-of-flight through the bolt and depends on the changes of the velocity of wave and theoretical elastic elongation under stress.

This system can be applied to the axial bolting force control for turbine bolts in a plant, when considering theoretical temperature dependence of time-of-flight through the bolt. The result is in good agreement with measurement on the elastic elongation with a gauge.

1. Introduction

The inner and outer casings of turbines in thermal power plants are so firmly bolted that they can be safely operated without any leakage of high-pressure steam. However, there is no definite method for the control of turbine bolt tightening force, except by measuring the elastic elongation of bolts after tightening. This tends to result in lower work efficiency and bolt tightening force imbalances. In addition, if bolts are tightened with excessive force for long periods of time, the life of the entire system, including the bolts, may be affected. Several instruments have been developed for measuring bolt force (axial bolting force) [1–3], and some of these are in practical use. There are, however, no measuring instruments applicable to controlling the axial bolting force for a large number of bolts over 1 m long, such as turbine casing bolts.

The authors have developed an ultrasonic axial bolting force instrument (QB1000), and applied this to non-destructive axial bolting force control and prevention of uneven or excessive tightening of turbine casing bolts. In general, outer casing turbine bolts are tightened with a bolt heater. In this paper the effect of temperature difference before and after tightening on time-of-flight of the ultrasonic wave propagation is calculated using both theoretical modelling and actual units. With this analysis the bolting force instrument was applied to measurement of thermal power plant outer turbine casing bolt tightening as a practical test.

2. Bolt Axial Force Measurement Principles and Analysis
2.1. Bolt Axial Force Measurement Principles

The principles of bolt axial force measurement with ultrasonic are illustrated in Fig. 1. The bolt force instrument propagates ultrasonic waves in the same direction as the axial force, and calculates the force from the difference in time-of-flight for untightened and tightened states. This change in time-of-flight is due to the change in the velocity of sound for longitudinal waves (velocity drops as tensile force increases), and bolt elongation. The change in sound waves due to axial force has been explained through acoustoelasticity [4]. The relation between stress and deformation in metals is assumed to be proportional in the elastic region, therefore the elasticity is treated as a constant. In retrospect, however, stress will cause a subtle change in the elastic constant, and this leads to a change in the velocity of sound propagation within the material.

2.2. Analysis with a Bolt Model

Ultrasonic wave propagation analysis as indicated in Fig. 2 was performed using a bolt model. As shown, ultrasonic waves were transmitted from the bolt head, and the total time to return after reflection at the opposite end is expressed in eqn (1), where V_o is the velocity of sound in a no-load state. Bolt tightening generates an axial force as indicated by the shaded portion of Fig. 2.

$$t_o = \frac{h + l_1 + l_2 + l_3 + l_4}{V_o} \times 2 \qquad (1)$$

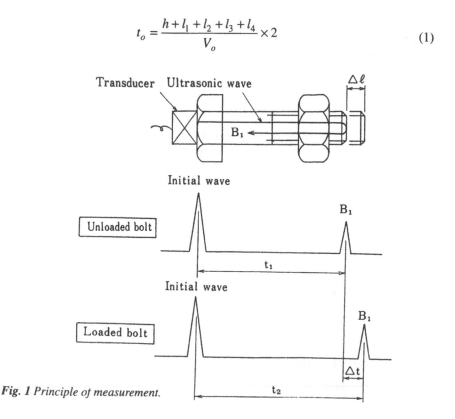

Fig. 1 Principle of measurement.

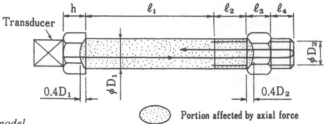

Fig. 2 *Bolt model.*

The generated axial force causes tensile elongation in the bolt and changes the velocity of sound within the bolt. The acoustoelastic constant, k, which represents the wave velocity change ratio with stress σ is defined in eqn (2).

$$V_\sigma = V_o (1 + k\sigma) \tag{2}$$

V_σ: wave velocity with applied stress
k : acoustoelastic constant.
The ultrasonic wave propagation time t_F for applied axial force is given by eqn (3).

$$t_F = 2\left[\frac{h - 0.4D_1 + l_3 - 0.4D_2 + l_4}{V_o} + \frac{0.4D_1 + \Delta h + l_1 + \Delta l_1}{V_o(1 + kF / A_1)} + \frac{l_2 + \Delta l_2 + \Delta l_3 + 0.4D_2}{V_o(1 + kF / A_2)}\right] \tag{3}$$

where
$A_1 = D_1^2 \pi/4$
$A_2 = D_2^2 \pi/4$
$D_3 = D_1 - 0.65P$
$\Delta h, \Delta l_1, \Delta l_2, \Delta l_3$: elastic elongation amount
F: bolt axial force
P: Thread pitch.
In eqn (3) t_F is calculated with the elongation and change in ultrasonic wave velocity for each portion of the bolt (Fig. 2). With Young's modulus E (N m^{-2}) and the acoustoelastic constant k (N m^{-2}), the relation between F and the difference (Δt) between unloaded time-of-flight t_o and loaded time-of-flight t_F is as indicated in eqn (4).

$$\Delta t = t_F - t_o = F(1 - kE) \times \frac{(0.4D_1 + l_1) / A_1 + (0.4D_2 + l_2) / A_2}{2V_o E} \tag{4}$$

Bolt axial force F can be determined from the difference in time-of-flight Δt from eqn (4), using the proportional constant $(1-kE)$.

3. Acoustoelastic Constant
3.1. Measurement of the Acoustoelastic Constant

The acoustoelastic constant was measured for representative bolts. Turbine bolts may be as large as 100 mm in diameter and 1 m in total length, so experimental bolts of appropriate size were fabricated.

The experimental bolt shape and block diagram of the experimental system are shown in Fig. 3. Bolt dimensions are: A = 105–20 mm, B = 53–204 mm, Ø D1 = 14–20mm and P = 1.5. The materials used were S25C, S35C, S45C and S50C stipulated in JIS, both as-manu-factured and heat-treated.

The experimental system consisted of compression load cells mounted on a hydraulic jack, and a nut for holding the bolt. In the experiment, axial bolting forces are applied to the bolt by the hydraulic jack, and are measured as load cell and bolt elongation with a dial gauge. The time for ultrasonic wave propagation inside the bolt (time-of-flight) was measured with the 'sing-around' technique using an ultrasonic probe set on the bolt head.

The acoustoelastic constant k for the experimental bolt was determined from eqn (4) by incorporating the applied axial tightening force F, the time-of-flight for one round trip t_F, and Young's modulus E.

3.2. Experimental Results and Review

Figure 4 shows the acoustoelastic constants k derived from the experiments. The mean of the measured values was -1.1×10^{-11} m^{-2} N^{-1}. Figure 5 shows a comparison of the calcu-

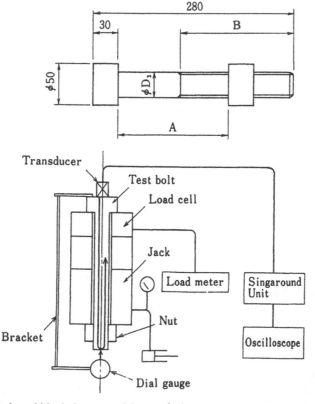

Fig. 3 *Test bolt and block diagram of the test device.*

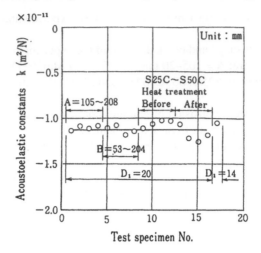

Fig. 4 *Acoustoelastic constants for various test pieces.*

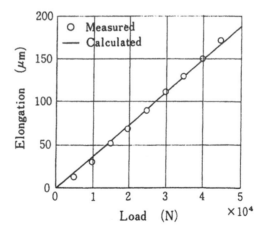

Fig. 5 *Calculated and measured values of elastic elongation (S45C).*

lated elongation for the bolt from the model with the actual value. There is good agreement, indicating that the model analysis closely approximates actual elongation. From the above results it can be seen that the proportional coefficient $(1 - kE)$ is nearly constant for steel. This makes it possible to calculate the axial tightening force from the dimensions of the bolt section and the difference in time-of-flight Δt between unloaded and loaded state.

These results were built into software and combined with an ultrasonic instrument to develop the axial tightening force instrument. Photo 1 shows the appearance of the developed instrument (QB1000), and Table 1 indicates its major specifications. The value for the acoustoelastic constant k was the -1.1×10^{-11} m^{-2} N^{-1} value determined from these experiments.

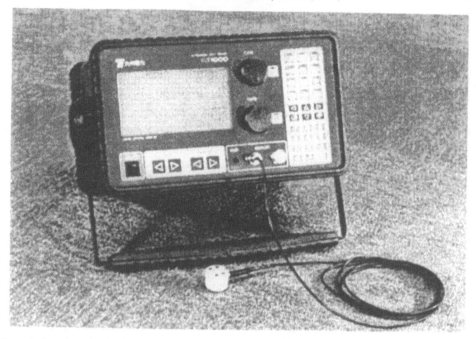

Photo 1. *Developed bolt axial force instrument*

Table 1. QB1000 specification

Applicable bolt length	50–9000 mm (steel)
Ultrasonic wave frequency	2–10 MHz
Display screen	60 × 120 mm liquid crystal display
Measurement result	Bolt axial force, stress, elongation
Printer	Data print, wave form copy
Stored measurements	50 types, 30 of each total 1500
External output	RS232C
Dimensions	H170 × W285 × D245 mm

4. Effect of Bolt Temperature on Axial Force

As a next step, the effect of temperature differences before and after bolt tightening on the difference in time-of-flight was investigated using QB1000. When temperature changes, the bolt dimensions and velocity of sound change accordingly. If the bolt temperature changes before and after tightening, measurement of the tightening force must take these changes into account.

In general, for a steel test piece at 239K with length L_o and speed V_o, length L and speed V at temperature T can be given by eqns (5) and (6):

$$L = L_o (1 + \alpha\Delta T) \tag{5}$$

$$V = V_o (1 + \beta\Delta T) \tag{6}$$

where

$\Delta T = T - 293$

α (coefficient of linear expansion) $= 1.2 \times 10^{-4}$ K^{-1}

β (temperature coefficient of velocity of sound) $= -1.1 \times 10^{-4}$ K^{-1}.

If the bolt temperatures are different both before and after tightening, the difference Δt between flight time in a no-load state t_o and at tightening t_F is calculated using the model shown in Fig. 6. Bolt temperatures before tightening were 273, 283, 29, 303 and 313K, and after tightening ranged from 273 to 373K. This is determined by eqn (7):

$$\Delta t = t_F \frac{(1 + \alpha\Delta T_F)}{(1 + \beta\Delta T_F)} - t_o \frac{(1 + \alpha\Delta T_o)}{(1 + \beta\Delta T_o)} \tag{7}$$

$\Delta T_F = T_F - 293$

$\Delta T_o = T_o - 293$

T_F: temperature after tightening.

T_o: temperature before tightening.

The calculation results for a bolt tightened at 1960 kN are shown in Fig. 7. Based on these results, the difference of time-of-flight Δt between a no-loaded state and the tightening state is found to vary with temperature difference between each state. Variations of Δt to T_F, T_o and axial force F are, respectively, highly linear, as expressed by eqn (8).

$$\Delta t = \varsigma F + \eta T_F - \xi T_o \tag{8}$$

where ς, η and ξ are proportional coefficients. From these, F can be determined in eqn (9).

$$F = (\Delta t - \eta T_F + \xi T_o)/\varsigma \tag{9}$$

From these calculations, results for Δt based on axial forces of 980, 1960 and 2940 kN, ζ, η and ξ were determined. Then these values are incorporated as shown in eqn (10).

$$F = \{\Delta t + 5.49 \times 10^{-8} T_F - 5.45 \times 10^{-8} T_o\}/3.79 \times 10^{-9} \tag{10}$$

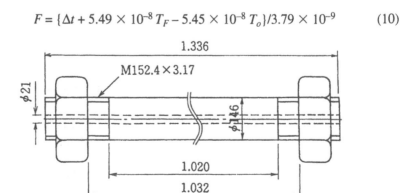

Fig. 6 *Bolt model.*

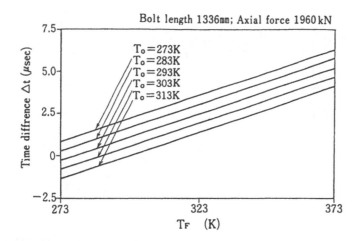

Fig. 7 *Time difference of a round trip between at unloaded temperature* T_o *and at 1960 kN loaded temperature* T_F.

Specifically, the parameters in this equation are not completely linear with respect to Δt, but the results closely match those in Fig. 7. From eqn (10), even if the calculation of Δt is accurate, a deviation of 1 K in T_F or T_o will cause an offset of *ca.* 15% in axial force at 980 kN, and *ca.* 7% at 1960 kN.

5. Axial Force Measurement at Thermal Power Plant

5.1. Measurement Method

With the objective of evaluating usage of the QB1000 ultrasonic axial bolting force instrument, we measured the axial force of tightened outer bolts on the medium- and high-pressure casings of the Shinkokura No. 5 turbine. Bolt specifications are given in Table 2, and bolt positions are shown in Fig. 8.

Table 2. *Bolt specification*

Bolt No.	L_1	L_2	L_3	D_1	D_2	P Thread pitch	d
1 ~ 6	1308	1016	1010	121	126.3	3.18	22
7 ~ 18	1367	1016	1010	146	155	3.18	22
19 ~ 22	1607	1228	1224	159	164	3.18	22
23 ~ 32	1636	1228	1224	171	177.5	3.18	22
33, 34	1266	916	909	146	151.8	3.18	22
35 ~ 52	1146	916	909	96	101	3.18	22
53 ~ 60	1122	916	909	83	88.4	3.18	22

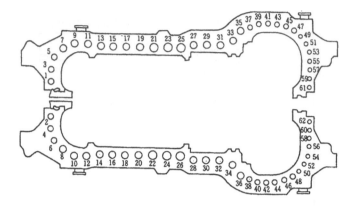

Fig. 8 *Position of turbine casing bolts.*

The ultrasonic probe used was a 5C12.7N with a magnetic mount, and axial force was determined from time-of-flight on measurements of the bolt before and after tightening.

Bolt temperature measurement was performed for both the head surface temperature and the median temperature of the central portion in the bolt heater hole. The surface temperature was used as the mean bolt temperature. The surface temperature was measured with a magnetic surface temperature sensor (manufactured by Hitachi Construction Machinery), and the median temperature with a thermocouple (Ø1.6). After thermocouple setting, the top of the heater hole was sealed to minimise air flow within the bolt hole.

5.2. Trial Results

5.2.1. Effect of measurement position

The position for measurement before tightening (initial measurement), as indicated in Fig. 9, is a representative position (angle 0 degrees). The effect of using a different position for measurement after tightening was investigated. Bolt stress is indicated in Fig. 10. Measurements were made on bolt numbers 5, 31 and 39, with the following results:

1. The variation due to change in the post-tightening measurement position was 49 MPa at maximum stress.

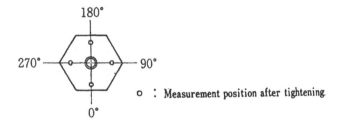

Fig. 9 *Measurement position of bolt axial force.*

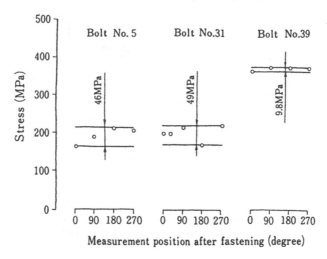

Fig. 10 *Effect of measuring position.*

2. This variation was larger than the management width for stress, which is 31–40 MPa.

This change, due to measurement position, is caused by non-uniform microstructure within the bolt, resulting from the effects of heat treatment and machining.
Consequently axial forces were measured at the same position.

5.2.2. Effects of surface temperature
In the ultrasonic measurement of bolt axial force it is assumed that the internal bolt temperature is uniform, and that the surface temperature is representative of bolt temperature. The effects of this representative temperature on axial force measurement were investigated, varying input data of surface temperature intentionally, although the actual surface temperature of 294 K. The results are indicated in Fig. 11, as follows:

(i) An error of 1 K in surface temperature reduced stress by 8 MPa (bolt number 38).

(ii) The permitted variation in surface temperature ranged from 3.8 to 4.9 K, compared to a stress management width of 31–40 MPa (using bolt number 38 as standard).

From the above results, an accuracy of ± 1 K is essential for surface temperature measurement to maintain the requisite precision for bolt management specifications.

5.2.3. Effects of backing off
If a bolt is over-tightened, then it must be backed off. The results of an investigation of the effect of such backing off with the QB1000 are shown in Fig. 12. Sufficient time for bolt cooling after backing off could not be allowed, but again the surface temperature was taken as representative of bolt temperature, with the following results:

1. Backing off was carried out at bolt number 52, which had exceeded the management

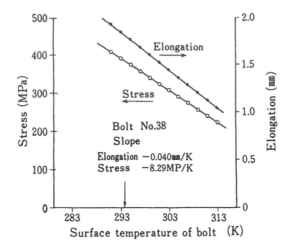

Fig. 11 *Effect of error in surface temperature.*

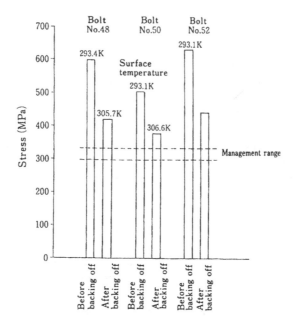

Fig. 12 *Change before and after backing off.*

value by a considerable margin (stress maximum 616 MPa), and axial force measurement clearly indicated the effect of backing off, even when recorded 1–2 h after backing off.

2. The cause of measurement error with QB1000 is considered to result from a sharper temperature inclination inside the bolt after, rather than before, backing off.

The effects of backing off were then confirmed with QB1000.

5.2.4. Axial force measurement results
The results of axial force measurements for the target bolts are given in Fig. 13.

(i) Results of bolt axial force measurement with the QB1000 show that 25% of the total (14 bolts) were within the management range.

(ii) Excessive tightening force detected by QB1000 was adjusted for bolt numbers 48, 50 and 52 through backing off.

(iii) Comparing values from QB1000 with the conventional gauge mesurement control, it was found that there was a trend for bolt numbers 1–34 to be lower, and for numbers 35–60 to be higher.

The possible factor for error in measurement of bolt axial force by QB1000 is thought to be the temperature distribution within the bolt.
From the above, it was confirmed that measurement of bolt axial force using QB1000 closely corresponds to the controlled values (or the conventional measurement method).

6. Conclusions

Acoustoelastic theory was applied to measurement of bolt axial force with ultrasonic waves. Analysis and measurement of wave propagation yielded the following results:

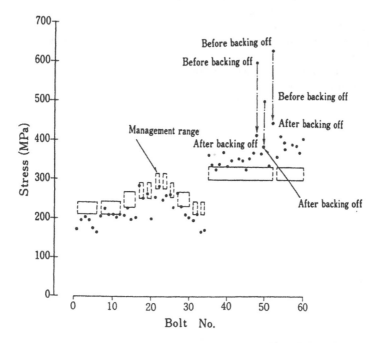

Fig. 13 *Stress after bolt tightening.*

1. The bolt material constant $(1 - kE)$ was calculated using the acoustoelastic constant k ($\text{m}^{-2}\,\text{N}^{-1}$), and it was confirmed that this constant is effective in measurement of bolt axial force.

2. A bolt axial force measurement system using the constant $(1 - kE)$ was developed and applied to the measurement of axial force for turbine bolts in a thermal power plant. A comparison with values obtained through conventional measurement methods was carried out, and exhibited a slight tendency to measure excessive values.

3. It is possible to control tightening force with a bolt force instrument during tightening.

4. A simulation of the effects of change in bolt temperature before and after tightening showed that the change in time-of-flight Δt between tightening and no-load states is strongly dependent on the no-load temperature T_o and the tightened temperature T_F. It is possible to express axial force F with three parameters, namely T_o, T_F and Δt.

References

1. T. Mori, Y. Takekoshi, T. Yagisawa, M. Ohyabu and T. Nakamura, Measurement of Bolt Axial Force by Magnetic Technique, *Trans. Japan. Soc. Mech. Eng.*, 1980, **47**, No. 413, A, 102 (in Japanese).
2. T. Makino, T. Sakai and H. Toriyama, Development of the Bolt Clamping Force Measurement Device with Ultrasonic Waves, *Toyota Engineering*, **25**, No. 1, 11 (in Japanese).
3. E. Yamamoto and R. Motegi, Ultrasonic Sensors for Measuring Bolt Axial Stress, *J. Japan. Soc. Mech. Eng.*, 1979, **82**, No. 731, 40 (in Japanese).
4. H. Fukuoka, Introduction to Acoustoelasticity – Part I. Theory of Acoustoelasticity, *J. Japan. Soc. NDI*, 1984, **33**, No. 9, 633 (in Japanese).
5. T. Takishita and Y. Suzuki, Management of Bolt Axial Force with Ultrasonic Sensor, *ISME Int. J.*, 1989 (August), p.91 (in Japanese).
6. T. Takishita, Principles and Usage of Bolt Axial Force Measurement, *Maintenance*, 1992 (October) (in Japanese).

SESSION 7

Materials Supply and Control

Chairman: RD Conroy
Parsons Power Generation Systems Ltd,
Newcastle-upon-Tyne, UK

Developments in Processing of High Temperature Bolting Alloys — Part I. Steels

H. EVERSON

UES Steels, Stocksbridge, Sheffield, UK

ABSTRACT

High temperature bolting steels are critical components that entail careful manufacturing control to ensure that the range of demanding properties required of them are consistently achieved. Properties such as toughness, stress rupture strength, stress relaxation characteristics and long terrn rupture ductility are all influenced by manufacturing control.

The majority of specialised high temperature bolting steels have a unique composition used in relatively small order quantities therefore production is concentrated on the very flexible electric arc steelmaking route.

Developments in electric arc furnace melting, secondary refining techniques and casting practices have led to improved analytical control, better homogeneity and cleaner steels.

Improved chemical analysis methods have ensured that better raw material control and narrower alloying element ranges can be achieved which, together with better heat treatment furnaces incorporating modern temperature control and monitoring techniques, result in optimised consistent properties.

1. Introduction

If we consider the bolt, stud or nut that is used for high temperature bolting applications it is quite a complex article in terms of shape, thread profile and dimensions. The normal method of production is to machine the fastener from round bar.

The bar for machining must have the correct composition, microstructure and properties to suit the end application. These features are governed by steelmaking, subsequent processing to size and final heat treatment.

The majority of specialised high temperature bolting steels have a unique composition needed in relatively small order quantities therefore production is predominantly via the very flexible electric arc steelmaking and ingot cast route.

2. Compositional Control

The alloy design of high temperature bolting steels is complex. Many contain a number of alloying elements carefully balanced to develop the optimum elevated temperature combination of strength and ductility dictated by service. It is therefore important that close analytical control over the the principal alloying elements is exercised to ensure that every cast made meets the desired alloy balance and that variability from cast to cast is minimised.

Residual elements can also affect properties and it is important that these are also controlled.

Modern methods of instrumented chemical analysis provide a rapid method of accurately analysing raw materials, in process and final chemical analysis.

3. Raw Materials

The principal raw material for the electric arc furnace is scrap.

For economic reasons it is preferable to obtain as much of the alloy content as possible from scrap. Careful selection and segregation of scrap with known analysis is necessary. This is particularly true of specific harmful residual elements as many elements cannot be removed during the steelmaking process. The introduction of portable, accurate instrumented analysis equipment has aided scrap segregation.

Alloying additions and other steelmaking materials such as fluxing compounds also have to be similarly controlled.

4. Melting

The electric arc furnace is now used primarily as a fast, efficient method of melting the charge. Developments have concentrated on speeding up melting rate by high electrical power inputs augmented by oxy fuel burners.

Whereas previously the arc furnace was used as both a melting and a refining unit the majority of arc furnace practices now use a secondary refilling unit. However it is normal for dephosphorisation to be carried out as a refining step in the arc furnace.

In order to prevent slag carry over on tapping from the furnace into the ladle submerged tap holes or eccentric bottom tap holes are common. Improved control of the tapping process is now possible due to the introduction of sliding gate valve tap holes. The introduction of these measures assists cleanness improvement [1–3].

Figure 1 shows a submerged tap hole, and Fig. 2 shows an eccentric bottom tap hole.

5. Secondary Refining

5.1. Ladle Furnaces

On tapping from the arc furnace into a ladle (slag free due to the submerged or slide gate tap hole) it is now common to refine the steel in a ladle furnace. Having removed the oxidising furnace slag a clean synthetic slag is added to maintain low oxygen levels to improve cleanness. The composition of the slag is also controlled to achieve the desired sulphur level.

Heating is provided by three electrodes as in the arc furnace but these are much smaller and fed by much lower power than the primary arc melting unit (see Fig. 3).

Most ladle furnaces are equipped with automatic, conveyor fed, metered alloying equipment. The fact that the weight of the molten metal can be obtained on tapping from the arc

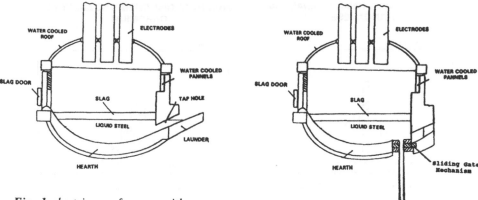

Fig. 1 *electric arc furnace with submerged tap hole.*

Fig. 2 *electric arc furnace with eccentric bottom tapping.*

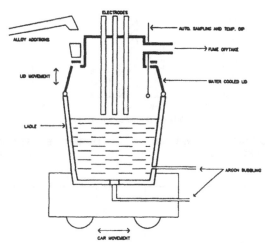

Fig. 3 *Ladle furnace.*

furnace into the ladle and the availability of metered alloying additions give much greater accuracy to compositional control and hence more consistency [4].

Figure 4 shows the control of principal alloying elements in an alloy steel, high temperature bolting material. The range achieved is well within the specified ranges.

Where the ladle furnace also incorporates a vacuum system, known as a vacuum arc degassing (VAD) unit, the dual benefits are realised.

5.2. Degassing

The simplest degassing process is tank degassing in which a ladle of molten steel is placed inside a vacuum chamber. There may be facilities to stir the molten metal by either inert gas bubbling or electromagnetic means (see Fig. 5).

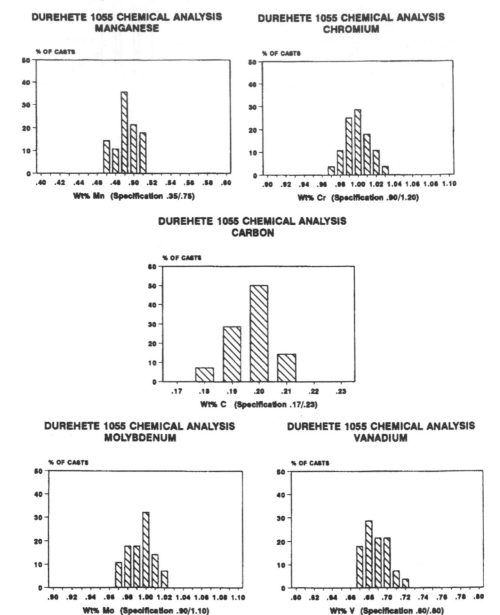

Fig. 4 *Durehete 1055 — chemical analyses.*

Stirring improves the efficiency of degassing, mixes the molten metal promoting homogeneity and assists the flotation of non-metallic inclusions.

There are other designs of degassing plant where the molten metal is recirculated through a separate chamber — these are commonly known as R–H units (see Fig. 6 [5]).

Degassing is the principal means of hydrogen removal. In many steels hydrogen control measures are needed to prevent hairline cracking. A disadvantage of several of the degassing

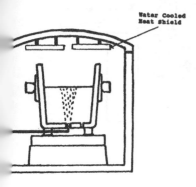

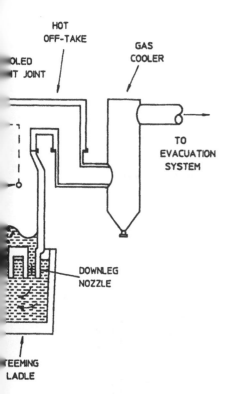

heat source and degassing time is limited before the
is disadvantage is removed if the vacuum system is
hich has a reheating facility (Fig. 7).

3. Stainless Steels

w sulphur stainless steels there are two principal types

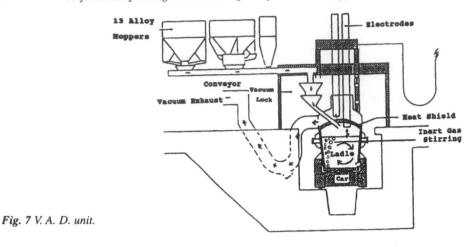

Fig. 7 *V. A. D. unit.*

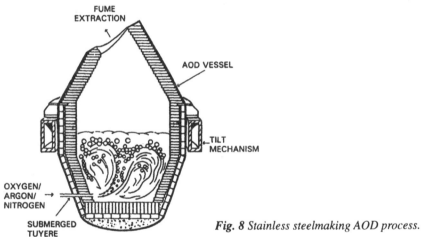

Fig. 8 *Stainless steelmaking AOD process.*

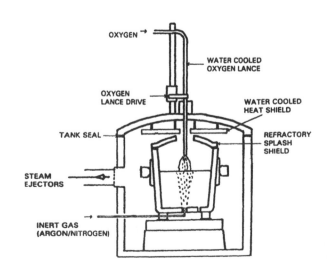

Fig. 9 *Stainless steelmaking VOD process.*

Primary melting of the scrap charge is by electric arc melting but refining of high chromium, low carbon, low sulphur steels in arc furnaces is not practicable as the conditions necessary for reducing sulphur content cause carbon to be picked up from carbon electrodes.

Reactions in the AOD and VOD process are given in Fig. 10.

The types of martensitic stainless steel used in high temperature bolting applications have a carbon content sufficiently high for electric arc furnace/ladle refining to be practicable and the use of AOD or VOD vessels is not essential.

Phosphorus removal cannot be effected in high chromium melts therefore very low phosphorus stainless grades have to be made using low phosphorus scraps or base mix practices [8].

6. Ingot Casting

The specialised high temperature bolting steels are invariably cast uphill, molten metal being teemed from the ladle through a sliding gate valve into a refractory runner system that feeds through a refractory distribution system into the bottom of a number of moulds, typically between four and eight moulds [9].

Protection of the liquid stream by inert gas shrouding systems, the development of high alumina erosion resistant refractory holloware maintain the cleanness of the steel on teeming (see Fig. 11).

TWO PROCESSES • VOD — vacuum oxygen decarburisation
• AOD — argon oxygen decarburisation

PRINCIPLE • removal of carbon to very low levels by oxidation to carbon monoxide whilst minimising oxidation of chromium

STEELMAKING REACTIONS

1. $[C] + \frac{1}{2}O_2 = CO\uparrow$

2. $2[Cr] + 1\frac{1}{2}O_2 = (Cr_2O_3)$ [ ] dissolved in steel
( ) dissolved in slag phase

3. $3[C] + (Cr_2O_3) = 2[Cr] + 3CO\uparrow$ ↑ gaseous product

Lowering of partial pressure of CO promotes reactions 1 and 3 by:

(i) blowing oxygen under vacuum in the VOD process; and

(ii) blowing argon with oxygen in the AOD process.

Fig. 10 *Stainless steelmaking.*

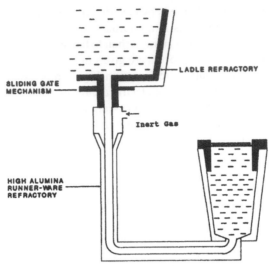

Fig. 11 Ingot teeming.

7. Reheating, Rolling and Cooling

High temperature bolting steels are required in small quantities in many grades and different sizes. The majority have high hardenability and are therefore prone to cracking if heated or cooled incorrectly.

Ideally it would be more economic to direct hot charge the ingots but as 20 or 30 ingots from a large cast may be destined for as many different sizes, slow cooling, ingot annealing and careful reheating of ingots is necessary.

The majority of grades can be direct rolled to a primary size — either a finished large diameter (typically above 75 mm dia.) or an intermediate size for subsequent re-rolling.

Cooling from primary rolling has to be carefully controlled to prevent cracking.

Secondary rolling to smaller sizes is usually straightforward but high hardenability grades need care on reheating and cooling to prevent cracking.

Reheating and soaking regimes prior to rolling can influence microstructure. For example, martensitic stainless steels can be prone to delta-ferrite formation at high temperature. This can cause problems on hot processing and if the phase persists after cooling from final heat treatment can give rise to spurious indications on magnetic particle inspection of finished fasteners.

8. Heat Treatment

Modern heat treatment furnaces utilising improved thermal insulation, intelligent temperature controls in conjunction with computerised burner controls give very uniform temperature distribution and freedom from overshoot conditions.

These result in much greater consistency of properties.

9. Bar Inspection

In-line inspection for surface and internal defects has made significant strides over the past decade.

A variety of techniques are available. Thermal imaging, Eddy current and magnetic methods are used for surface inspection. In-line ultrasonic inspection has also been improved in both speed and sensitivity [10].

These developments in inspection procedures result in greater product assurance.

10. Summary and Conclusions

Modern secondary steelmaking techniques have given rise to significant quality improvements in relation to steels for high temperature bolting purposes.

Better chemical composition control due to improved automated analytical techniques and secondary steelmaking have enabled manufacturers to minimise cast to cast variation and to work to optimised analysis ranges leading to greater consistency of properties.

The availability of portable computensed spectrographic chemical analysis equipment capable of analysing a wide range of elements, including undesirable residuals has improved scrap control and contributed to the greater consistency noted above. The ability to work to lower residual levels has enabled improvements to be made to the long term rupture ductility and notch strengthening characteristics of high performance bolting steels.

Degassing, secondary reflning and improved ingot casting techniques plus the development of specialised casting refractories have made a significant improvement in steel cleanness leading to better ductility and freedom from indications due to non metallic inclusions on inspection of the finished fastener.

Improvements in process thermal operations and final heat treatment equipment have also contributed to greater material control and consistency.

11. Acknowledgements

The author wishes to thank his colleagues at UES Steels (in particular Mr J. Beardwood) for help in preparation of the paper.

References

1. K. A. Broome, *SMEA Conf. on Quality Steel — Advances in Secondary Steelmaking and Casting*, 9–10 April 1992, Sheffield, Paper No. 1.
2. A. Shelbourne, Submerged Taphole on the Electric Arc Furnace. *Ibid.*, Paper No. 4.
3. F. Marsh, Use of the Slidegate Taphole Valve on the Electric Arc Furnace. *Ibid.*, Paper No. 2.
4. I. G. Davies, K. A. Broome and K. Thomas, Major Improvements in Steel Cleanness. Process Route Modifications and the Introduction of Ladle Furnace Operation. *Clean Steel III*, June 1986, Balalonfused, Hungary.

5. J. Rushe, An Overview of the Designs and Practices used in Modern Vacuum Degassers. *SMEA Conf.*, 9–10 April 1992, Sheffield, Paper No. 20.

6. R. J. Choulet and S. K. Mehlman, Status of Stainless Steel Refining. *Metal Bulletin International Stainless Steel Conf.*, 12 November 1984.

7. K. A. Broome, J. Beardwood and M. Berry, The Production of Carbon, Low Alloy and Stainless Steels using VAD, VOD and LF Secondary Steelmaking Facilities at Stocksbridge Engineering Steels, *Int. Conf. on Secondary Metallurgy*, September 1987, Aachen, Germany.

8. H. Everson and M. A. Clarke, Influence of Steelmaking and Primary Processing Factors on Availability and Properties of Stainless Steels. *Stainless Steel '87*, September 1987, Institute of Metals, York.

9. Clean Steels for Aerospace Applications Seminar, Control of Oxygen During Steelmaking and the Production of Ultra-Clean Steel. Institute of Metals, London, April 1988.

10. A. D. Cope, I. G. Davies, I. Fretwell and A. Hardman, The Ultrasonic Inspection of Clean Steels, *ATS Steelmaking Days*, December 1988, Paris.

Developments in Processing of High Temperature Bolting Alloys — Part II. Superalloys

G. OAKES

Special Melted Products Limited, Atlas House, Sheffield, UK

ABSTRACT

Improvements in the properties of steel bolts can be further enhanced by the use of consumable electrode remelting, Vacuum Arc (VAR) and Electroslag (ESR). This results in an improvement in both structure and cleanliness. The use of material produced in this manner is dependent on both commercial (price) and technical considerations.

However, for a number of applications the designers have felt the need to use nickel base and semi nickel base alloys. These materials have generally been used where the properties of the conventional bolting steels have been deemed inadequate.

The hot strength, corrosion/oxidation resistance of the creep resistant nickel based and semi nickel based alloys is unquestioned. These alloys, however, are far less defect tolerant than steels, are generally less ductile and are often notch sensitive.

The production of such alloys therefore calls for extreme vigilance at all stages of manufacture if bolts are to be produced fit for purpose.

The methods available for the production of such materials are vacuum induction melting (VIM) followed by either vacuum arc remelting or electroslag remelting.

Processing of the ingot to bloom billet and eventually bar also calls for close control of the thermal mechanical processing route involved. This control is necessary due to (a) the inability of the materials to be grain refined by simple thermal treatment and (b) the presence of harmful intermetallic phases.

1. Introduction

The advances made in the processing of steel for critical high temperature bolting applications have been presented by Hugh Everson of Stocksbridge Engineering Steels. If, however, one looks at the high temperature bolting materials used in both aerospace (Table 1) and industrial (Table 2) gas turbines, it is apparent that these industries utilise a wide variety of high temperature materials varying from low alloy steels through nickel base alloys to high cobalt bearing alloys.

Unlike the steel alloys, nickel base bolting alloys obtain their strength through the precipitation of intermetallic phases. In general terms the higher the volume fraction of intermetallics the greater the resistance to deformation at high temperatures (Fig. 1). This progressive strengthening of the 'matrix' by an increase in the volume fraction of intermetallics results in the need for clean ductile grain boundaries capable of accommodating strain. The presence of very small amounts of impurities can lead to a significant reduction of ductility and notch weakening. Thus the processing of such alloys requires attention to detail at all

Table 1. Nominal composition of some bolting alloys used in aerospace applications

Alloy Grade	C	Si	Mn	Cr	Ni	Mo	Co	Fe	Ti	Al	Nb	V
HCM3	0.40	0.30	0.50	3.0	0.20	1.0	–	Bal.	–	–	–	0.25
HCM5	0.30	0.30	0.45	3.0	0.20	0.40	–	Bal.	–	–	–	–
M152	0.12	0.40	0.80	11.5	2.5	1.75	–	Bal.	–	–	–	0.3
FV448	0.13	0.35	0.75	10.5	0.75	0.65	–	Bal.	–	–	–	0.25
A286	0.04	0.55	1.4	15.0	26.0	1.3	–	Bal.	2.2	0.2	–	
Alloy 718	0.04	0.1	0.05	18.5	52.0	3.0	–	Bal.	0.9	0.45	5.2	
Nim 105	0.20	0.7	0.5	14.5	53.5	5.0	20.0	Bal.	4.5	1.2	–	
Alloy 80A	0.07	0.02	0.01	19.5	Bal.	0.01	–	0.1	2.35	1.3	–	
Alloy 90	0.08	0.02	0.01	19.5	Bal.	0.01	16.5	0.1	2.45	1.35	–	
Waspaloy	0.03	0.02	0.02	19.0	Bal.	4.2	13.2	0.2	3.1	1.4	–	
MP159	0.02	–	–	19.0	25.5	7.0	35.7	9.0	3.0	0.2	0.6	

Table 2. Nominal composition of some bolting alloys used in industrial gas turbines

Alloy Grade	C	Si	Mn	Cr	Ni	Mo	Co	Fe	Ti	Al	Nb
A286	0.04	0.55	1.4	15.0	26.0	1.3	–	Bal	2.2	0.2	–
Alloy 718	0.04	0.1	0.05	18.5	52.0	3.0	–	Bal	0.9	0.45	5.2
Alloy 80A	0.07	0.02	0.01	19.5	Bal.	0.01	–	0.1	2.35	1.3	–
Alloy X750	0.05	0.02	0.04	15.25	Bal.	–	–	7	2.58	0.95	1.1

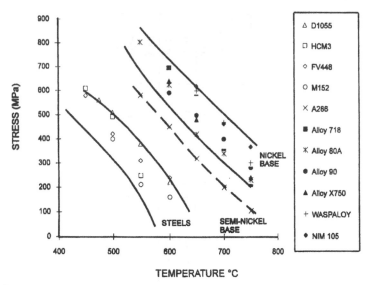

Fig. 1 *Comparison of stress rupture properties of bolting alloys at 1000 h life.*

stages of manufacture from selection of raw materials through melting, remelting and thermal mechanical working.

2. Compositional Control

2.1. Tramp Elements

The rupture ductility of nickel base alloys is far more sensitive to the presence of minute amounts of tramp elements, such as sulphur, tellurium, selenium, bismuth, lead, silver, tin, etc. Segregation of these elements to grain boundaries causes weakening. With the exception of the volatile elements of lead and bismuth, control of these elements can only be achieved by selection of the raw materials. To this end most suppliers of super alloys strictly control the sources of both their 'virgin' materials Co, Ni, Fe, Mo etc. and also control and monitor their sources of scrap. For the user, however, the quest for increased purity results in increased cost. An indication of the demands made by the user (Table 3) against sources of iron, nickel and cobalt available illustrates the difficulty the supplier has meeting the customer demands.

2.2. Finishing Elements

The deleterious effect of some of the tramp elements can be partially overcome by the use of finishing elements. These elements are largely used to tie up the final traces of sulphur and oxygen and to modify the carbides, etc. Examples of elements used in this manner are Mg, Ca, Zr, B etc. Whilst these elements have a beneficial effect, overdoping can also be deleterious.

Examples of the levels of both tramp and finishing elements specified by users are shown

Table 3. Typical residual analyses of nickel, cobalt and iron raw materials

			Cu	As	Pb	S	Bi	Sb	Sn
Nickel	Electrolytic	99.9% Ni	<0.0001	<0.00005	0.00005	<0.0003	0.0003	0.0003	0.0003
	Russian origin	99.6% Ni	0.01	0.001	0.0002	0.001	0.0005	0.0005	0.0005
Cobalt	Electrolytic	99.8% Co	<0.0001	0.0005	0.0001	0.001	0.0003	0.0003	0.0003
	Russian ingot	99.3% Co	0.02	0.002	0.0005	0.004	0.0005	0.001	0.001
Iron	Electrolytic		0.004	0.003	<0.0001	0.001	<0.00003	0.0001	0.001
	Armco		0.03	0.007	0.0001	0.006	<0.00003	0.0005	0.006

in Table 4. This demonstrates quite clearly the different approach taken, albeit on different materials, by two different users. User A leaves the use and level of finishing additions purely to the discretion of the supplier whereas user B attempts to control both the use and levels of the finishing elements and all known harmful tramps.

3. Primary Melting

The quest for purity and the need to add in a controlled manner reactive elements (Mg, Ca and Zr) for finishing the heats plus the high levels of reactive elements used to achieve the properties (Ti and Al) results in the use of vacuum induction melting as the preferred method of primary melting.

S.M.P. operate two vacuum induction furnaces for the production of nickel based super alloys. A conventional one of 10 tonne capacity (Fig. 2(a)). In the 10 tonne furnace the furnace body is completely enclosed inside a vacuum chamber. The whole melting cycle is carried out under high vacuum with the exception of the addition of the finishing element. These are added under a slight partial pressure of argon. Unlike air melting no refining of sulphur and phosphorus can be achieved by the use of active slag. The only refining processes avilable in the vacuum induction furnace are

(a) Removal of volatile gaseous elements by vacuum treatment, e.g. N_2, H_2, Pb, Mn, Bi etc.

Table 4. Limits of tramp and finishing elements specified

	Specification A (Alloy 80A)		Specification B (Waspaloy)	
	Min.	Max.	Min.	Max.
S	–	0.015	–	0.0015
P	–	0.002	–	0.015
Ag	–	0.0005	–	–
B	–	0.008	0.003	0.010
Bi	–	0.0001	–	0.00003
Cu	–	0.2	–	0.10
Pb	–	0.002	–	0.0005
Zr		0.02	0.12	
Te		–	0.00005	
Th		–	0.00005	
Se		–	0.00005	
Mg		0.003	0.010	
N_2		–	0.0040	
O_2		–	0.0030	
Ca		To be reported		

Fig. 2(a) *Conventional vacuum induction furnace 10 tonne capacity.*

(b) Removal of oxygen by driving the carbon oxygen reaction to completion using stirring in conjunction with the application of vacuum.

(c) The use of finishing elements to form solid removable deoxidation, desulphurisation products by floatation or during subsequent remelting.

As mentioned earlier S.M.P. operate a second smaller vacuum induction furnace of 4.5 tonne capacity (Fig. 2(b)). This furnace purchased in 1991 is of a modular construction and unlike the larger conventional furnace is not totally enclosed within a vacuum chamber. This design results in easier maintenance, smaller volume and reduced surface area. This results in the achievement of lower, ulffmate pressures and hence increased refining potential. In addition, this furnace readily lends itself to the use of ceramic foam filters. These are used in the process to filter out solid deoxidation/desulphurisation products prior to pouring (Fig. 3).

4. Secondary Melting

The high levels of hardening elements coupled with, in some cases, high level of molybdenum, mean that the materials are very prone to macro/micro alloy segregation. To reduce

Fig. 2 (b) *Vacuum induction degassing and pouring furnace (VIDP) 4.5 tonne capacity.*

this segregation to acceptable levels it has become common practise to consumable remelt nickel base alloys intended for critical high temperature service. Remelting can be carried out by either the vacuum arc remelting (VAR) or electroslag remelting (ESR) process. S.M.P. operate four VAR furnaces and two ESR furnaces and are currently installing one new VAR and one new ESR furnace (Fig. 4).

The choice of furnace is either left to the discretion of the supplier or specified by the user. In both cases the effect of consumable remelting is to refine the structure, reduce segregation by control of the solidification and improve cleanness by the removal of non metallic inclusions by flotation and dissociation.

The preferred melting route for the production of nickel based super alloys has thus largely gravitated to vacuum induction primary melting followed by either vacuum arc remelting or electroslag remelting, e.g. VIM–VAR or VIM–ESR. In a number of instances, particularly for the production of A286 material, AOD and VOD are acceptable alternatives to VIM melting.

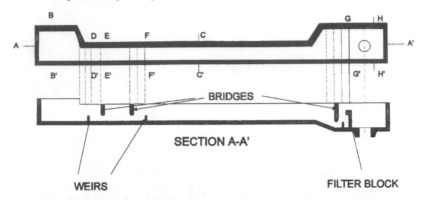

Fig. 3 *Schematic of VIDP launder system.*

Fig. 4 *Vacuum arc and electroslag furnaces at S.M.P.*

5. Post Melting Processes

The high levels of alloying elements can result in the formation of undesirable phases and unacceptable levels of segregation even in material which has been consumable remelted.

S.M.P.'s practise is to subject the leaner alloys to a prolonged soak forging at temperature and to homogenise the richer alloys at temperatures close to the solidus. In some special cases a double temperature homogenisation is carried out. The use of such treatments is instrumental in eliminating harmful phases such as laves phase in Alloy 718 and reducing segregation in other alloys. 'Homogenisation' is also important in raising the local solidus temperature and thus improving/enhancing the forgeability of these materials.

6. Forging to Bloom/Billet Ingot Breakdown

The nickel base alloys, unlike the steels, do not undergo phase transformation. Grain refinement by thermal cycling alone is thus not feasible. Processing of nickel base alloys is therefore of necessity slightly more complex than that of steels. The generally accepted procedure is to:

(a) Carry out ingot breakdown above γ/δ solvus.

(b) Complete forging at a temperature below or straddling the γ/δ solvus temperatures.

The differences in alloy content means that it is necessary to define accurately the thermal mechanical process for each alloy if a fine consistent grain structure is to be achieved.

Once achieved the fine grain size can be destroyed by heating and holding above the δ solvus temperature and thus care must be taken in the rolling of billet to bar. This, however, is relatively easily controlled compared to the achievement of fine grains during forging.

S.M.P. break down their nickel alloys on a large rotary forge (Fig. 5) which is very versatile as regards forging pass sequence temperature control etc. Bar rolling is carried out on a jobbing mill.

Fig. 5 Rotary forge GFM SXP 65.

The need for a fine grained product is twofold:

(a) It improves inspectability by ultrasonic techniques.

(b) It increases rupture ductility (at the expense of rupture life), reduces notch sensitivity and reduces crack propagation rates. The trade off between strength and ductility is shown schematically in Fig. 6.

It is interesting to compare the structural requirements specified by the various users. Some of the older specifications pay little attention to grain size whilst the majority of modern users request grain sizes in the region of ASTM 4-6 with the occasional user requesting finer grain sizes still. It is not uncommon for the same alloys used for disc applications where fatigue and crack propagation are of extreme importance for the user to stipulate grain size in the order of ASTM12. The final part of the processing of nickel base bolting materials is heat treatment. The simple alloys are normally supplied in the solution treated and aged condition whilst some alloys can be given a two stage ageing cycle, an initial cycle to optimise the formation of discrete grain boundary carbides followed by a second age to intensify the hardening. These treatments are aimed at producing material having the correct strength whilst retaining adequate ductility.

One aspect of processing not discussed in the present paper is the use of cold work after hot rolling. This can, and is, used in some alloys to improve gauge control (cold sized hexagons) and also to improve properties, particularly ambient temperature strength and rupture ductility through additional grain refinement.

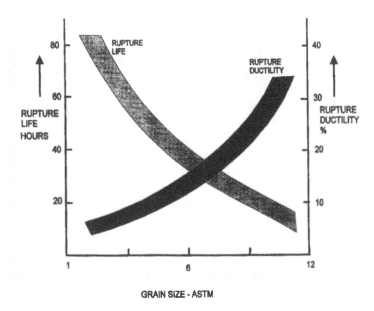

GRAIN SIZE - ASTM

Fig. 6 *Effect of grain size on rupture properties of nickel base alloys.*

7. Summary

To summarise, the production of high integrity nickel base materials for elevated temperature bolting and applications requires a slightly modified approach to that taken for steel bolts. Particular attention must be paid to:

(a) Use of high purity changes.

(b) Use of finishing elements to improve properties and processability.

(c) Use of consumable remelting to reduce segregation.

(d) Use of control thermal mechanical processing to achieve structural control.

(e) Use of controlled multistage heat treatment cycles.

All the above are essential if one is to obtain a satisfactory product which combines the excellent high temperature strength capable from these alloys whilst retaining adequate ductility.

Finally, it is worthwhile to point out that the application of such a process route to steel materials can result in significant improvements in ductility at the expense of costs. For example the rupture ductility of the CrMoV bolting steel can be enhanced by the use of high purity materials coupled with consummable electrode remelting.

32

Typical Grades of Steels and Alloys Used for High Temperature Bolting and their Methods of Production

H. SCHAFF

Aubert & Duval, 41, rue de Villiers, 92200 Neuilly Sur Seine, France

1. Introduction

For many decades, Aubert & Duval has melted, converted and heat treated numerous steel and superalloy grades destined for high temperature bolting in steam turbines, gas turbines, aircraft turbines and aerospace applications.

The aim of this presentation is to give some examples of grades in current production and some information on the manufacturing methods used by Aubert & Duval.

2. Examples of Grades Used for Bolting

Table 1 lists the chemical composition of some bolting grades, classified into the following families:

- low alloy chrome–moly–vanadium steels
- semi stainless 9/12% chromium martensitic steels
- precipitation hardening austenitic steels
- nickel base alloys.

Table 2 gives examples of heat treatments and minimum mechanical properties for these grades, as defined by standards or 'classic' specifications.

3. Some Aspects of Melting Techniques

Dependent upon the steel grade and the application, the complete panoply of the different melting methods industrially available today is used for the manufacture of high temperature alloys. Aubert & Duval have great experience is the classical melting techniques in electric arc furnace with vacuum degassing (more than 40 years), vacuum arc remelting (VAR) (33 years), electroslag remelting (ESR) (26 years) and vacuum induction melting (VIM) (28 years).

Table 1. Chemical composition of some current grades used for high temperature bolting

Type	Designations AD	Designations Usual	C	Si	Mn	Ni	Cr	Mo	V	Nb	N	Al	B	Ti	W	Co	Fe
Low alloy Steels	56 TG	21CrMoV57	0,17 0,24	≤ 0,50	0,30 0,60	≤ 0,50	1,10 1,50	0,50 0,80	0,15 0,35	--	--	--	--	--	--	--	Base
Low alloy Steels	MOCV 2	40CrMoV47	0,36 0,47	0,15 0,35	0,45 0,70	--	0,80 1,15	0,50 0,65	0,25 0,35	--	--	--	--	--	--	--	Base
9-12 % Cr steels	X 13 VD	JETHETE M152	0,08 0,13	≤ 0,35	0,50 0,90	2,00 3,00	11,00 12,50	1,50 2,00	0,25 0,40	--	0,020 0,040	--	--	--	--	--	Base
9-12 % Cr steels	X 13 W3	GREEK ASCOLOY	0,15 0,20	≤ 0,50	≤ 0,50	1,80 2,20	12,00 14,00	≤ 0,50	--	--	≤ 0,080	≤ 0,150	--	≤ 0,05	2,50 3,50	--	Base
9-12 % Cr steels	56 GE	X22CrMoWV121	0,20 0,25	0,20 0,50	0,50 1,00	0,50 1,00	11,00 12,50	0,90 1,25	0,20 0,30	--	--	≤ 0,025	--	≤ 0,05	0,90 1,25	≤ 0,20	Base
9-12 % Cr steels	56 T5	X19CrMoVNbN111	0,16 0,22	0,10 0,50	0,30 0,80	0,30 0,80	10,00 11,50	0,50 1,00	0,10 0,30	0,30 0,60	0,050 0,100	≤ 0,020	≤ 0,005	--	--	--	Base
9-12 % Cr steels	56 CW	FV 535	0,06 0,11	0,10 0,70	0,60 1,15	0,20 0,80	9,80 11,20	0,50 1,00	0,10 0,35	0,20 0,45	0,010 0,035	--	0,005 0,012	--	≤ 0,70	5,00 7,00	Base
Austenitic Precipitation Hardening steels	XN 26 TW	A 286	≤ 0,08	0,40 1,00	1,00 2,00	24,00 27,00	13,50 16,00	1,00 1,50	0,10 0,50	--	--	≤ 0,35	0,001 0,010	1,90 2,30	--	--	Base
Austenitic Precipitation Hardening steels	XN 26 AW	A 286 Mod.	≤ 0,06	≤ 0,50	≤ 2,00	24,00 27,00	13,50 16,00	1,00 1,50	0,10 0,50	--	--	≤ 0,35	0,003 0,010	1,70 2,00	--	--	Base
Nickel Base Alloys	PER 2X	NIMONIC 80A	0,04 0,10	≤ 1,00	≤ 1,00	Base	18,00 21,00	--	--	--	--	1,00 1,80	≤ 0,008	1,80 2,70	--	--	≤ 1,50
Nickel Base Alloys	PYRAD 53 NW	IN 718	0,03 0,10	≤ 0,35	≤ 0,35	Base	17,00 21,00	2,80 3,30	--	5,00 5,50	--	0,40 0,80	≤ 0,006	0,65 1,15	--	≤ 1,00	17,00 20,00
Nickel Base Alloys	PER 3	WASPALOY	0,03 0,10	≤ 0,15	≤ 0,10	Base	18,00 21,00	3,50 5,00	--	--	--	1,20 1,60	0,003 0,010	2,75 3,25	--	12,00 15,00	≤ 2,00

Table 2. Heat treatments and mechanical properties

Type	Designations AD	Designations Usual	Heat Treatment	Tensile Properties UTS (MPa)	YS (MPa)	A (%)	Z (%)	Creep θ (°C)	σ (MPa)	Life (h)	A (%)
Low alloy Steels	56 TG	21CrMoV57	890-940°C/ A or Oil + 670-720°C	700-850	≥ 550	≥ 16	≥ 60	- -	- -	- -	- -
Low alloy Steels	MOCV 2	40CrMoV47	880-930°C/ A or Oil + 670-730°C	850-1000	≥ 700	≥ 16	≥ 45	- -	- -	- -	- -
9-12 % Cr steels	X 13 VD	JETHETE M152	1030-1060°C/Oil + 620-700°C	930-1130	≥ 750	≥ 14	- -	- -	- -	- -	- -
9-12 % Cr steels	X 13 W3	GREEK ASCOLOY	950-1040°C/A or Oil + 560-600°C	970-1200	≥ 810	≥ 13	≥ 40	- -	- -	- -	- -
9-12 % Cr steels	56 GE	X22CrMoWV121	1030°C/Oil + 670-700°C	≥ 970	≥ 720	≥ 13	≥ 30	650	180	- -	- -
9-12 % Cr steels	56 T5	X19CrMoVNbN111	1100°C/A or Oil + 670-700°C	900-1050	≥ 780	≥ 10	≥ 40	- -	- -	- -	- -
9-12 % Cr steels	56 CW	FV 535	1120-1170°C/Oil + 600-650°C	1000-1140	≥ 880	≥ 8	≥ 30	550	325	100	≤ 0,15
Austenitic Precipitation Hardening steels	XN 26 TW	A 286	900-980°C/A + 720°C 16 h	960-1150	≥ 660	≥ 10	- -	650	480	≥ 23	≥ 4
Austenitic Precipitation Hardening steels	XN 26 AW	A 286 Mod.	900-980°C/A + 720°C 16 h	≥ 850	≥ 550	≥ 20	- -	650	410	≥ 30	≥ 3,5
Nickel Base Alloys	PER 2X	NIMONIC 80A	1050-1080°C/A (+ 850°C 24 h) + 700°C 16 h	≥ 1000	≥ 600	≥ 12	≥ 12	750	305	≥ 75	- -
Nickel Base Alloys	PYRAD 53 NW	IN 718	955-990°C/A + 720°C 8 h + 620°C 8 h	≥ 1240	≥ 1030	≥ 12	≥ 15	650	680	≥ 30	≥ 5
Nickel Base Alloys	PER3	WASPALOY	1020°C/A + 850°C 4 h + 760°C 16 h	≥ 1080	≥ 770	≥ 12	≥ 18	730	530	≥ 30	≥ 5

3.1. Classical Electric Arc Melting

The first operation, fundamental to the quality of the final product, is the choice of raw material. The mastery and reproductibility of the melting procedure depend upon adherence to fixed technical controls of raw materials, regulated by approval of each type of product and of each supplier, and verified by statistical process control. It is particularly important to limit the levels of residual elements which cannot be extracted during melting, like certain 'tramp elements' such as Arsenic, Tin, Antimony, etc.; these being responsible in the majority of grades, when not controlled, for scatter in impact values, toughness and for risk of embrittlement in service.

The melting itself is directed towards two objectives: for the one part to obtain the most precise final chemical composition conforming to our own limits and for the other to obtain the metallurgical cleanness compatible with the desired properties.

Reproducibility of chemical composition from one cast to another is an important feature in the mastery of the subsequent manufacturing processes, in particular the thermo-mechanical transformations and heat treatment operations.

Equally, the reproductibility allows the minimalisation of scatter of material properties: those which are verified on each delivery batch, such as tensile and impact values, but also those which are only determined on more detailed studies or qualification tests such as fatigue, creep resistance, long term relaxation, embrittlement in service, corrosion, etc.

Our melting techniques allow us to achieve low sulphur levels (typically 20 ppm) thereby almost completely avoiding the presence of sulphide particles which work against mechanical property isotropy and also resistance to pitting corrosion of stainless and semi-stainless steels, and significantly increase the risk of cracking in service by corrosion fatigue of by stress corrosion. Ladle degassing under vacuum permits both additional decantation and deoxidation of the liquid metal thereby minimising the levels of oxide type inclusions, also a source of mechanical heterogeneity of the microstructure.

Special care during the ingot pouring also plays a role in achieving the desired metallurgical cleanness and controlled solidification structures.

3.2. Remelting by Consumable Electrode

Two processes are used:

• Vacuum Arc Remelting (VAR);

• Electro-Slag Remelting (ESR).

The improvement in properties is linked to two different factors:

• reduction in the inclusion level;

• a much finer structure on solidification than that of the classical ingot, allowing a more thorough homogeneisation of the chemical composition in the transverse section of the product.

The almost total absence of longitudinal inclusions in the direction of the grain flow and the reduction of transverse chemical heterogeneity ('banding') leads to an often spectacular improvement in the macroscopic isotropy of the properties of the final product.

The choice of remelting method depends on the material type. Vacuum Arc Remelting (VAR) leads to an almost total removal of hydrogen, a large part of the nitrogen, a certain loss of manganese and a significant reduction in the oxygen content. This process is used for a large number of steel grades and practically all of the nickel base alloys. Electro-Slag Remelting is carried out in a technologically simpler installation, because there is no vacuum system, but the physio–chemical reactions are more complex due to the exchanges between the liquid metal and the slag. It allows the control of materials with high nitrogen and/or manganese contents and the obtention of satisfactory solidifications structures in the presence of other volatile elements.

3.4. Vacuum Induction Melting (VIM)

Electric arc melting ('air melting') does not allow sufficient mastery of the composition and cleanness for those grades containing a high level of very reactive elements like for example aluminium and titanium and a low level of carbon, silicon and manganese because these latter elements lead, when they are present, to the deoxydation of the liquid metal.
For this type of chemistry, a Vacuum Induction Furnace is used, this allowing the melting, processing, adjustments to composition, and pouring to be performed under vacuum.
An electrode is cast under vacuum and it is then remelted under vacuum. This VIM + VAR sequence is the classic route for nickel base alloys.

Double vacuum melting is also used today for precipitation hardening steels containing titanium and aluminium (classic or stainless maraging steels), for a particular version of an A 286 type (our grade XN 26 AW) and, for the moment exclusively for aerospace applications, certain bearing steels, nitriding and carburising steels, for which exceptional fatigue properties are requ ired.

4. Melting Techniques Relative to Bolting Materials

Returning to the examples in Table 1, the differents families are given below.

4.1. Low Alloy Steels and 9/12% Chromium Steels

For the majority of land based turbine bolting applications, these steels are not remelted whereas they sometimes are (in particular the 9/12% Cr steels) for turbine blades and often are for aerospace bolting.

One particular point merits mention with regard to those grades containing carbide or nitride forming elements, such as niobium, titanium or boron, added to improve creep and relaxation properties. These elements are extremely reactive, and only work effectively in the finished product if they do not form parasitic clusters on solidification.

With these types of grades, it is particularly important to look to fix the manufacturing parameters as much as possible, particularly the inclusion level, to allow consistent in service properties to be obtained. A certain prudence is equally desirable in relation to the results obtained from small laboratory melts, which are rarely representative of production heats in this field of study. The in depth evaluation of truly industrial products is distinctly preferable.

4.2. A 286-Type Steels

Our grade XN 26 TW is first electrically arc melted in air and then systematically remelted under vacuum. Aubert & Duval have developed a variant, XN 26 AW, melted by the VIM + VAR route, which has very low levels of silicon and manganese and reduced levels of phosphorus and carbon. These modifications give improved weldability. Its behaviour after exposure to temperature for very long periods warrants investigations and comparison with that of the classic grade.

4.3. Nickel Base Alloys

Melting is by VIM + VAR. In these types of alloys, the selection of raw materials and the control of trace elements play equally important roles with respect to the control of and the reproductibility of properties in service.

The metallographic structure being austenitic, that is with no possibility of grain refinement simply by heat treatment, it is necessary to research the greatest precision possible for the thermo-mechanical transformations and solution treatments, to achieve an homogeneous and reproducible grain size. This allows meaningful non-destructive testing and consistent mechanical properties.

5. Conclusions

The various aspects of the production methods presented give an idea of the complex subject with which we are confronted. Mastery of these manufacturing methods necessitates the technological mastery of the industrial processes and the best possible understanding of the metallurgical phenomena associated with each of these processes. The tools for advancement are the study by statistical methods of parameters related both to the process and to the product at each manufacturing stage and process modelling which helps to identify those parameters which must be under control.

We look to use these techniques at each stage of manufacture. Today's tools are parameters acquisition, data bases, statistics, guidance through previous results and comparison with models, these allowing us to deepen our understanding.

In this way, we can increasingly reduce the dispersion of properties in service and ensure that we maintain from on fabrication batch to another, those properties which cannot be verified systematically on each production batch, thereby giving the maximum security to our customers and permitting them to better their design margins.

SESSION 8

Future Prospects

Chairman: A Strang
GEC ALSTHOM Large Steam Turbines, Rugby, UK

33

Bolting Requirements for Advanced Turbine Plant

D. V. THORNTON

GEC ALSTHOM Large Steam Turbines, Rugby, UK

ABSTRACT

The higher pressure and temperature associated with advanced turbine plant will make the design duty of bolted joints more onerous. Bolting materials capable of improved performance will be required to match the increased temperature capabilities of the castings and pipework flange materials. Bolting requirements for advanced steam turbine plant are reviewed by consideration of the material characteristics needed for design of bolted joints and needed to confer resistance to bolt fracture. The properties of the various classes of bolting material are surveyed reylative to the steam conditions currently identified as possible for advanced turbine plant.

1. Introduction

Bolting requirements for advanced turbine plant, in principle, remain very similar to requirements for any other machine, viz. to maintain pressure tightness of metal/metal joints such as casing flanges and steam chest covers. In advanced plant the steam pressure and temperature at turbine inlet is increased necessitating the use of materials with improved creep resistance for the casings and chest components in the high temperature cylinders and the use of bolting materials with adequate relaxation strength for these more arduous conditions.

A few high temperature supercritical turbine plants were built circa 1960, e.g. Eddystone in the USA (372 bar/654°C/565°C/565°C) and Drakelow C in the UK (250 bar/593°C/565°C) (Table 1). Ferritic steels available at this time were limited to a maximum temperature of around 565°C. For Drakelow C, with the aid of steam cooling in the inlet region, ferritic steels were used for HP cylinder components such as rotors and casings whereas the HP steam chests and nozzle box were manufactured from austenitic steels. However, significant materials' development programmes conducted in Europe (COST) [1, 2], USA (EPRI RP1403) [3] and in Japan (EPDC) [4] since the early 1980s have led to the establishment of several new ferritic alloys. At this time ferritic materials have been developed for rotors, castings and pipework applications at temperatures up to 600°C. Currently, programmes are in hand to optimise ferritic alloy steel systems to improve their high temperature performance by means of alloy design to confer improved microstructural stability. The aim is to extend the useful service temperature of ferritic steels to at least 625°C and possibly to 650°C.

Turbines for advanced steam power plant have recently been ordered in Japan with a progression of steam conditions from 246 bar/538°C/593°C to 246 bar/593°C/593°C with a further step to 255 bar/600°C/610°C planned. In Europe machines have been ordered with

Table 1. Steam conditions of advanced turbine plant

	Pressure (bar)	Main Steam (°C)	Reheat (°C)	
Eddystone	372	654	565	565
Drakelow	250	593	565	
Kawagoe	310	566	566	566
Hekinan	246	538	593	
Matsuura	246	593	593	
Japan future	314	593	593	593
Japan future	343	649	593	593
Skaerbaek	285	580	580	580
Europe future	275	580	600	
Europe future	325	610	630	630

285 bar/580°C/580°C/580°C steam conditions and others are planned to operate at about 266 bar/580°C/600°C. The plan for the medium term future is to increase efficiencies to ~50% using steam at 325 bar/630°C with double reheat to 630°C.

2. Bolting Requirements

The design engineer requires a knowledge of certain properties of bolting materials, and the associated flange materials, in order to develop a satisfactory bolted joint design. This includes the definition of the loading parameters to maintain a pressure-tight joint from the cold condition up to the maximum operating temperature for the desired time interval to the next retightening.

The various design considerations have been reviewed in the opening keynote paper to this conference by J. L. Bolton [5]. The manner in which the available data on bolting materials is manipulated and used for high temperature joint design is illustrated and potential problem areas are highlighted. In addition to this specific data required for calculation to ensure joint tightness there is a need to consider other factors to provide an adequate margin against failure of the bolt.

The prime property of interest to the design engineer for a bolting material is stress relaxation resistance. Testing is conducted either as a constant strain uniaxial test or with the aid of model tests. In the former case the data is material specific whereas in the case of the model test the relaxed stress measured in the bolt material is dependent upon the associated flange material and their relative areas. A new series of model tests are required for changes in joint design or material combinations. In both cases residual strength after various times on load for the temperature range of application for the alloy type has to be determined. Retightening intervals have traditionally been after 30 000 h but it is now common for operators to wish to extend the interval to 50/70 000 h to reduce overhaul costs.

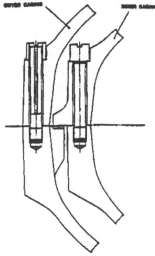

Fig. 1 *Horizontal joint bolting.*

The bolted joints are made at room temperature with the bolt extended to a specific strain value. If the bolt and flange material are the same material and hence have the same thermal expansion characteristics and the same elastic modulus over the temperature range of interest then the total applied strain will remain essentially constant. However, where different alloy combinations are used account must be taken of the temperature dependent changes in these characteristics. The elastic limit of bolting materials is required to be adequate to ensure that the initial loading strain is elastic. It is obviously damaging to the bolt life if significant amounts of plastic strain are absorbed during loading as this will reduce the ability to absorb creep strain during service.

In addition to property information which is utilised by the engineer to design the bolted joint and to define the loading conditions and the retightening interval to maintain a leaktight assembly the bolting material should also display a high resistance to potential failure mechanisms.

Experience has shown that failures can generally be ascribed to low notch ductility or a susceptibility to stress corrosion cracking. Notch ductility includes both the ability to accommodate the accumulation of plastic creep strain during service and to tolerate any service cracks which may form.

A notch strengthening behaviour in creep rupture testing would appear to be a necessary credential for a high temperature bolting material. In reality, the materials which have sufficient high temperature deformation resistance to confer a high relaxation strength are not particularly creep ductile. The most commonly used bolting materials, such as 1CrMoVNbTiB, 12CrMoVNb and Nimonic 80A, may all tend to notch weakening in standard creep rupture tests. Nevertheless by paying attention to the detail of bolt design to restrict the stress level in the threaded region by the use of waisted bolts these materials have been successfully used for many turbine bolt applications.

In modern power plant with the use of bolt heaters or hydraulic equipment to stretch the bolts the loading strain can be applied 'quietly' which is in contrast to the severe shock loading imparted by the former method of 'flogging' bolts using large wrenches, with large levers and large hammers. A high impact strength is therefore not required although there is a significant benefit if bolting materials exhibit a high fracture toughness. This confers a tolerance to the presence of small cracks, whether formed as a result of creep strain accumulation (Fig. 2) or stress corrosion (Fig. 3), and therefore increases the likelihood of cracks being found prior to failure during the routine inspection of bolts when the joint is opened for routine maintenance.

Since the effective bolt life is some 250 000 h and certain bolts will experience almost full steam temperature it is a significant advantage if the chosen bolting alloys exhibit microstructural stability. This is normally assessed by creep rupture tests to confirm absence of severe creep ductility troughs and by use of ageing 'embrittlement' tests to investigate any significant deterioration in toughness behaviour.

The historical observation of susceptibility of bolts in certain machines to stress corrosion cracking has often been ascribed to the use of molybdenum disulphide containing lubricants. In general the performance of low alloy steel, 12Cr and Nimonic alloys has been satisfactory since high temperature bolts operate dry. In certain circumstances where there are prolonged shutdowns (the bolts remaining fully loaded) and the bolting configuration permits condensate contaminated with molybdenum disulphide to build up around the bottom threads stress corrosion may occur. There have been reported incidents of stress corrosion cracking of low alloy steel bolts in wet steam turbines and also of 12% Chromium steel and Nimonic bolting in high temperature machines. With the banning of molybdenum disulphide containing lubricants which have now been replaced with graphite based or graphite/nickel powder anti-seize compounds the incidence of stress corrosion cracking in bolts has disappeared. Nonetheless immunity of bolting alloys to stress corrosion cracking in turbine quality steam or condensate is normally tested.

3. Bolting Alloys

The above review of bolting characteristics is appropriate for all high temperature turbines. For advanced plant it may be necessary to find new bolting materials for the high temperature components where new materials are employed. Nonetheless, lower temperature joints in the advanced machine will use established bolting materials. It is therefore appropriate in this review of bolting requirements for advanced steam turbine plant to consider current bolting materials, highlighting certain characteristics, and also to look at planned future steam conditions and possible new bolted joint materials which may be required.

3.1. Low Alloy Steels

Creep resistant low alloy steels have been used for several decades for high temperature turbine construction. In particular, several bolting steels based on the 1%CrMo, 1%CrMoV and the 1%CrMoVTiB alloys have been in service for extended periods at temperatures from 300 to 565°C. Their relative relaxation strength is shown in Fig. 4. Some problems have been encountered particularly with the higher strength 1%CrMoVTiB alloy as a result of accumulation of creep strain and cracking at the first unsupported thread.

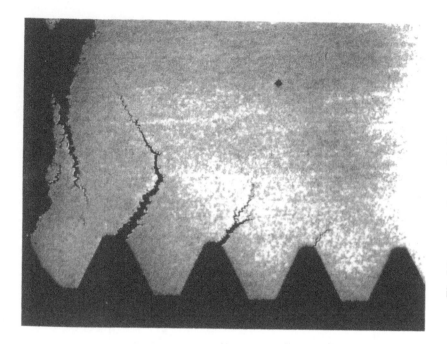

Fig. 3 Stress corrosion attack in Nimonic 80A bolt.

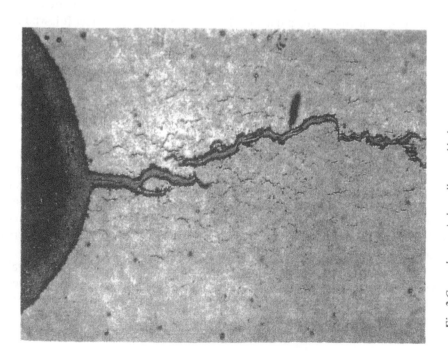

Fig. 2 Creep damage in low alloy steel bolt.

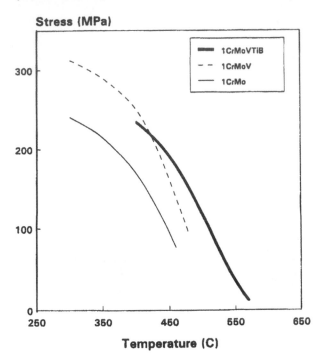

Fig. 4 *Relaxation strength of low alloy bolting steels.*

Failures were noted to occur in material which had high hardness, high creep strength and generally low (2–3% in 104 h) creep ductility. Following the work of Tipler [6], who had demonstrated that the removal of residual elements led to significant improvement in rupture ductility enquiries were made around 1980 with steelmakers to reduce residual levels. The effect was demonstrated [7] by making a cast by the VIM/VAR melting route using a Japanese electrolytic iron base. There was no notch sensitivity and the rupture ductility was improved by a factor of 5 for this high purity material with $R = 0.026$ ($R = P + 2.43$ As $+ 3.57$ Sn $+ 8.165$Sb) in comparison with commercial purity material with $R \cong 0.20$ as shown in Fig. 5. The extra cost of double vacuum remelted steels has prevented this melting route becoming the norm. By attention to bolt design, bolt operating procedures and to the material manufacture bolt failures are now rare. Of particular note is the general improvement in quality of electric arc furnace ladle refined steel, especially the reduction in the levels of residual elements. Low levels can now be routinely achieved as evidenced by the acceptance of a maximum R value in low alloy steel purchase specifications. Similar consideration of residual elements and their limitation to low levels have enabled improvement of material ductility in 12% Chromium steel and Nimonic 80A bolting materials.

For the commonly employed bolting materials test data out to 10 000 h is readily available, with some longer term tests to 30 000 h. The requirement for longer operation between overhauls leads to the need for more, even longer term, data or a reliable method to extrapolation of existing data.

Materials data generated from high temperature testing exhibit a degree of scatter. Expe-

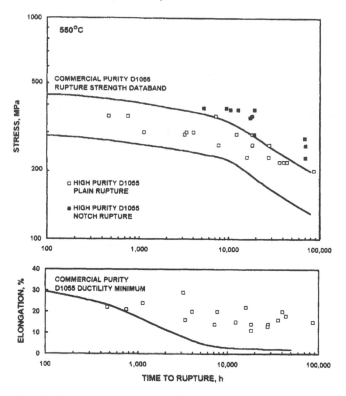

Fig. 5 *Effect of residual elements on creep rupture data for 1%CrMoVTiB steel at 550°C [7].*

rience has shown that for creep rupture strength and for stress to specific creep strain values over a time of 105 h there is a scatter of about 20% around the mean value. Such scatter results from variation between heats of similar material and from testing variations, even if the testing is conducted in 'approved' laboratories to a standardised procedure. Figure 6 shows the results of creep rupture tests and relaxation tests conducted on a number of casts of a low alloy steel all at 550°C and tested in the same laboratory. The scatter in strength values is +18% for rupture and +42% for the relaxation tests. If this is typical and that with the use of current test methods there is significantly more scatter in the determination of stress relaxation values than there is for creep strain or creep rupture values attention should be paid to standardisation of procedures. This is particularly important as we move to collaborative European assessment of materials data [8] on the activities of the European Creep Collaborative Committee Working Group on high temperature bolting materials. In addition, the design engineers who use the data need to be aware of the accuracy of the data in order to be able to design a reliable bolted joint.

3.2. 9–12% Chromium Steels

The search for improved bolting steels for low alloy steel joints in the 1960s at steam temperatures of 565°C resulted in the use in the UK of the highly creep resistant alloy 12%CrMoVNb. This material showed a significant advantage in relaxation strength being

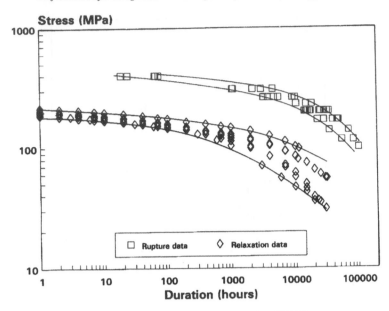

Fig. 6 *Rupture and relaxation data for 1%CrMoVTiB steel at 550°C.*

almost twice as strong as low alloy steels at 550°C. Although the expansion coefficient was about 20% less than that of low alloy steel (Fig. 7) an appropriate allowance was made in the cold tightening strain so that the required initial hot strain value was attained. The high creep resistance of 12%CrMoVNb was accompanied by low strain accommodation resulting in a number of failures and loss of favour leading to decreased usage. A similar alloy with slightly modified composition and heat treatment, X19 CrMoVNbN11.1, was developed in Germany and has a long record of successful service and this alloy continues to be used.

New 9–12% Chromium steels which have superior long term creep properties to existing ferritic steels have been developed for use in advanced power plant [1–4]. High temperature casings, steam chests and pipework operating with inlet steam at 580–600°C will be made from these new steels and will employ bolted joints. A collaborative test programme was undertaken in COST 501/2 to assess the relaxation strength of a range of the new alloys. The materials included Grade 91, the boron alloyed TAF steel and the COST 10CrMoVNbN rotor steels with additions of boron, tungsten and high nitrogen. The materials were heat treated to a proof strength of about 750 MPa and subjected to a 1000 hscreening test at 600°C. The results were most disappointing (Fig. 8) in that only two of the new steels showed the same relaxation strength as X19 CrMoVNbN11.1, with the others as much as 35% lower. The poor relaxation performance of the new steels was also surprising as the alloy compositions had been optimised to develop high creep resistance and, indeed, these new steels have up to two times the creep rupture strength of X19 CrMoVNbN11.1 at 600°C which was not reflected in their relaxation strength.

In the continuing search for an improved ferritic bolting steel the COST 501/3 group are investigating the potential of a titanium nitride strengthened 9% chromium steel for bolting [9]. The initial development will produce a sample from a base alloy powder coated with chromium nitride through a process of mechanofusion. A compact of this powder is to be

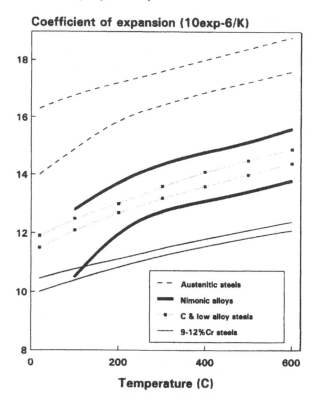

Fig. 7 *Coefficient of thermal expansion for various bolting materials.*

extruded at high temperature. During this process the chromium nitride dissociates and chromium and nitrogen diffuse into the base metal. The nitrogen reacts with titanium present in the base metal to form a fine dispersion of TiN particles. The alloy composition is to be carefully controlled to give an austenitic structure at the extrusion temperature, both before 'nitriding', to ensure a fine TiN dispersion, and afterwards, to ensure a martensitic transformation on cooling. The sample of TiN strengthened 9% chromium steel will initially be tested to determine its stress relaxation characteristics and if the results are sufficiently encouraging a larger sample of this new steel will be made for a more detailed assessment. In the longer term another production route shows promise. The 'Osprey' process involves co-deposition of molten steel droplets with the chromium nitride particles to ensure an even more uniform and intimate mixture prior to nitride dissociation.

This radical change to alloy design is being investigated after only limited experience with austenitic alloys but undoubtedly a new higher strength ferritic bolting steel would be of advantage in the design of advanced machines.

3.3. Austenitic Steels

Although some earlier machines were designed with 600/650°C main steam and employed austenitic materials, including bolting, the current endeavours to optimise the high tempera-

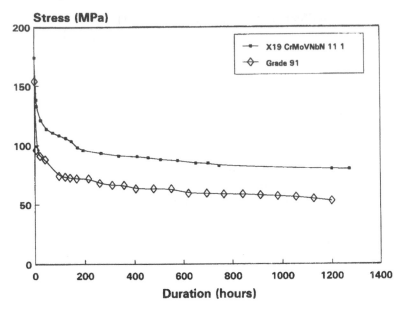

Fig. 8 *Comparative stress relaxation of modified 9%CrMo and X19 CrMoVNbN 11 1 at 600°C, 0.15% strain.*

ture creep strength of ferritic alloys are aimed at deferring the need for austenitic materials in modern supercritical turbines. Low proof strength, poor thermal conductivity and high thermal expansion coefficients of austenitic materials make it significantly more difficult to design components to withstand the thermal cycles associated with stop/start cycling demanded in modern machines. Experience at Drakelow and Eddystone has confirmed the susceptibility of such components to thermal cycling damage and, in particular, problems due to dimensional stability.

Drakelow, operating at 593°C, had wrought austenitic steam chests and the covers were bolted with austenitic material which has performed satisfactorily. Two types of austenitic materials were used, alloys which were warm worked to increase the proof strength and a precipitation hardening alloy (Table 2). An indication of the relaxation behaviour is given in Fig. 9. The relaxed stress of the austenitic bolting materials at 650°C is superior to the ferritic alloys at 540°C and therefore should be sufficient for the designers' requirements. However, when a positive interest in the application of austenitic materials develops additional data, particularly long term relaxation data, on austenitic bolting materials will be required.

3.4. Nickel Alloys

For many years Nimonic 80A has been used for bolting low alloy steel joints, the associated cost premium limiting its use to critical joints. The marked advantage in relaxation strength of Nimonic 80A is illustrated in Fig. 9 and the similarity of expansion coefficient with low alloy steel is shown in Fig. 7. The metallurgical problems associated with the use of Nimonic 80A have been reviewed in Session 5 of this conference. The overall performance of Nimonic 80A has been excellent, the reported [10] failure rate in a recent worldwide survey was <

Table 2. *Composition and properties of austenitic bolting steels*

Composition									
	C	Cr	Ni	Mo	V	Mn	Nb	Al	Ti
Esshete 1250	0.1	15	10	1	0.2	6.0	1	–	–
A286	0.05	15	25	1	0.3	1	–	0.25	2

Manufacture	
Esshete 1250	Solution treatment 1050 – 1150° AC Warm work ~10/15% at 700 – 600°C
A286	980 O.Q. 720°C AC

Mechanical Properties at 20°C				
	0.2% PS	UTS	EL	RA
Esshete 1250	550	710	38	64
A286	620	1080	29	55

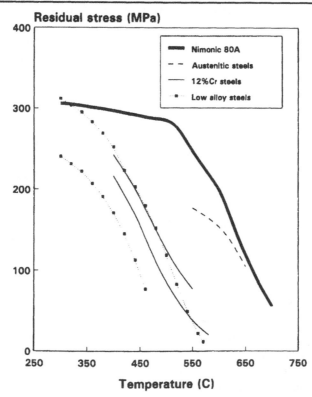

Fig. 9 *Comparative stress relaxation behaviour for various bolting materials.*

0.4% for a total of over 20 000 bolts operating in machines with inlet temperatures up to 566°C.

In the absence of an improved ferritic steel bolting material the high relaxation strength of Nimonic 80A at temperatures up to 650°C makes it a candidate for the bolting of 9% chromium steel flanges in advanced turbines. However, the difference in expansion coefficients means the bolt expands more than the flange and a proportion of the initial tightening strain is lost as the joint is heated to the operating temperature. The reduction is too much to be compensated for by increasing the initial tightening strain and therefore austenitic steel sleeves, having higher expansion coefficient than 9CrMo, are designed to fit between the flange and the nut to compensate for the differences in expansion. Accurate data for the value of the expansion coefficients is required by the designer for the three materials in the combination.

In an effort to minimise this problem the COST 501 group are currently reviewing other nickel base alloys searching for a material with a lower expansion coefficient. There are no obvious established alloys which have a significantly lower value at 600°C (see Fig. 7), but the alloy producers are continuing the investigation. Even alloys listed with lower expansion coefficients have, when measured, been found to be not significantly different to Nimonic 80A.

4. Future Advanced Machines

If a satisfactory solution using austenitic steels is not practical due to limitation on size of rotor forging, or difficulty in attaining the property requirements in castings, or problems in design, it would be possible, though expensive, to use nickel alloys at about 650°C.

For the more distant future for steam turbines operating in the region of 700°C it is most probable that multi-casing designs would be employed reducing bolt loading. Superalloy casings would be required at these temperatures and the bolting would be likely to be in matching materials. Nimonic 80A at 700°C shows equivalent relaxed strength to the ferritic steels at 540°C and is therefore a potential candidate.

Srivastava and Rothman [11] indicate that a nickel base alloy containing 25% molybdenum and 8% chromium is a potential bolting material for application at about 650°C. In addition, Buzalits and Newnham [12] discuss a CoNiCrMo alloy with the highest strength and creep resistance of any fastener up to 593°C and a NiCoCrMo alloy recently developed for gas turbine fastener duty with the highest creep resistance of any fastener material between 650 and 760°C.

If the time approaches when steam turbines are required to operate in the region of 650–700°C the experience available in the gas turbine field will need to be reviewed to identify if any of their established bolting materials are capable of operating over the longer overhaul periods required for steam turbine operation.

However, a broader perspective of materials for advanced machines, as shown in Table 3, indicates that the materials for the header and superheater of the boiler are required to operate at higher temperatures than turbine materials. For turbine steam temperatures of 580 and 600°C boiler and turbine materials are fully developed and, where necessary, codified. It is anticipated that development programmes in hand will establish improved ferritic materials

Table 3. *Material developments for advanced machines*

Steam X°	Header (X + 15)°C	Superheater (X + 35)°C	Turbine X°C
580	P91	Type 347 HFG	10% CrMoVNbN
600	NF616	Type 347 HFG	10% CrMoVNbN
610	HCM12A* NF12*	HR3C* NF709*	Improved * 10% CrMoVNbN
630	HCM12A* NF12*	HR3C* NF709*	Imrpoved * 10% CrMoVNbN
650	??	HR6N* IN617*	Forgings A286* Castings ??
700	??	IN617*	??

Composition

P91	9Cr	1Mo	V	Nb	N			
NF616	9Cr	0.5Mo	V	Nb	N	1.8W		
HCM12A	12Cr	0.4Mo	V	Nb	N	2W	1Cu	
NF12	11Cr	0.2Mo	V	Nb	N	2.6W	3Co	B
A286	15Cr	25Ni	1Mo	0.3V	2Ti	0.25Al	B	
347H.FG	18Cr	10Ni	1Nb					
HR3C	25Cr	20Ni	0.4Nb					
NF709	20Cr	25Ni	1.5 Mo	0.25Nb	0.05Ti			
HR6W	23Cr	43Ni	1Mo					
IN617	22Cr	52Ni	9Mo	12Co	2Al			

*Not fully established.
?? Not known.

to permit turbine steam temperatures up to 630°C. Materials for advanced plant at 650°C and above are still undecided and hence the selection of the associated bolting materials is not an immediate requirement.

5. Summary

A number of commercial steam turbines to operate with advanced steam conditions have now been ordered and some are already in operation. Further development of ultra-supercritical machines will be dependent upon the evaluation of the economics of new designs, which employ new high temperature materials and will be more expensive to build, relative to the increase in efficiency resulting from the advance in steam conditions.

For the near future with steam conditions up to about 310 bar and 630°C where ferritic materials will be used for construction the following requirements have been identified:

- to standardise the methods of relaxation testing of materials to reduce scatter in results and to facilitate the exchange of data in a collaborative international framework;

- to establish and verify methods for the extrapolation of stress relaxation data to longer durations to provide the design data to meet the plant operators' requirements for extended periods between overhauls; and

- to develop a new material for bolting of the new 9/12% chromium steels which has a matching thermal expansion coefficient to avoid the complications of incorporating a compensating sleeve to accommodate mis-match in expansion coefficients of the joint materials.

In the longer term there may be further advance in steam conditions to 650°C or even 700°C and the use of austenitic steels or nickel alloys as materials of construction. Although austenitic steam chests with austenitic bolting have been in successful operation on 'old' supercritical machines there is little reliable relaxation data available and additional testing of austenitic bolting steels would be necessary. In the case of nickel alloys the obvious starting point to identify potential materials for steam turbine bolting is a review of experience gained in gas turbine applications, bearing in mind that there may be a difference in bolt size and service lifetime for steam turbine duty.

References

1. C. Berger *et al.*, 'Steam turbine materials: high temperature forgings', *5th Int. Conf. Materials for Advanced Power Engineering*, Liège, Belgium, October 1994.
2. R. B. Scarlin *et al.*, 'Steam turbine materials: high temperature castings', *ibid.*
3. *Third EPRI International Conference on Improved Coal-fired Power Plants*, San Francisco, USA, April 1991.
4. Y. Nakabayashi, 'Development of advanced steam plant in Japan', *COST 501 Conf. Materials for Power Engineering Components*, Jülich, Germany, October 1992.
5. J. Bolton, 'Design considerations for high temperature bolting', this volume, pp. 1–14.
6. H. R. Tipler, *Conf. on Residuals, Additives and Materials Properties*, London, May 1978.
7. A. Strang *et al.*, 'Rupture ductility of 1CrMoV high temperature bolting steels', *Conf. on Rupture Ductility of Creep Resistant Steels*, York, UK, December 1990.
8. J. Orr, 'Activities of ECCC Working Group high temperature bolting materials', these proceedings.
9. H. G. Hamerton *et al.*, 'Titanium nitride strengthened steels for bolting applications', this volume, pp. 220–225.
10. K. H. Mayer *et al.*, 'High temperature bolting of steam turbines for improved coal-fired plants', *Conf. on Improved CoalFired Power Plants*, San Francisco, USA, April 1991.
11. S. K. Srivastava, 'A Ni–25Mo–8Cr alloy for high temperature fastener application', this volume, pp. 284–293.
12. J. Newnham and S. Buzolits, 'High strength, high temperature bolting alloys for turbine applications', this volume, pp. 318–328.

34

Conference Overview

F. B. PICKERING

Sheffield Hallam University, Sheffield, UK

During the two days of the conference well over 30 papers have been presented. The amount of information in those papers has been enormous. In some respects, an observer of the conference may well have had a distinct sense of 'deja vu' since many old 'friends' in terms of materials and subjects have been revisited. For example, there has been much talk on established materials such as Durehete 1055, Esshete 1250, Nimonic 80A, A286 (to mention just a few), and well covered topics such as extrapolation (which produced animated discussion), high purity materials, embrittlement effects, environmental cracking and testing techniques. Another feature of the conference, which may have surprised someone not immersed in the field of high temperature bolting materials, was the overall paucity of papers on newly developing materials. Moreover, another impression, which may be more apparent than real, is that the metallurgists and engineers involved in the field form a close knit community, the members of which talk mainly with one another, attend the same conferences and sit together on the many committees — both national and international. There may be a tendency for concepts and developments to become fossilised and channelled into an accepted consensus. Perhaps an influx of new blood and new ideas would be especially beneficial.

There were several papers on what may loosely be called engineering design. Apparently the engineer never really gets the data he requires; it was ever thus! The use of bolting models, some simplified and others more complex were mentioned, but never the marked microstructural changes which can occur in service. This will be returned to. It is very necessary to differentiate between design problems and metallurgical problems, and apparently only about one third of the factors leading to bolting failures are materials based. The rest are largely engineering design and operating problems. Clearly embrittlement effects are of paramount importance and fortunately there is now much detailed understanding of the causes and remedies for these. Tightening and bolt removal was dealt with, together with various methods for bolt strain control. Some of the damage inducing operating processes seem quite horrific, and there is a need to match high temperature creep, stress relaxation and rupture ductility with adequate cold fracture toughness. These different properties are not always controlled by the same metallurgical phenomena.

Stress relaxation testing, perhaps the most difficult of all mechanical property testing procedures, occupied considerable time. Reproducibility, as shown by the scatter in results, is clearly a problem, as well as temperature control over a long specimen. Can an improvement be made in the sensitivity of extension measurement and control? Equally important is calibration drift of temperature measuring devices, especially for 5–10 year testing periods. The use of uniaxial testing may give an incomplete guide to the complex interaction of bolt and flange, and particularly to the effect of bending and torsional stresses. Hence there is the tendency to use model bolt/flange assemblies or even tests on full sized bolts which are often of massive dimensions. The latter often indicates a lack of understanding of what is happening in the real world. The need to extrapolate data to 30 years or more, for obvious maintenance, replacement and re-tightening reasons, as well as the economics of power generation, is still a major problem. It has been discussed and studied for very many years, but yet can still raise fierce passions as the discussions revealed. We resort to curve fitting exercises. Surely no one can doubt that the relaxation or creep strength levels out at the longest times as the material reaches a coarse grained, large particle size, minimum solid solution strengthened condition. But extrapolation poses peculiarly difficult problems as the stress/temperature/time conditions change during service — that is the test or service conditions are represented by a moving point which passes over the landscape of the deformation or fracture map, the regime boundaries of which are moving as the microstructure alters during testing or service. Extrapolation is therefore only likely to be valid if the microstructures and the deformation mechanics involved do not alter, which is definitely not the case at higher temperatures, or longer times. With regard to techniques of testing, property collection and standardisation, there seems to be a plethora of national or international committees looking into these problems. Whilst recognising the need for such technique evaluations, it can be questioned whether committee based programmes are the best way forward. Understanding is more important than anything else, especially the routine collection of data.

Turning to embrittlement effects, there have been identified creep embrittlement (often associated with stoichiometry in carbide hardened steels), temper embrittlement, impurity embrittlement, ordering embrittlement and negative creep effects (not always associated with ordering), irradiation embrittlement and various environmentally induced embrittlements sometimes associated with thread lubrication, i.e. stress corrosion due to the lubricants. Purity is clearly a prime requisite in modern materials, but high purity does not necessarily increase creep or relaxation strength, although it does have a very beneficial influence on creep ductility and results in much less notch sensitivity. Hence we have, at a cost, the availability of high purity very clean steel produced by the ESR or VIM/VAR process routes. This has spurred some manufacturers to improve Electric Arc Steelmaking by control of and attention to deoxidation, secondary steelmaking, scrap selection for low residuals, etc., to try to produce at much lower cost the equivalent of the high purity vacuum processed steels. There will be fierce arguments about this approach, which is a real step forward, and about the balance of cost against properties. May we see a sliding cost scale for progressive residual element reductions?

But the question must be asked as to whether one can accept less than the highest purity if one produces a very fine austenite grain size and thus a very low solute concentration per unit of grain boundary surface area (S_v)? Do we ever plot ductility against solute/S_v? No such plot was shown in the conference. Also, how best can we achieve very fine grain sizes? Possibly by TiN technology as in the HSLA steels. Is it possible to do even better by grain refinement by Ti oxides? Also what are the dangers of very low residual steels and very fine austenite grain sizes? Will there be heat treatment problems, more tendency for grain growth and less than optimum machinability in very clean steels? Will these new and important developments require the heat treatment processes to be re-optimised to overcome potential problems? We saw in the conference how important austenitising temperature can be in the 12%Cr–Mo–V bolting steels. But we must differentiate between ultra pure and super clean steels because they are very different in terms of their effects on different properties. We have also seen intriguing effects of increased Al contents during deoxidation in terms of lower rupture ductilities, but the precise reasons for this is not known, i.e. it could be due to large AlN particles or the effect of Al affecting the solubility of MnS. It is known that re-precipitation of MnS can produce a form of creep embrittlement and the question may be asked whether the current use of Ca treatment could be very beneficial. There is no doubt that the newer high purity materials have a further and marked advantage in producing a greater reproducibility of properties — which is of the greatest importance as shown by the scatter in the stress relaxation data for steels of conventional purity and cleanness, which can vary by half an order of magnitude on the time scale and up to 150% on the stress scale. This leads to the thought that much of the current long time data has been accumulated on steels made long before these new processing and manufacturing techniques were developed. It will be interesting to compare the properties of these newer steels, and their actual service performance, in years to come with those of steels made by the old conventional methods.

Unfortunately, the conference heard little about mechanisms and the optimisation of the design of alloy steels. The metallurgical rationale for alloy design is most important, which is why the paper dealing with the use of solubility products and stoichiometry in explaining the behaviour of 12%Cr steels containing Nb and N was so impressive. This is surely a methodology which needs to be expanded, as much may be learnt from its application in the HSLA microalloyed steels. Also, there should be the application of the well established principles for decreasing particle growth in order to produce a more stable microstructure which degrades much more slowly. This would help improve the accuracy of extrapolation techniques.

Equally we have heard relatively little about very novel materials, except perhaps for the Ni–Co–Cr–Mo alloys and the TiN strengthened austenitic and ferritic steels. But there must be doubts at present on the manufacture of these alloys particularly for massive bolts. The approach seems largely to have been one of incremental development, as shown by the work reported on D1055, Esshete 1250 and Nimonic 80A. It is encouraging however to see clearly that there is a better understanding of current materials which helps these steels to be used more closely to their optimum potential. Much has been discussed on the subjects of negative creep effects which can be due to carbide precipitation as well as ordering in Ni base alloys, the limit of Ti in A286 type alloys in terms of chi (sigma) phase and c.p.hex Ni_3Ti formation, the potential for warm working in Esshete 1250 which also involves a stoichiometric Nb:C effect, etc. As working temperatures increase, surely there must be concern about intermetallic compound formation especially in some of the more highly alloyed 12%Cr steels. Most of these effects have been discussed over many years, and in some respects various areas covered by the conference simply reiterated what was already known and understood.

Finally, and inevitably, in conferences dealing nowadays with high temperature plant applications, there was discussion of life management and remnant life evaluation. The use of models is commonplace, but it is necessary to be aware that such models may simply be supplying an aura of respectability to predictions of long term properties. It is important to be very aware that the models only give predictions, which need to be validated by long term testing. This experimental work is time consuming and expensive, whilst mathematical model building is cheap. One may ask whether models can be developed which take into consideration the microstructural changes that occur during service; this would be a real advance.

The overriding impression of the conference was that we are hardly, if ever, seeing absolutely novel materials, but rather a more acute understanding and awareness of the underlying processes which occur in current materials. This is however solid progress, and one wonders whether there is a real need for novel materials. Of particular note is the development of the improved purity materials, applying as it does to steels and Ni base alloys, and the efforts now being made to improve and optimise alloy manufacture and the thermal and mechanical processing. If one may conclude with what may well be an idiosyncratic observation, it is that a little more understanding is worth quite a lot of extra data accumulation. If we can get the metallurgy right, the engineers will have far fewer problems.

Author Index

List of Delegates

J. Abbott	Hydra-Tight Ltd, Walsall, UK	K.H. Mayer	MAN Energie GMBH, Nürnberg, Germany
O. Andersen	Statoil, Stavanger, Norway		
I. Artinger	Tech University of Budapest, Budapest, Hungary	P.R. McCarthy	ERA Techonology Ltd, Leatherhead, UK
D.C. Barker	K.S. Paul Products Ltd, London, UK	Mr Micallef	Hydra-Tight Ltd, Oldham, UK
W. Barr	Scottish Nuclear Ltd, East Kilbride, UK	W. Mihoub	Imphy SA, Imphy, France
M. Baxter	North Bridge Fasteners, Leicester, UK	C.W. Miles	Hydra-Tight Ltd, Oldham, UK
J. Beardwood	UES Steels, Stocksbridge, UK	N. More	Hedley Purvis Ltd, Northumberland, UK
S.M. Beech	Int'l Research & Develop Ltd, Newcastle upon Tyne, UK	D. Morrison	Electricity Corp of NZ Ltd, Wellington, New Zealand
J.L. Bolton	GEC ALSTHOM, Rugby, UK		
P. Bontempi	Enel, Milan, Italy	B. Nath	National Power PLC, Swindon, UK
A.G. Callagy	Electricity Supply Board, Dublin, UK	J.A. Newnham	SPS Technologies Ltd, Leicester, UK
J. Cawley	Materials Research Inst., Sheffield, UK	P. O'Hara	Metal Improvement Co, Newbury, UK
D.R. Columbine	Dartec Limited, Stourbridge, UK	G. Oakes	Special Melted Products Ltd, Sheffield, UK
J.M. Cooper	GEC ALSTHOM, Rugby, UK		
J.H. Davidson	Imphy SA, Imphy, France	J. Orr	British Steel Technical, Rotherham, UK
M.J. Durbin	Aubert et Duval, Sheffield, UK	S. Osgerby	National Physical Laboratory, Teddington, UK
A.J. Dykes	Enpar Ltd, Rotherham, UK		
H. Everson	UES Steels, Stocksbridge, UK	P.A.M. Palij	North Bridge Fasterners, Leicester, UK
M. Eyckmans	Laborelec, Linkbeek, Belgium	F.B. Pickering	Sheffield Hallam University, Sheffield, UK
N.C. Farr	Inco Alloys Ltd, Hereford, UK		
C.J. Flavell	Dartec Limited, Stourbridge, UK	M.O. Pitkanen	Imatran Voima Oy, Ivo, Finland
B.M. Fletcher	Met Ltd. Lancaster, UK	H. Purper	MPA Stuttgart, Stuttgart, Germany
V. Foldyna	Vitkovice Research Institute,, Ostrava, Czech Republic	B.W. Roberts	GEC ALSTHOM, Rugby, UK
		R.H. Ryder	General Atomics, San Diego, Ca, USA
P. Foster	National Power PLC, Telford, UK	H. Schaff	Aubert & Duval, Neuilly sur Seine, France
R.J.N. Gommans	DSM Research, Geleen, The Netherlands		
		N.B. Shaw	National Power PLC, Knottingley, UK
D.J. Gooch	National Power, Swindon, UK	D. Smith	Hydra-Tight Ltd, Oldham, UK
R.G. Hamerton	AEA Techonlogy, Warrington, UK	M.W. Spindler	Nuclear Electric Plc, Berkeley, UK
R. Holinski	Molykote, Munich, GermAny	R.A. Stevens	Nuclear Electric Plc, Berkeley, UK
A.D. Hope	Shell Research BV, Arnhem, The Netherlands	A. Strang	GEC ALSTHOM, Rugby, UK
		A.J. Tack	Industrial Research Ltd, Auckland, New Zealand
F. Hunter	National Power PLC, Telford, UK		
G.T. Jones	ERA Techonlogy Ltd, Leatherhead, UK	E.F. Tate	Parsons Turbine Generators Ltd, Newcastle-upon-Tyne, UK
B. Jouan	Aubert & Duval, Neuilly sur Seine, France		
		D.V. Thornton	GEC ALSTHOM, Rugby, UK
S. Kawakami	Kyushu Electric Power Co. Fukuoka, Japan	R.D. Townsend	ERA Technology Ltd, Leatherhead, UK
		F. Vee	Statoil, Stavanger, Norway
H. König	MAN Energie GmbH, Nürnberg, Germany	G. Walton	National Power PLC, Didcot, UK
		G.J. White	Scottish Power PLC, East Kilbride, UK
I. Lennox	Prosper Engineering Ltd, Irivine, UK	A. Whitton	Aubert & Duval, Hemel Hempstead, UK
C. Li	Materials Research Inst., Sheffield, UK		
J.G. MacDonald	ICI Engineering, Billingham, UK	B. Wilshire	University of Wales, Swansea, UK
A. Marucco	CNR ITM, Milan, Italy	J.W. Woolley	GEC ALSTHOM, Rugby, UK
I. Matsuura	Mitsubishi Heavy Ind., Madrid, Spain	M. Zannoni	Ansaldo Gie SRL, Genova, Italy

Printed and bound by CPI Group (UK) Ltd, Croydon, CR0 4YY

23/10/2024

01778230-0001